SAFETY AND REGULATORY ISSUES OF NANOENCAPSULATED FOOD INGREDIENTS

Nanoencapsulation in the Food Industry

SAFETY AND REGULATORY ISSUES OF NANOENCAPSULATED FOOD INGREDIENTS

VOLUME 7

Edited by

SEID MAHDI JAFARI

Department of Food Materials and Process Design Engineering, Gorgan University of Agricultural Sciences and Natural Resources, Gorgan, Iran

ELSEVIER

ACADEMIC PRESS

An imprint of Elsevier

Academic Press is an imprint of Elsevier
125 London Wall, London EC2Y 5AS, United Kingdom
525 B Street, Suite 1650, San Diego, CA 92101, United States
50 Hampshire Street, 5th Floor, Cambridge, MA 02139, United States
The Boulevard, Langford Lane, Kidlington, Oxford OX5 1GB, United Kingdom

Library of Congress Cataloging-in-Publication Data
A catalog record for this book is available from the Library of Congress

British Library Cataloguing-in-Publication Data
A catalogue record for this book is available from the British Library

ISBN: 978-0-12-815725-1

For information on all Academic Press publications
visit our website at https://www.elsevier.com/books-and-journals

Publisher: Charlotte Cockle
Acquisitions Editor: Nina Bandeira
Editorial Project Manager: Lena Sparks
Production Project Manager: Joy Christel Neumarin Honest Thangiah
Designer: Christian J. Bilbow

Typeset by Thomson Digital

This book is dedicated to Mohammad-Reza Shajarian (23 September 1940–8 October 2020), “Iranian vocalist and master (Ostad) of Persian traditional music.”

Contents

Contributors

Safoura Akbari-Alavijeh
Department of Food Science and Technology, Faculty of Agriculture and Natural Resources, University of Mohaghegh Ardabili, Ardabil, Iran; Department of Food Science and Technology, College of Agriculture, Isfahan University of Technology, Isfahan, Iran

Sareh Boostani
Department of Food Science and Technology, School of Agriculture, Shiraz University, Shiraz, Iran

Esra Capanoglu
Department of Food Engineering, Faculty of Chemical and Metallurgical Engineering, Istanbul Technical University, Maslak, Istanbul, Turkey

Marjan Daeihamed
Department of Pharmaceutics, School of Pharmacy, Guilan University of Medical Sciences, Rasht, Iran

Ali Ebrahimi Pure
Department of Food Science and Technology, College of Agriculture and Natural Resources, University of Tehran, Tehran, Iran

Monir Ebrahimi Pure
Department of Chemistry, Alzahra University, Tehran, Iran

Zahra Emam-Djomeh
Transfer Phenomena Laboratory (TPL), Functional Food Research Core, Department of Food Science and Technology, College of Agriculture and Natural Resources, University of Tehran, Tehran, Iran

Masoud Faghih Akhlaghi
Department of Medicinal Chemistry, School of Pharmacy, Guilan University of Medical Sciences, Rasht, Iran

Seid Reza Falsafi
Department of Food Materials and Process Design Engineering, Gorgan University of Agricultural Sciences and Natural Resources, Gorgan, Iran

Seid Mahdi Jafari
Department of Food Materials and Process Design Engineering, Gorgan University of Agricultural Sciences and Natural Resources, Gorgan, Iran

Sara Khoshnoudi-Nia
Seafood Processing Research Group, School of Agriculture, Shiraz University, Shiraz, Iran

Zahra Mohammadi
Department of Food Materials and Process Design Engineering, Gorgan University of Agricultural Sciences and Natural Resources, Gorgan, Iran

Atefe Rezaei
Department of Food Science and Technology, School of Nutrition and Food Science, Food Security Research Center, Isfahan University of Medical Sciences, Isfahan, Iran

Hadis Rostamabadi
Department of Food Materials and Process Design Engineering, Gorgan University of Agricultural Sciences and Natural Resources, Gorgan, Iran

Rezvan Shaddel
Department of Food Science and Technology, Faculty of Agriculture and Natural Resources, University of Mohaghegh Ardabili, Ardabil, Iran

Özgür Tarhan
Department of Food Engineering, Engineering Faculty, Uşak University, Uşak, Turkey

Merve Tomas
Department of Food Engineering, Faculty of Engineering and Natural Sciences, Istanbul Sabahattin Zaim University, Halkali, Istanbul, Turkey

Atike Ince Yardımcı
Technology Transfer Office, Uşak University, Uşak, Turkey

Alireza Zibaei
Faculty of Medicine, Islamic Azad University of Mashhad, Mashhad, Iran

Preface to Volume 7

Nanotechnology has significant impacts on many industrial areas including agriculture and food at various stages from production to sale. Agricultural raw materials continuously feeding food industry, food formulations and production methodologies, and chemical polymeric materials essentially associated with food packaging are naturally involved in the whole process and can be designed, improved, and fabricated through innovative technological principles. The benefits of nanotechnology-based food products for consumers and manufacturers are widely reported in terms of bioavailability, food safety, target, and controlled delivery systems. Inorganic nanoparticles such as titanium oxide and silver, and organic nanoparticles such as carbohydrates, proteins, and lipid-based nanostructures incorporated in food materials can be uptaken into body continuously through consumption. Unforseseen consequences may arise from extended and uncontrolled exposure to them. Safety issues associated with the use of nanomaterials in food and food contact products have raised public concerns. In fact, in normal daily diet, human can intake various tiny particles ranging from large macromolecules to nanoscale structures with no harm. However, when deliberately incorporated organic or inorganic nanoparticles are present in food and their contact materials, it can lead to overexposure, allergenity, and toxicity at different levels since they are readily absorbed through digestion system and deposited in tissues. Those potential risks may lead to public concerns and unwillingness to pay for and consume nanofood products.

This book entitled "*Safety and Regulatory Issues of Nanoencapsulated Food Ingredients,*" as volume 7 in the book series "*Nanoencapsulation in the Food Industry*" presents an overview of the safety evaluation and legislation of nanostructures in different countries dealing with food products. It covers recent and applied research in all categories of food nanostructures.

After presenting the safety and toxicity aspects of food nanoparticles in Chapter 1, the book covers possible health risks associated with food nanoparticles in Chapter 2. Chapters 3 and 4 deal with in vitro and in vivo assays for predicting the safety of food-based nanomaterials, respectively. Another important topic, that is, regulatory aspects and legislation of nanostructures in different countries has been discussed in Chapters 5–8; including European countries, (Chapter 5), Asian and Oceania countries

(Chapter 6), North and South American countries (Chapter 7), and International Organizations and Authorities (Chapter 8).

All who are engaged in micro/nano-encapsulation of food, nutraceutical, pharmaceutical, and cosmetic ingredients world-wide can use this book as either a textbook or a reference which will give the readers a good and recent knowledge of health risks and safety of nanocarriers and nanoencapsulated ingredients, as well as their toxicology, safety analysis, and legislation. We hope this book will stimulate further research in this rapidly growing area, and will enable scientists to get familiar with the safety of different nanocarriers loaded with bioactive ingredients.

I really appreciate the great cooperation of all authors of the chapters for taking time from their busy schedules to contribute to this project. Also, it is necessary to express my sincere thanks to all the editorial staff at Elsevier for their help and support throughout the project. Finally, special acknowledgment is to my family for their understanding and encouragement during the editing of this great project.

Seid Mahdi Jafari
February, 2021
Gorgan, Iran

CHAPTER 1

Safety and toxicity aspects of food nanoparticles

Alireza Zibaei[a], Seid Mahdi Jafari[b]
[a]Faculty of Medicine, Islamic Azad University of Mashhad, Mashhad, Iran
[b]Department of Food Materials and Process Design Engineering, Gorgan University of Agricultural Sciences and Natural Resources, Gorgan, Iran

1.1 Introduction

Safety has been considered as one of the most important human necessities before accomplishing anything in the world. Potential harmful effects from novel discoveries, and their volatilities raise the question: How to provide safety in new products? In 1959, Nobel Prize winner Richard Feynman introduced the concept of nanotechnology, which was later on defined as the science of studying materials of very small sizes, between 1 and 100 nm (10^{-9} m) (Bayda, Adeel, Tuccinardi, Cordani, & Rizzolio, 2019). This cutting-edge branch of science has revolutionized several industries- including food- in four areas: (1) quality and safety, (2) packaging, (3) ingredients and delivery systems, and (4) methods of food processing (Jafari & McClements, 2017).

Many nanomaterials of diverse origins have secured their position in the main 12 types of applications related to the final food chain (Peters et al., 2014). Fig. 1.1 illustrates the nanomaterials classification based on their chemical composition and application in the food industry: inorganic, organic, and composite materials (Khan, 2020). As far as this concept is an accepted practice that humans have been eating nano-particles made during digestion of natural foods like milk for ages, and that such nanoparticles are considered relatively safe, in this book we focus on Engineered Nanoscale Materials (ENMs) because related research demonstrates potential adverse effects on the human body originating from some of them.

Wide-range applications of nanoparticles in the food industry raise this question "is manufactured nano-food safe?" To answer this recently growing concern, first, we need to identify how nano-based products affect our body when creating conventional or toxic materials, and how we can minimize the hazards of their possible toxicity features and regulatory principles so

Safety and Regulatory Issues of Nanoencapsulated Food Ingredients
http://dx.doi.org/10.1016/B978-0-12-815725-1.00001-X

Figure 1.1 ***Nanomaterial classification based on their chemical position.***

as to ultimately ensure that the nano-food products released to the market are safe. "You are what you eat" is a famous quote, highlighting the critical position of food-safety both in the eyes of the public and the organizations supervising food products.

Downsizing the bulk form of food has several benefits (e.g., higher bioavailability, targeted delivery). Moreover, safe production practices and appropriate consumption instructions can ensure the safety of nanosized-digested particles (Chun, 2009; Assadpour, Dima, & Jafari, 2020).

1.2 Nano-food in the human body: exposure and toxicity routs

Bioavailability is one of the most important considerations for the majority of health-promoting nutrients when assessing their degree of efficacy on the human body. It comprises: (1) releasing and mixing of food particles with body fluids (bioaccessibility); (2) letting the favorable parts into the mucus layer (absorption); and (3) chemical as well as physical changes of the food materials starting in the mouth and ending in different parts of the body (transforming) (McClements, 2019). Unfortunately, low bioavailability of many substances is a big barrier in the food and medicine industry due to factors such as size, shape (to be discussed later in this chapter); gastrointestinal tract (GIT) effects on things such as stomach acidity and/or enzymes, human conditions such as old/young ages as well as illnesses (e.g., Crohn's disease). Nanotechnology is seen as one of the most novel sciences applied in combating food scarcity through nanoencapsulation, nanodelivery as well as other alternative ever-growing methods (Assadpour et al., 2020). As an example, iron and zinc shortage are reported as the most prevalent nutrition deficiency and the least bioavailable, respectively all over the world (World Health, 2000). Nano-iron and the application of nanotechnology in zinc delivery are believed to be effective in large populations, especially developing countries, where nutrient deficiencies threaten human life (Chaudhry & Castle, 2011).

Nanoparticles arrive in our body through five different paths: (1) blood circulation, (2) inhalation, (3) dermal absorption, (4) ingestion, and (5) implantation; nevertheless, they may penetrate through other parts as well, such as nervous system, urinary tract, liver, spleen (Buzea, Pacheco, & Robbie, 2007a). This chapter aims to study "nano-food"; as such, we will discuss the digestion route of downsized particles in this section. Nano-food is found to enter the GIT not only via swallowing, but also through inhaling nanosized particles (De Matteis, 2017).

Compared to stand-alone versus combination scenarios, in living systems, originality or final effect does not just depend on particular components as each plays a different role and presents distinct effects upon presence in the system. This is caused by the influence of each compound unit on other parts. Safety/toxicity survey of nano-based products without considering what the human body will do with such particles is a big mistake occurring in some reviews conducted on nanosized particles in various areas (Bellmann et al., 2015). Therefore, downsized particles should be discussed in view of the GIT effect and the food matrix (McClements & Xiao, 2017).

Ingestion of nanomaterials could occur through food/drug consumption and packaging as well as cosmetics utilization. Nanomaterials reduction from the final food-products is a formidable task as they often have been found in the food-chain, and as such, continue to remain in the human body thereafter (Dasgupta & Ranjan, 2018). Foods including nanoparticles or bulk food, which eventually turns into nano-sized particles, start the GIT journey from mouth, which contains teeth, enzymes, and electrolytes all contributing to the structural/chemical changes in the food. Long esophagus (~25 cm length) is the next pathway in which foods are mixed up and downsized through the peristalsis movement before they are delivered into the stomach. High acidic environment (pH~ 2), powerful digestive enzymes (lipase and pepsin) and electrolytes cause many chemical transformations in the stomach. Digestive juices from liver, pancreas, gallbladder, and intestine wall meet in duodenum—the final part of the upper GIT—to make downsized food ready for absorption by intestine if it has not already (Blijdenstein, van der Linden, van Vliet, & van Aken, 2004). GIT microbiota impact nano-food metabolism and absorption by generating products (e.g., enzymes) that can change nanoparticle features (Clarke et al., 2014). Absorption of tiny food products is done while passing through the mucus layer and showing up in the blood circulation to feed all parts of the body.

1.2.1 Absorption-factors of food nanoparticles

Many studies have investigated the influence of nanoparticle properties on the process of passing through the intestine wall in the form of absorption. These studies show:

Smaller goes faster: The size of particles is one of the determining factors in downsized food absorption. Higher surface area, improved water dispersibility, and faster penetration provides opportunities more in favor of nanoparticles than microparticles to move and reach mucus surface (Kulkarni & Feng, 2013; Katouzian & Jafari, 2016; Jafari & McClements, 2017). As an example in nano-scale, absorption of polystyrene microspheres was faster and larger when it was fed to cats in sizes of 50 nm versus 100 nm; moreover,

300 nm particles did not exist in this animal's blood—hence zero absorption (Jani, Halbert, Langridge, & Florence, 1990). In some studies, size was not seen to have any effects; however, in the majority of such studies, particle size usually affects transferring of the nano-food into the circulatory system (Magnuson & Bouwmeester, 2011).

Anionic (-) particle more absorption: Surface charge of nanomaterial is another important factor when passing the mucus layer. GIT usually has (+) surface charge in the mucus layer; thus, much more likely to transport particles with anionic (−) surface charge in contrast to cationic (+) ones. For instance, polystyrene latex nanomaterials adhere to the mucosal surface rapidly; therefore, they cannot enter the epithelial tissue, and as a result, they usually disappear over time (Szentkuti, 1997).

Dissolution/solubility rate, interfacial chemistry, biochemical barriers (like influx transporter, enzymes), and presence/absences of a ligand are other factors impacting absorption rate, where more research needs to be done in order to measure which one exerts a higher influence rate (Martins et al., 2015).

Illnesses such as ulcerative colitis, Crohn's disease, and colon cancer affect the regular GIT absorption process, which results in constant presence of nanoparticles in the patient's colon tissue as opposed to its absence in healthy people. The bulk form of particles mentioned below have been considered as toxin-free; nevertheless, various nano-forms of the following particles were found through spectroscopy analysis of pathological human colon tissue having cancer or Crohn's disease: calcium, carbon, ceramic, gypsum, silicon, sulfur, stainless steel, filosilicates, zirconium, and silver. Absence of debris within the normal colon tissue of the control sample suggests a possible link between accumulation of particular debris (known as biocompatibles) within the pathologic tissue on the one hand and the underlying disease in the study on the other (Gatti, 2004). Also, several controversial studies indicate that nanoparticles might be the cause of Crohn's disease on account of abnormal response to nanosized dietary intake followed by tissue inflammation (Powell et al., 1996; Lomer et al., 2004; Buzea et al., 2007; Riasat, Guangjun, & Tariq, 2016).

As indicated, GIT has the potential to influence nano-food properties such as dimension, aggregation state, composition, and surface characterization, which can, in turn, lead to conversions of nano-particle safety/toxicity properties (des Rieux, Fievez, Garinot, Schneider, & Préat, 2006).

Toxicokinetics of nano-food, which is the study of how particles get into and what changes do they receive from the body as discussed briefly earlier.

This subject studies such topics as the five main routes of entry (dermal, intravenous, inhalation, ingestion, implantation); the GIT journey of nano-food; the impact of organs on the chemical/structural composition of these particles; absorption; and factors influencing the entrance of downsized particles into the blood.

1.2.2 Toxic characteristics of food nanoparticles

In this section, *toxicodynamic* of nano-food, otherwise known as the interactions and biological effects of such particles on the human body will be discussed. For this purpose, first, toxic features of nanoparticles will be briefly described; second, routs of toxicology along food nanoparticles will be discussed. For the past two decades, many studies have investigated several factors contributing to nano-food toxicology; nevertheless, there is still a long way before we discover how each factor specifically reacts in different mixtures so that we can find a suitable approach to particular nanomaterial toxicological factors.

1.2.2.1 Size

Needless to say reduction in nanoparticle size results in higher surface-to-volume ratio, which could in turn increase penetration into and interaction with the target organs; in addition, more surface molecules lead to more toxicity. An important example of this occurs in inhaled TiO_2, which induces pulmonary responses like inflammation (Oberdorster et al., 1998; Tran et al., 2000), which this particle finally found in organs such as kidney, liver, brain, spleen, and GIT (Borm & Kreyling, 2004; Elder & Oberdörster, 2006; Mills et al., 2006; Xu, Liu, Bai, & Chen, 2010). Several studies indicate that the risen inflammatory response is size-dependent (Kyung et al., 2009), and that tumor rate increases in ultrafine *vs.* fine TiO_2 (Höhr et al., 2002). However, a survey comparing the toxicity of TiO_2 fine particles versus nanoscale materials (nanorod, nanodots) declined the size and surface area relation with toxicity activation of TiO_2 (Warheit, Webb, Sayes, Colvin, & Reed, 2006). It is worth noting that TiO_2 particles did not appear to be toxic during in vivo situations whereas at the very same size, their toxic effect was clear in renal cells (i.e., in vitro situations) (L'azou et al., 2008). In the case of GIT, TiO_2 accumulates in the small intestine (Peyer's patches), induces an immunity system response within local cells, which results in aggravation of inflammation responses in earlier inflamed intestine (Powell, Faria, Thomas-McKay, & Pele, 2010). The reviewed study entitled "The role of titanium dioxide in the gut" emphasizes that both accumulation and size reduction of TiO_2 could contribute to inflammatory bowel diseases (Achtschin & Sipahi, 2017). A recent in vitro study (Bettencourt et al., 2020) on the effects of several TiO_2 nanomaterials (NM-103, NM-103, NM-105) on GIT cells reports more obvious toxicity in NM-105 form compared to undigested nanomaterial; however, with subtle changes within the last particle.

1.2.2.2 Surface structure/charge

Diverse structures (both organic and inorganic) are used to significantly improve biocompatibility, absorption and activity of nanoparticles, leading to changes in safety/toxicity characteristics through these structures. For example, iron oxide nanoparticles *coated* with pullulan (Pn-SPION) is proven to be non-toxic while the uncoated iron oxide nanoparticles is toxic (Gupta & Gupta, 2005).

Another key factor having a roll in nanoparticles safety is the surface charge. Evidence shows cationic (+) nanoparticles surfaces tend to be more toxic compared with anionic (−) ones, while neutral surfaces are at the top of biocompatible properties (Xu et al., 2010). Some reasons for this include interaction-tendency from negative (−) charge cell molecules (e.g., glycosaminoglycans) with positively (+) charged particles. These higher electrostatic interactions with cells result in greater uptake, which could finally lead to cytotoxic effects (Huang, Cambre, & Lee, 2017). For instance, ZnO nanoparticles produce more cytotoxic effects in the human lung epithelial (A549) cells when positively charged than when negatively), with no difference in size and shape between the two groups (Baek et al., 2011).

1.2.2.3 Solubility

Based on several studies investigating the relationship between solubility and toxicity of nanoparticles, both solubility and insolubility could play a crucial role in the toxicity/safety of nanoparticles. Low solubility material could accumulate in GIT for a longer period of time so one can observe more toxicity activity in the human body by them (Renwick, Brown, Clouter, & Donaldson, 2004). On the other hand, (Brunner et al., 2006) exposed human/rodent cells to soluble NPs like ZnO nanoparticles in 15 ppm concentrations and higher, which resulted in the death of all cells under experiment (MSTO/CT3 cells); nevertheless, some cells still remained alive in high concentration (30 ppm) of insoluble nanoparticles. From these studies, one can understand the high impact of solubility as a determining factor in nanotoxicology studies.

1.2.2.4 Chemical composition

One of the most important factors of nanoparticles toxicity is their chemical composition (Donaldson, Stone, Tran, Kreyling, & Borm, 2004). Lung and its epithelial were inflamed/injured to a greater extent by nano/micro particles than by their larger bulks (Xia et al., 2004). A large group of surface-active substances called surfactants have diverse applications in nanotechnology. Decreasing the surface tension is the most important function of these substances. Another action of surfactants is encapsulating the dirt; therefore, the soiling would not be redeposited on surfaces (Lee, Shih, Hsu, Hung, & Lin, 2008). (Song et al., 2012) investigated the cellular toxicity of surfactants through assessing their effect on metabolic/mitochondrial activity, membrane integrity as well as the total amount of protein produced in the cell culture. Sodium lauryl ether and polyoxyethylene lauryl ether decreased the mitochondrial/cellular metabolic activity; however, they profoundly hurt the membrane. Polyethylene glycol, propylene glycol, and monoethylene glycol had no cellular toxicity even in high concentration (100 ppm). Polyoxyethylene tallow amine, polyoxyethylene lauryl amine ether and polyoxypropylene glycol block copolymer exhibited severe toxicity. The cytotoxic dose of the surfactants named was 3.13–100 ppm. These studies show that in order to choose optimal composition for formulation purposes, toxicity and their concentration, which are distinct in chemical composition, should be considered.

1.2.2.5 Shape

The shape could also influence the toxicity grade of food nanoparticles in different activity/inhibitory mechanisms. Three particular shapes of TiO_2 generate different toxicity levels in this order: Amorphous> anatase > rutile (Jiang et al., 2008; Wu, Sun, & Xue, 2010). The possible cause of this is that high surface defects on amorphous TiO_2 leads to active sites that are more capable of producing reactive oxygen site (ROS). Fe_2O_3 is another example of nanoparticles with possible toxicity, which has been studied in its two shapes: rod and sphere. Rod-shape Fe_2O_3 developed a higher level of cytotoxic response than the sphered-shape one in a macrophage cell line of murine with these increased activity/inhibitory responses: leakage of lactate dehydrogenase (LDH), inflammation, developing ROS, necrosis (Lee et al., 2014). Octahedron and cubic shapes of CeO_2 had less-toxic effects on RAW 264.7 cells, despite the rod-shaped forms, which developed significant LDH, tumor necrosis factor alpha-TNF, and caused toxic-effect (Forest et al., 2017).

1.2.3 Toxic mechanisms of food nanoparticles

1.2.3.1 Reactive oxygen species (ROS)

ROSs are the group of vital signaling molecules which could be both protective and, at the same time, injurious depending on the amount of ROS-production and targeted cell type (normal or cancerous). The balanced level of ROS is generally produced by mitochondrial electron transport-chain, peroxisomes, inflammatory responses, and other external agents. Excessive ROS-production could participate in cell death (apoptosis), which is why novel studies find this mechanism helpful for combating cancerous cells (Chattopadhyay et al., 2014; Guo et al., 2015; Perillo et al., 2020). Nevertheless, by generating oxidative stress, the extremely produced ROS is one of the most common toxicity pathways of nanoparticles (Jain, Ranjan, Dasgupta, & Ramalingam, 2018). The different nanoparticles toxicity characteristics discussed earlier (Section 2.2) might influence the rate of ROS. These include size, surface characteristics, chemical composition, as well as cellular activity (Jafari, Katouzian, & Akhavan, 2017b). Electron-presence in nanoparticles boundaries is also a key element in generating ROS, which could contribute to certain intracellular-damage ways like: protein denaturation, peroxidation of lipid, strands breaking, DNA/mitochondrial damage (Jafari, Katouzian, & Akhavan, 2017c; Jain et al., 2018). Several nanoparticles -ROS studies show toxicity particularly in inorganic ones such as: silver nanoparticles (Stensberg et al., 2011; Budama-Kilinc et al., 2018; Mao, Chen, Wang, & Yan, 2018; Ferdous & Nemmar, 2020), ZnO nanoparticles (Fu, Xia, Hwang, Ray, & Yu, 2014; Singh, Cheng, & Singh, 2020), silica nanoparticles (Lehman et al., 2016; Murugadoss et al., 2017; Kustov et al., 2018), and cobalt oxide nanoparticles (Chattopadhyay et al., 2015). As mentioned earlier, most of the nanoparticles which tend to be toxic due to oxidative stress are inorganic as organic nanoparticles have been digested before absorption; however, the toxicity scope of both categories of nanoparticles, particularly the inorganic ones, must be examined in more details.

1.2.3.2 Immune response

As several studies have shown (Liu, Hardie, Zhang, & Rotello, 2017; Muhammad et al., 2020), in contrast to the desirable immune-modulations that nanoparticles induce like boosting vaccine efficacy, antitumor properties-undesirable immune responses of nanoparticles are significant too. These responses could be categorized into two mechanisms: *immunostimulation* and *immunosuppression*, wherein hypersensitivity reactions, anaphylaxis, and inflammation are examples of the first, while decrease in immune response to infection or diseases, suppression of bone marrow and thymus are instances of the second (Zolnik, González-Fernández, Sadrieh, & Dobrovolskaia, 2010; Jafari et al., 2017b). The solution stabilizers mostly consist of amphiphilic molecules (having both hydrophobic and hydrophilic parts) called emulsifiers such as polycinylpyrrolidone, sodium docusate, and polyethylene glycol (PEG). Nowadays, nanosized oral-delivery drugs include emulsifiers to enhance the bioavailability of water-insoluble compounds (Merisko-Liversidge & Liversidge, 2011). The human body recognizes many emulsifiers as foreign-body leading to *hypersensitivity reactions*, which can be serious or even fatal. Moreover, depending on the exposure time of the nanoparticles, the immune system might respond differently (Őrfi & Szebeni, 2016); for example, i.v. injection, which entails sudden exposure, as opposed to oral administration, which requires gradual exposure. Moreover, several studies investigated immunotoxicity of diverse nanoparticles such as latex nanomaterial (Inoue, Takano, Yanagisawa, Koike, & Shimada, 2009), gold nanomaterial (Cho, Kim, Han, Son, & Jeong, 2009; Brandenberger et al., 2010), TiO_2 (Morishige et al., 2010; Müller et al., 2010), and cerium oxide (Xia et al., 2008) through different mechanisms such as apoptosis, acute inflammation, necrosis, etc.

1.2.3.3 Protein corona

There is a dynamic interplay between circulatory system biomolecules (lipid, DNA, protein) and nanoparticles in the human veins. Protein corona, named as a group of proteins, attaches to nanoparticles, changing so many of their physicochemical characteristics such as size, aggregation state, surface charge, etc., which leads to modifying pharmacokinetic responses (Mahmoudi, Bertrand, Zope, & Farokhzad, 2016; Jafari et al., 2017c). The critical role of protein corona in introducing nanoparticles to the immune system was established recently. This could lead to immune responses discussed in the former Section (2.3.2) called immunological toxicity. Furthermore, quick clearance of nanoparticles by organs like liver, spleen, GIT through phagocyte cells leads to inappropriate nanoparticles response (Rampado, Crotti, Caliceti, Pucciarelli, & Agostini, 2020). For example, surface coating of gold nanoparticles with poly(acrylic) acid induces adsorbed fibrinogen to unfold, leading to interaction with leukocyte receptor (MAC-1) triggering inflammation (Deng, Liang, Monteiro, Toth, & Minchin, 2011).

Interestingly, protein corona could decrease nanoparticles toxicity by coating potential toxic particles, which results in masking interactions with negatively charged cells (Corbo et al., 2016). For instance, zinc oxide (ZnO) nanoparticles coated with silica (Ramasamy, Das, An, & Yi, 2014), or with a protein corona (Yin et al., 2015) reduce their toxicity through human skin fibroblasts and hepatocarcinoma cells, respectively by preventing surface contact between nanoparticles and cells.

1.2.3.4 Genotoxicity

Unlike the bulk form of food, nanosized particles could easily enter the cell and interact with different parts like mitochondria and/or chromosome, which may later on cause genotoxicity (Jafari, 2017). Such genotoxicity could be primary when nanoparticles enter the cells and influence their toxicity effect directly, developing ROS, inhibiting the chromosome repair, interrupting proteins, and regulating cycle checkpoints. It could also be secondary as caused by inflammation due to high level of ROS (Kumar & Dhawan, 2013; Patel et al., 2016). DNA hypermethylation, histone alteration, miRNAs (a gene expression modifiers) alterations are the main examples of nanoparticles genotoxicity effects, which are comprehensively discussed by (Jafari, 2017).

Boudreau et al. (2016) discovered the average intake of silver nanoparticles (AgNPs) through human diet is 0.4–27 μg/day, while another recent study (Li, Tang, & Xue, 2019) indicates the daily silver intake could be as high as 90 μg. A recent review article (Rodriguez-Garraus, Azqueta, Vettorazzi, & López de Cerain, 2020) collected AgNPs genotoxicity levels from 43 articles all using high-established standards, and concluded that AgNPs generate genotoxicity at all DNA damage levels (i.e., gene mutations, primary DNA damage) in both in-vitro/in-vivo studies, with high positive results in in-vitro assays. Observations conclude that size and coating have a genotoxic-effect on both in-vitro/in-vivo situations; however, complete genotoxicological properties of AgNPs are needed for more precision.

1.2.3.5 GIT normal function impediment

Foods in nano-scale containing high surface area have the potential to absorb every surface-active biomolecule present in GIT, which results in the modification of their function/structure. For instance, globular proteins are denatured after adsorption on surfaces of nanoparticles as a result of optimal environmental changes. Therefore lipid, carbohydrate, protein digestion might be reduced due to the high level of nano-foods (Bergin & Witzmann, 2013; Liu et al., 2017). Some studies (McClements & Xiao, 2017) show a smaller effect of this mechanism for these reasons (1) aggregation of inorganic particles may reduce their exposed surface area; (2) less adsorption of a molecule such as a lipid to nanoparticles than target adsorption structures. However, *GIT microbiome disruption by nanoparticles* is generally an accepted theory. About 100 trillion bacteria from 1000 diverse species living in GIT known as "gut microbiome" have many beneficial health effects in the human body such as: protection against intestinal infections, contributing to developing vitamin k, and serving as a probiotic (Shreiner, Kao, & Young, 2015). Consequently, any changes in the balance of these microbiomes could cause health problems. Several nanoparticles have been studied in this way, resulting in the possible negative effects on the gut microbiome composition (Bouwmeester, van der Zande, & Jepson, 2018).

It is worth noting that the adsorption of nanoparticles to bacteria reduces the toxicity of nanomaterials in some situations. Nanoparticles–bacteria complex impact on toxicity was reviewed in silica nanoparticles exposition to human gastric epithelial cells. The toxicity of silica nanoparticles and bacterial complexes was significantly reduced compared with the silica nanoparticles alone (Siemer et al., 2018). Further studies should be conducted on different nano-foods with diverse concentrations through each biomolecule to explain in which concentration each specific surface area of nanoparticles could be harmful to biomolecules.

Therefore, it can be concluded for this section that:

1. Changes due to GIT effect on nano-food as well as studies reviewed here, convince us of the importance of in vitro studies in situations similar to food-GIT systems and in vivo evaluations of nanoparticles toxicity.
2. As far as different subtypes of each nanoparticles have particular properties, the toxicity of each subtype should be assessed in several conditions to find which one has more safety/toxicity potential.
3. For each nanomaterial, toxicity properties profile such as size, aggregation, and surface characterization should be evaluated to distinguish the influence of each property on safety/toxicity of nanoparticles

1.3 Evaluating nano-food safety

Nanoparticles related risks of identification/evaluation have a vital influence and can include safety, health, environmental, societal and regulatory issues. For these possible harmful effects of nano-food, legal agencies in Europe and United States have introduced multiple route maps and guidelines for risk-assessment and safety-evaluation in order to safely use nanoparticles in foods (EFSA, 2011; Hamburg, 2012). Multiple criteria have been suggested for nanotoxicology assessments such as toxicity evaluation, determining transportation, and nanomaterial sustainability (Dima, Assadpour, Dima, & Jafari, 2020). Nonetheless, validation and characterization of methods need more attention for improving toxic-assessment (OECD, 2013; Amenta et al., 2015). Also, there is limited knowledge about nanotoxicology and risk assessment of nanoparticles, which creates a big barrier for legal authors to introduce a clarified evaluating route for the safety of nanoparticles (Maksimović, Omanović-Mikličanin, & Badnjević, 2019).

Current toxic methodologies utilized for risk assessment of conventional and bulk forms of material need to be modified in order to evaluate foods in the nano-scale form. Considering the issue of assessment of nanomaterials with diverse properties like chemical composition and presenting these particles in food matrix, makes toxic-assessment much more complicated compared to bulk form of food. Thus, *Case-by-case* toxic assessment besides multiple tests in one case by *different situations* is crucial for nanoparticles. Chapters 3 and 4 of the current book describe multiple in-vivo/in-vitro studies carried out on the safety/toxicity assessment of nano-foods. One of the best approaches for *preventing* nano-food potential hazards before the occurrence of the risk is the safety-by-design method, with the aim of reducing adverse effects during design and manufacturing of nanoparticle incorporated products (Iavicoli, Leso, Beezhold, & Shvedova, 2017). This approach first collects all data and other requirements about nano-food toxicity, stability, and toxic/safety characteristics of nanoparticles. Second, risk assessment of nanomaterials is performed on all stages of their life cycle including: raw material, process stage, manufacturing, usage, transportation, and last step disposal or recycling (Maksimović et al., 2019).

When nanoparticles are proposed to be toxic-analyzed individually, first they should be separated from the food matrix carefully with a minimum amount of treatment. High reactivity of nanoparticles may change their shape, surface properties, and even composition, if environmental conditions are not provided so well (Szakal et al., 2014). The next step is to analyze the potential toxicity via several methods. Spectroscopy, chromatography, and high-resolution microscopy are the main powerful tools for this purpose (Fig. 1.2). Both organic (proteins, carbohydrates, lipids) and inorganic (metal) nanoparticles can be identified quantitatively and qualitatively by these methods (Jafari, Faridi Esfanjani, Katouzian, & Assadpour, 2017a).

Electron microscopy analyzes potential morphologic toxic features such as shape, size, and surface properties. Other useful instruments for nano-food assessment are mass spectroscopy, liquid chromatography, and gas chromatography. Specific hazard analysis of nano-food effects could be

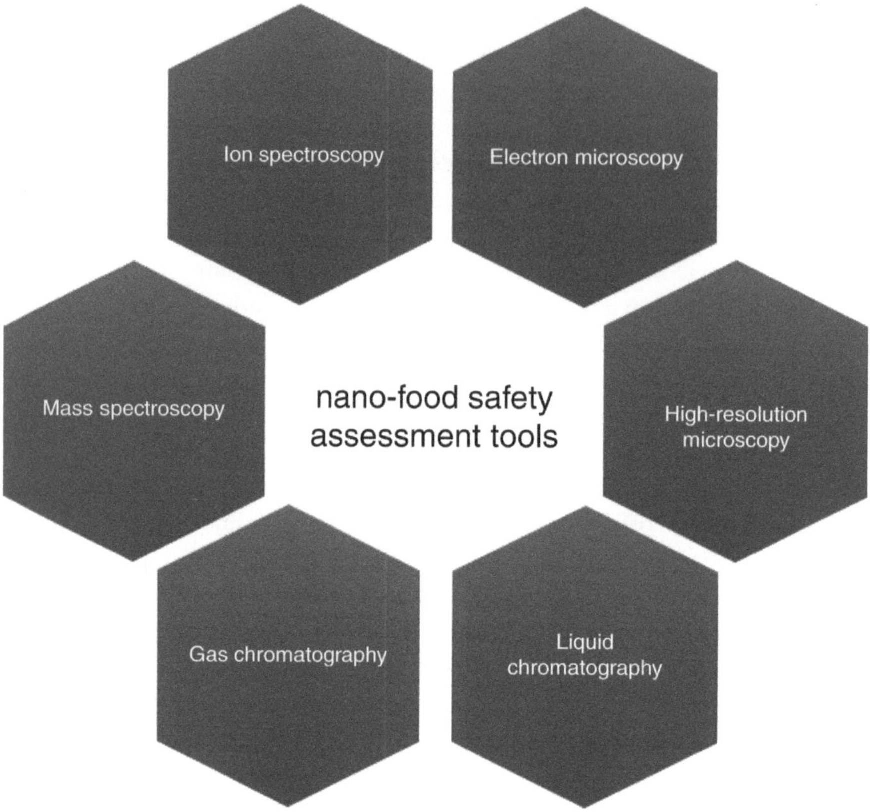

Figure 1.2 ***Nano-food safety assessment tools.***

additionally tested with particular tests. As noted in Section 2.3.4, DNA could be a target of nanoparticles cellular toxicity. DNA damage caused by nanoparticles could be tested in several cell cultures like thymidine kinase (TK) assay, hypoxanthine phosphoribosyltransferase (HPRT) assay, phosphotidylinositol glycan, as well as chromosomal aberration test (He, Yuan, Liu, & Zhang, 2008; Chen et al., 2014). Chinese hamster cells were the target of the study investigating HPRT forward mutation (Finette, Kendall, & Vacek, 2002). Phosphoribosylation of guanine/hypoxanthine is performed by HPRT gen, which exists in the X chromosome of hamster cells. Media having 6-thioguanine, which is guanines analogous but could be toxic in nature, was the grown culture of cells. While DNA is busy replicating via the HPRT enzyme, 6-thioguanine combines with the DNA duplex, which results in the cell death. Nevertheless, if nanoparticles-exposure leads to the HPRT gen mutation, deathly-pathway does not occur because

6-thioguanine couldn't combine with DNA duplex. This deleterious mutation caused by nanoparticles was visible through a number of colonies on the medium; however, some nanoparticles have been negatively responded (Chen et al., 2014; Dasgupta & Ranjan, 2018).

True and more evaluation of nano-food toxic/safety potentials is not only one of the regulatory powerful arms, but also leads to consumer acceptance of nanoproducts.

1.4 Public and food organizations concerned with the potential toxicity of nano-food

Novel food-technologies such as cultured meat, gen technology, food irradiation, and nanotechnology have revolutionized the food-industry in its various sectors like packaging, bioavailability, and sustainability (Siegrist & Hartmann, 2020). Two important consumer orientations trending nowadays in the market are (1) rising demand for natural-foods such as organic, traditional foods, and local products and (2) emergence of modern-foods like functional foods, nutrient-rich, and health foods (e.g., low-sodium/calorie foods) (Vidigal et al., 2015). Rising concerns about novel-food poses this question to public and food-organizations: "is the novel-food safe and beneficial?" and "how should we improve novel-foods safety along with its benefits?"

The rapidly growing application of nanotechnology, which holds a very superior position in the food-technology, can influence such most important food values as security, taste, shelf life, and nutrition in consumers' view as discovered in studies like (Yang & Hobbs, 2020). However, there is little knowledge about public perception, attitude and acceptance by consumers toward the usage of nanotechnology particularly in the food industry (Sanguansri & Augustin, 2006, Gruère, 2012). People are concerned, or sometimes have food neophobia, about novel technologies utilization in the food industry, of which "nano-food" is an example (Matin et al., 2012).

A survey (Currall, King, Lane, Madera, & Turner, 2006) investigated the benefits/risks perception of nanotechnology as well as 43 other food technologies among 503 responders to a national telephone Questioner. Their study concluded that nanotechnology is viewed as an average benefit, and that it has a moderate risk level; however, it was perceived as less risky with more benefits compared to other growing technologies such as human genetic engineering, genetically modified organisms, pesticides, and chemical disinfectants. At present, consumers have a wide range of novel-food choices to select from. This includes genetically modified (GM) foods, irradiated food, and nano-food (Rollin, Kennedy, & Wills, 2011). Extensive research has proven that food sustainability improves through diverse technologies applied in the food sector; however, a study found higher percentage of food technology-neophobia among consumers concerned about food-sustainability compared to consumers with less concern about this factor (Cavaliere & Ventura, 2018). Lack of enough *facts-Release* via media or other community networks could be a reason for this. Despite the innovation that some technologies have brought to food products, their feedback from consumers was negative. For instance, the food irradiation and preservation technique is considered safe and effective by the scientific community even though it is not accepted generally by consumers (Ronteltap, Van Trijp, Renes, & Frewer, 2007). The importance of multiple measuring studies of consumer attitude, finding positive/negative aspects of novel technology, covering the potential holes of this technology before broadcasting, it is more sensible than risking a negative response from the public.

Some studies demonstrate nanotechnology has a higher rate of acceptance among other technologies. As an example, Yue, Zhao, and Kuzma (2015) surveyed 1117 US consumers and their preferences of technology use in food products. The results indicated lower negative reactions toward nano-food than GM food. This negative reaction is not limited to high-income countries. Studies have revealed similar reactions to neophobia of acceptance in some low-income countries like Uganda (De Steur, Odongo, & Gellynck, 2016) as well. However, people's attitude toward nano-food is different from country to country. An experiment (Bieberstein, Roosen, Marette, Blanchemanche, & Vandermoere, 2012) measuring consumers willingness to pay (WTP) for food nanotechnology in Germany and France revealed (1) high percentage of acceptance of nano-food among consumers of both countries, (2) nano-packaging and nano-fortification were less attractive among French and German consumers. This study shows the importance of several determining factors like media, culture, lifestyle, as well as regulatory issues, which could be different from one country to another; thus, investigations of nano-food acceptance should not only be worldwide, they should also be society specific.

Another study of a European country (Sodano, Gorgitano Maria, Verneau, & Vitale Cosimo, 2016) performed in Italy shows a significant reluctance of people toward nano-food in this society. Assessment of the economic model with regard to the six nano-food studies in this article revealed that WTP is affected through (1) *risks and benefits* appreciation of product; (2) the level of food novel-technology neophobia; (3) the level of *trust* in the food industry. Benefits, trust, health, organicity are particular concepts that influence consumer's attitude toward foods produced by nanotechnology in yet another study (Stampfli, Siegrist, & Kastenholz, 2010); moreover, the results show nanotechnology utilization in food and food-packaging have different acceptance rates, with higher ranges for the latter. Somehow near to this is the result of a *review* study (Giles, Kuznesof, Clark, Hubbard, & Frewer, 2015), which maintains that application of nanotechnology in agri-food industry directly has received less consumer interest than in novel food packaging.

1.4.1 Factors affecting consumer acceptance of food nanotechnology

The main influencing factors in consumer acceptance toward nano-food are introduced in several studies, which will be presented further.

1.4.1.1 Risks and benefits of nano-food

Consideration of risks and benefits is one of the most crucial factors influencing people's and experts' choice to pay for nano-food or to subsequently judge its quality (Giles et al., 2015). The major risks and benefits of nano-food identified in multiple studies include: *safety*, freshness, enhanced shelf-life, reducing adverse properties (e.g., high calories/sodium), improved taste, supplemental effects (Giles et al., 2015; Zhao, Yue, & Kuzma, 2020). For instance, increasing foods shelf life through novel technology is proven to be effective with regard to the consumer attitude toward novel-food (Hosseini, Tajiani, & Jafari, 2019); furthermore, nano-packaging has the highest rank of food packaging when it comes to boosting shelf life (Azeredo, 2009).

1.4.1.2 Socio-economic characteristics of consumers

As noted earlier, several studies indicate that certain populations having close characteristics like culture, beliefs, ethnicity and/or religious rituals might accept nano-food more than others (Mintz & Du Bois, 2002; Coveney, 2000). A survey even discovered acceptance differences based on gender and age, which revealed that females are more interested in health-related benefits of nano-food in comparison to males; and that younger customers have better knowledge, and as such, more acceptance of nanotechnology in food (McCarron, 2016).

1.4.1.3 Information and knowledge about nano-food

Generally, many studies demonstrate consumers with high level of knowledge tend to be less food technology neophobic compared to consumers having less knowledge (Cavaliere & Ventura, 2018). Limited nutrition information, incorrect perceptions about nano-food safety/toxicity or environmental impact and negative image of food technology are the critical barriers to marketing nano-food worldwide (Wardle, Parmenter, & Waller, 2000; Satterfield, Kandlikar, Beaudrie, Conti, & Harthorn, 2009). Additionally, the presence of the "hungry for information about nano-food" factor among the vast majority of consumers results in the urge for knowledge broadcasting via several means, such as media, labeling, community devices (Brown & Kuzma, 2013).

1.4.1.4 Marketing and advertising

There is no doubt about the *power* of marketing/advertising especially through media, which can even incorporate a lie in their coverage of news! Marketing has a critical role in three important dimensions of nano-food acceptance: knowledge, encouragement and trust. As far as studies have shown, application of nanotechnology in the food industry was not familiar to most people (Gaskell, Eyck, Jackson, & Veltri, 2005), transferring knowledge is vital to nanotechnology and the food industry. In order to encourage people to pay for nano-food, marketing could have a critical role in uniting scientists and people by first collecting impressive articles about different aspects of nano-food like safety/toxicity, health-benefits, targeted nano-food groups; second, presenting the studies findings in the best possible way to the customer in several broadcasting-tools; and third collecting the responses from different groups of consumers, which surely is of great value to the food industry and nanotechnology. Fig. 1.3 demonstrates how market encouragement could lead to a better product–hence the best advertising tool for novel products. More encouragement leads to more positive attitude toward nanoproducts, which then results in more WTP. Money rotation between the food industry and consumer causes more public investigations into products, which will ultimately lead to higher quality products in all aspects of *safety*, price, health-benefit, etc. Beside the *knowledge* and *encouragement* that marketing could provide in the nano-food industry, *trust* is another decisive acceptance factor in which marketing plays a key role (Zhao et al., 2020). Trust is also known as a critical WTP factor for people having less knowledge about novel food technology (Siegrist & Cvetkovich, 2000).

Figure 1.3 ***Marketing circle: how marketing-encouragement could lead to a better product.***

1.4.1.5 Appropriate legislation

With the rapidly growing influence of nanotechnology on the food industry and its novel product nano-food, and in view of the public concern about the safety of novel food, food organizations show their concern by establishing several definitions and regulations to promote nano-food safety and quality. Also, governments could help nanotechnology and the food industry with good regulations, which in turn can result in the progression and propagation of both systems (Chau & Wu, 2007). Food and health organizations of several countries have made particular legislations (Table 1.1) on nano-food, like FDA (Food and Drug Administration) in USA, EU (European Union) in some European countries, and FDO (Food and Drug Organization) and INIC (Iran Nanotechnology Initiative Council) in Iran (Jafari et al., 2017b). However, many countries do not have any particular regulations for ensuring the safety of nano-foods. Moreover, there is a long way to go before we can tap all the definitions, classifications, and regulations already introduced by different countries as far as implementation of

Table 1.1 Web links of nano-food related regulatory frameworks in different countries.

Country	Legislation authorities	Link
Australia and New Zealand	Food standards Australia New Zealand (FSANZ)	www.foodstandards.gov.au/consumer/foodtech/nanotech/Pages/default.aspx
Brazil	National agency of sanitary surveillance (ANVISA) Ministry of Agriculture, livestock, and food supply (MAPA), Ministry of health (MS)	www.portal.anvisa.gov.br/wps/portal/anvisaingls
Canada	Health Canada	www.lawslois.justice.gc.ca/eng/regulations/C.R.C.%2C_c._870/
European Union (also applies to France, Germany, Netherlands, Norway, and Poland)	European Food Safety Authority (EFSA)	www.eurlex.europa.eu/LexUriServ/LexUriServ.do?uri=CONSLEG:1997R0258:20090120:EN:PDF
Iran	The Iran Nanotechnology Initiative Council (INIC)	www.nanohealth.ir/pages/static_page.php?id=14&site=1&lang=2
Japan	Ministry of Health, Labor and Welfare	www.mhlw.go.jp/english/policy/healthmedical/food/index.html
Korea	Ministry of Food and Drug Safety (MFDS)	www.kfda.go.kr/eng/index.do?nMenuCode=61
South Africa		www.sani.org.za/
USA	Food and Drug Administration (FDA)	www.fda.gov/RegulatoryInformation/Legislation/ucm148717.htm

nano-food legislations are concerned (Linsinger et al., 2013; Babick, Mielke, Wohlleben, Weigel, & Hodoroaba, 2016; Miernicki, Hofmann, Eisenberger, von der Kammer, & Praetorius, 2019).

Generally, many principles could create opportunities for promotion of nano-food. For instance, some studies argue for mandatory labelling of nanotechnology utilization in food products; however, optimistically speaking, labeling could raise this question for a consumer: "what is nano-food?" Responding to such potential questions and transferring *knowledge* of the nano-food to consumers can be done in two ways; briefly explaining nanotechnology on the product or asking the consumer to go on the Internet to obtain more detailed information about the nano-food product.

Possible challenges facing organizations for successful regulation are (1) lack of enough studies in every nanoparticle to investigate their short-term and especially long-term harmful adverse effects on the human body; (2) the wide diversity of nanoparticles as well as the extensive nano-food complexes they form, each of which calls for specific toxicity research which could result in specific regulations; (3) diverse toxicity-characteristics and mechanisms of nanoparticles, which, as noted earlier, demand deeper and more extensive studies in every nanosized particle under different situations and/or shapes. Chapters 5–8 of the current book provide more detailed regulatory principles of nano-food in several countries.

1.5 Future directions

With regard to nanoparticle toxicity/safety, there is a broad *influencing line*, which needs to be interpreted in every point with respect to the final goal, that is, "enhancing nano-food safety." Attentions to nano-food toxicity studies have increased over the past decade although lack of studies on nano-food exposure routes, toxic characteristics (e.g., size, surface charge, solubility, shape), and diverse potential toxic mechanisms is the first point of this influencing line, which motivates scientists to design more experiments in order to obtain a wider view in the field of nanotoxicology. It is worth noting that every toxic property/mechanism could be more or less important in each nanoparticle. For example, in inorganic nanoparticles chemical composition, accumulation in tissues, and absorption properties are of paramount importance; while in organic nanoparticles, solubility, level of toxicity, and enhancing the bioavailability of other potential toxic particles are more salient. The second point of the influence line is evaluating nano-food toxicity/safety as well as assessing possible health risks in the short, mediate, and long run. Some nanoparticles safety/toxicity have been well established; however, confusion over controversial toxicity studies of nanoparticles, where some result in the nanoparticle toxicity while others claim its safety, bring scientists to highlighting possible toxicity or safety of diverse nanoparticles in *different situations*. For this purpose, improvement in evaluating instruments is essential, which could analyze food safety in the nanosized form in view of particular objects. The final critical point of the toxic influencing line is to resolve the public concern through good regulations and trust marketing. The huge current market of nano-food ($20 billion a year) can serve as an incentive for food regulatory systems to emulate their regulations and define clear guidelines that are based on ample studies in order to enhance nano-food quality/safety. Moreover, all key food regulators like WHO, FAO, FDA, EFSA, OECD better tap their regulations into a single tabulation to provide clear choices of toxicity assessment as well as legislations with the purpose of nanotechnology development in the food industry.

References

Achtschin, C. G., & Sipahi, A. M. (2017). The role of titanium dioxide in the gut. *Nutrition & Food Science, 47*(3), 432–442.

Amenta, V., Aschberger, K., Arena, M., Bouwmeester, H., Moniz, F. B., Brandhoff, P., et al. (2015). Regulatory aspects of nanotechnology in the agri/feed/food sector in EU and non-EU countries. *Regulatory Toxicology and Pharmacology, 73*(1), 463–476.

Assadpour, E., Dima, C., & Jafari, S. M. (2020). Fundamentals of food nanotechnology. In S. M. Jafari (Ed.), *Handbook of food nanotechnology* (pp. 1–35). Academic Press.

Azeredo, H. M. C. d. (2009). Nanocomposites for food packaging applications. *Food Research International, 42*(9), 1240–1253.

Babick, F., Mielke, J., Wohlleben, W., Weigel, S., & Hodoroaba, V. -D. (2016). How reliably can a material be classified as a nanomaterial? Available particle-sizing techniques at work. *Journal of Nanoparticle Research, 18*(6), 158.

Baek, M., Kim, M., Cho, H., Lee, J., Yu, J., Chung, H., et al. (2011). Factors influencing the cytotoxicity of zinc oxide nanoparticles: particle size and surface charge. *Journal of Physics: Conference Series, IOP Publishing*.

Bayda, S., Adeel, M., Tuccinardi, T., Cordani, M., & Rizzolio, F. (2019). The history of nanoscience and nanotechnology: from chemical-physical applications to nanomedicine. *Molecules (Basel, Switzerland), 25*(1), 112.

Bellmann, S., Carlander, D., Fasano, A., Momcilovic, D., Scimeca, J. A., Waldman, W. J., et al. (2015). Mammalian gastrointestinal tract parameters modulating the integrity, surface properties, and absorption of food-relevant nanomaterials. *Wiley Interdisciplinary Reviews Nanomedicine and Nanobiotechnology*, 7(5), 609–622.

Bergin, I. L., & Witzmann, F. A. (2013). Nanoparticle toxicity by the gastrointestinal route: evidence and knowledge gaps. *International Journal of Biomedical Nanoscience and Nanotechnology, 3*(1–2), doi: 10.1504/IJBNN.2013.054515.

Bettencourt, A., Gonçalves, L. M., Gramacho, A. C., Vieira, A., Rolo, D., Martins, C., Assunção, R., Alvito, P., Silva, M. J., & Louro, H. (2020). Analysis of the characteristics and cytotoxicity of titanium dioxide nanomaterials following simulated In Vitro digestion. *Nanomaterials, 10*(8), 1516.

Bieberstein, A., Roosen, J., Marette, S., Blanchemanche, S., & Vandermoere, F. (2012). Consumer choices for nano-food and nano-packaging in France and Germany. *European Review of Agricultural Economics, 40*(1), 73–94.

Blijdenstein, T. B. J., van der Linden, E., van Vliet, T., & van Aken, G. A. (2004). Scaling behavior of delayed demixing, rheology, and microstructure of emulsions flocculated

by depletion and bridging. *Langmuir: The ACS Journal of Surfaces and Colloids*, *20*(26), 11321–11328.
Borm, P. J., & Kreyling, W. (2004). Toxicological hazards of inhaled nanoparticles—potential implications for drug delivery. *Journal of Nanoscience and Nanotechnology*, *4*(5), 521–531.
Boudreau, M. D., Imam, M. S., Paredes, A. M., Bryant, M. S., Cunningham, C. K., Felton, R. P., et al. (2016). Differential effects of silver nanoparticles and silver ions on tissue accumulation, distribution, and toxicity in the Sprague Dawley rat following daily oral gavage administration for 13 weeks. *Toxicological Sciences*, *150*(1), 131–160.
Bouwmeester, H., van der Zande, M., & Jepson, M. A. (2018). Effects of food-borne nanomaterials on gastrointestinal tissues and microbiota. *Wiley Interdisciplinary Reviews: Nanomedicine and Nanobiotechnology*, *10*(1), e1481.
Brandenberger, C., Rothen-Rutishauser, B., Mühlfeld, C., Schmid, O., Ferron, G. A., Maier, K. L., et al. (2010). Effects and uptake of gold nanoparticles deposited at the air–liquid interface of a human epithelial airway model. *Toxicology and Applied Pharmacology*, *242*(1), 56–65.
Brown, J., & Kuzma, J. (2013). Hungry for Information: Public Attitudes Toward Food Nanotechnology and Labeling. *Review of Policy Research*, *30*(5), 512–548.
Brunner, T. J., Wick, P., Manser, P., Spohn, P., Grass, R. N., Limbach, L. K., et al. (2006). In vitro cytotoxicity of oxide nanoparticles: comparison to asbestos, silica, and the effect of particle solubility. *Environmental Science & Technology*, *40*(14), 4374–4381.
Budama-Kilinc, Y., Cakir-Koc, R., Zorlu, T., Ozdemir, B., Karavelioglu, Z., Egil, A. C., et al. (2018). Assessment of nano-toxicity and safety profiles of silver nanoparticles. In M. Khan (Ed.), *Silver nanoparticles-fabrication, characterization and applications* (pp. 185). London: Intechopen.
Buzea, C., Pacheco, I. I., & Robbie, K. (2007). Nanomaterials and nanoparticles: sources and toxicity. *Biointerphases*, *2*(4), Mr17–71.
Cavaliere, A., & Ventura, V. (2018). Mismatch between food sustainability and consumer acceptance toward innovation technologies among Millennial students: The case of Shelf Life Extension. *Journal of Cleaner Production*, *175*, 641–650.
Chattopadhyay, S., Dash, S. K., Mahapatra, S. K., Tripathy, S., Ghosh, T., Das, B., et al. (2014). Chitosan-modified cobalt oxide nanoparticles stimulate TNF-α-mediated apoptosis in human leukemic cells. *JBIC Journal of Biological Inorganic Chemistry*, *19*(3), 399–414.
Chattopadhyay, S., Dash, S. K., Tripathy, S., Das, B., Mandal, D., Pramanik, P., et al. (2015). Toxicity of cobalt oxide nanoparticles to normal cells; an in vitro and in vivo study. *Chemico-Biological Interactions*, *226*, 58–71.
Chau, C. -F., Wu, S. -H., & Yen, G. -C. (2007). The development of regulations for food nanotechnology. *Trends in Food Science & Technology*, *18*(5), 269–280.
Chaudhry, Q., & Castle, L. (2011). Food applications of nanotechnologies: An overview of opportunities and challenges for developing countries. *Trends in Food Science & Technology*, *22*(11), 595–603.
Chen, Z., Wang, Y., Ba, T., Li, Y., Pu, J., Chen, T., et al. (2014). Genotoxic evaluation of titanium dioxide nanoparticles in vivo and in vitro. *Toxicology Letters*, *226*(3), 314–319.
Cho, W. -S., Kim, S., Han, B. S., Son, W. C., & Jeong, J. (2009). Comparison of gene expression profiles in mice liver following intravenous injection of 4 and 100 nm-sized PEG-coated gold nanoparticles. *Toxicology Letters*, *191*(1), 96–102.
Chun, A. L. (2009). Will the public swallow nanofood? *Nature Nanotechnology*, *4*(12), 790–791.
Clarke, G., Stilling, R. M., Kennedy, P. J., Stanton, C., Cryan, J. F., & Dinan, T. G. (2014). Minireview: Gut microbiota: the neglected endocrine organ. *Molecular Endocrinology (Baltimore, Md.)*, *28*(8), 1221–1238.
Corbo, C., Molinaro, R., Parodi, A., Toledano Furman, N. E., Salvatore, F., & Tasciotti, E. (2016). The impact of nanoparticle protein corona on cytotoxicity, immunotoxicity and target drug delivery. *Nanomedicine (Lond)*, *11*(1), 81–100.

Coveney, J. (2000). Food, morals, and meaning: The pleasure and anxiety of eating, Psychology Press.

Currall, S. C., King, E. B., Lane, N., Madera, J., & Turner, S. (2006). What drives public acceptance of nanotechnology? *Nature Nanotechnology*, *1*(3), 153–155.

Dasgupta, N., & Ranjan, S. (2018). Nano-food toxicity and regulations. In N. Dasgupta, & S. Ranjan (Eds.), *An introduction to food grade nanoemulsions* (pp. 151–179). Singapore: Springer.

De Matteis, V. (2017). Exposure to inorganic nanoparticles: routes of entry, immune response, biodistribution and in vitro/in vivo toxicity evaluation. *Toxics*, *5*(4), 29.

De Steur, H., Odongo, W., & Gellynck, X. (2016). Applying the food technology neophobia scale in a developing country context. A case-study on processed matooke (cooking banana) flour in Central Uganda. *Appetite*, *96*, 391–398.

Deng, Z. J., Liang, M., Monteiro, M., Toth, I., & Minchin, R. F. (2011). Nanoparticle-induced unfolding of fibrinogen promotes Mac-1 receptor activation and inflammation. *Nature Nanotechnology*, *6*(1), 39–44.

des Rieux, A., Fievez, V., Garinot, M., Schneider, Y. -J., & Préat, V. (2006). Nanoparticles as potential oral delivery systems of proteins and vaccines: A mechanistic approach. *Journal of Controlled Release*, *116*(1), 1–27.

Dima, C., Assadpour, E., Dima, S., & Jafari, S. M. (2020). 15 - Characterization and analysis of nanomaterials in foods. In S. M. Jafari (Ed.), *Handbook of Food Nanotechnology* (pp. 577–653). Academic Press.

Donaldson, K., Stone, V., Tran, C., Kreyling, W., & Borm, P. J. (2004). *Nanotoxicology*. BMJ Publishing Group Ltd.

EFSA (2011). Guidance on the risk assessment of the application of nanoscience and nanotechnologies in the food and feed chain. *EFSA Journal*, *9*(5), 2140.

Elder, A., & Oberdörster, G. (2006). Translocation and effects of ultrafine particles outside of the lung. *Clinics in Occupational and Environmental Medicine*, *5*(4), 785–796.

Ferdous, Z., & Nemmar, A. (2020). Health impact of silver nanoparticles: a review of the biodistribution and toxicity following various routes of exposure. *International Journal of Molecular Sciences*, *21*(7), 2375.

Finette, B. A., Kendall, H., & Vacek, P. M. (2002). Mutational spectral analysis at the HPRT locus in healthy children. *Mutation Research/Fundamental and Molecular Mechanisms of Mutagenesis*, *505*(1–2), 27–41.

Forest, V., Leclerc, L., Hochepied, J. -F., Trouvé, A., Sarry, G., & Pourchez, J. (2017). Impact of cerium oxide nanoparticles shape on their in vitro cellular toxicity. *Toxicology in Vitro*, *38*, 136–141.

Fu, P. P., Xia, Q., Hwang, H. -M., Ray, P. C., & Yu, H. (2014). Mechanisms of nanotoxicity: Generation of reactive oxygen species. *Journal of Food and Drug Analysis*, *22*(1), 64–75.

Gaskell, G., Eyck, T. T., Jackson, J., & Veltri, G. (2005). Imagining nanotechnology: cultural support for technological innovation in Europe and the United States. *Public Understanding of Science*, *14*(1), 81–90.

Gatti, A. M. (2004). Biocompatibility of micro-and nano-particles in the colon. Part II. *Biomaterials*, *25*(3), 385–392.

Giles, E. L., Kuznesof, S., Clark, B., Hubbard, C., & Frewer, L. J. (2015). Consumer acceptance of and willingness to pay for food nanotechnology: a systematic review. *Journal of Nanoparticle Research*, *17*(12), 467.

Gruère, G. P. (2012). Implications of nanotechnology growth in food and agriculture in OECD countries. *Food Policy*, *37*(2), 191–198.

Guo, C., Xia, Y., Niu, P., Jiang, L., Duan, J., Yu, Y., Li, Y. et al. (2015). Silica nanoparticles induce oxidative stress, inflammation, and endothelial dysfunction in vitro via activation of the MAPK/Nrf2 pathway and nuclear factor-κB signaling. *International Journal of Nanomedicine*, *10*, 1463.

Gupta, A. K., & Gupta, M. (2005). Cytotoxicity suppression and cellular uptake enhancement of surface modified magnetic nanoparticles. *Biomaterials*, *26*(13), 1565–1573.

Hamburg, M. A. (2012). FDA's approach to regulation of products of nanotechnology. *Science*, *336*(6079), 299–300.

He, Q., Yuan, W. M., Liu, J., & Zhang, Z. R. (2008). Study on in vivo distribution of liver-targeting nanopaticles encapsulating thymidine kinase gene (TK gene) in mice. *Journal of Materials Science: Materials in Medicine*, *19*(2), 559–565.

Höhr, D., Steinfartz, Y., Schins, R. P., Knaapen, A. M., Martra, G., Fubini, B., et al. (2002). The surface area rather than the surface coating determines the acute inflammatory response after instillation of fine and ultrafine TiO_2 in the rat. *International Journal of Hygiene and Environmental Health*, *205*(3), 239–244.

Hosseini, H., Tajiani, Z., & Jafari, S. M. (2019). 5 - Improving the shelf-life of food products by nano/micro-encapsulated ingredients. In C. M. Galanakis (Ed.), *Food Quality and Shelf Life* (pp. 159–200). Academic Press.

Huang, Y. -W., Cambre, M., & Lee, H. -J. (2017). The toxicity of nanoparticles depends on multiple molecular and physicochemical mechanisms. *International Journal of Molecular Sciences*, *18*(12), 2702.

Iavicoli, I., Leso, V., Beezhold, D. H., & Shvedova, A. A. (2017). Nanotechnology in agriculture: opportunities, toxicological implications, and occupational risks. *Toxicology and Applied Pharmacology*, *329*, 96–111.

Inoue, K. -i., Takano, H., Yanagisawa, R., Koike, E., & Shimada, A. (2009). Size effects of latex nanomaterials on lung inflammation in mice. *Toxicology and Applied Pharmacology*, *234*(1), 68–76.

Jafari, S. M. (2017). *Nanoencapsulation of food bioactive ingredients: Principles and applications*. Academic Press.

Jafari, S. M., & McClements, D. J. (2017). Chapter One - Nanotechnology approaches for increasing nutrient bioavailability. In F. Toldrá (Ed.), *Advances in food and nutrition research* (pp. 1–30). (81). Academic Press.

Jafari, S. M., Faridi Esfanjani, A., Katouzian, I., & Assadpour, E. (2017a). Chapter 10 - Release, characterization, and safety of nanoencapsulated food ingredients. In S. M. Jafari (Ed.), *Nanoencapsulation of food bioactive ingredients* (pp. 401–453). Academic Press.

Jafari, S. M., Katouzian, I., & Akhavan, S. (2017b). 15 - Safety and regulatory issues of nanocapsules. In S. M. Jafari (Ed.), *Nanoencapsulation technologies for the food and nutraceutical industries* (pp. 545–590). Academic Press.

Jafari, S. M., Katouzian, I., & Akhavan, S. (2017c). Safety and regulatory issues of nanocapsules. In *Nanoencapsulation technologies for the food and nutraceutical industries* (pp. 545–590). Elsevier.

Jain, A., Ranjan, S., Dasgupta, N., & Ramalingam, C. (2018). Nanomaterials in food and agriculture: an overview on their safety concerns and regulatory issues. *Critical Reviews in Food Science and Nutrition*, *58*(2), 297–317.

Jani, P., Halbert, G.W., Langridge, J., & Florence, A.T. (1990). Nanoparticle uptake by the rat gastrointestinal mucosa: quantitation and particle size dependency. *Journal of Pharmacy and Pharmacology*, *42*(12), 821–826.

Jiang, J., Oberdörster, G., Elder, A., Gelein, R., Mercer, P., & Biswas, P. (2008). Does nanoparticle activity depend upon size and crystal phase? *Nanotoxicology*, *2*(1), 33–42.

Katouzian, I., & Jafari, S. M. (2016). Nano-encapsulation as a promising approach for targeted delivery and controlled release of vitamins. *Trends in Food Science & Technology*, *53*, 34–48.

Khan, F. A. (2020). Nanomaterials: types, classifications, and sources. In F. A. Khan (Ed.), *Applications of nanomaterials in human health* (pp. 1–13). Singapore: Springer.

Kulkarni, S. A., & Feng, S. S. (2013). Effects of particle size and surface modification on cellular uptake and biodistribution of polymeric nanoparticles for drug delivery. *Pharmaceutical Research*, *30*(10), 2512–2522.

Kumar, A., & Dhawan, A. (2013). Genotoxic and carcinogenic potential of engineered nanoparticles: an update. *Archives of Toxicology*, *87*(11), 1883–1900.

Kustov, L. M., Mashkin, P.V., Zakharov, V. N., Abramenko, N. B., Krysanov, E.Y., Aslanov, L. A., et al. (2018). Silicon nanoparticles: characterization and toxicity studies. *Environmental Science: Nano*, *5*(12), 2945–2951.

Kyung, O. Y., Grabinski, C. M., Schrand, A. M., Murdock, R. C., Wang, W., Gu, B., et al. (2009). Toxicity of amorphous silica nanoparticles in mouse keratinocytes. *Journal of Nanoparticle Research*, *11*(1), 15–24.

L'azou, B., Jorly, J., On, D., Sellier, E., Moisan, F., Fleury-Feith, J., et al. (2008). In vitro effects of nanoparticles on renal cells. *Particle and Fibre Toxicology*, *5*(1), 1–14.

Lee, C. -H., Shih, C. -P., Hsu, K. -H., Hung, D. -Z., & Lin, C. -C. (2008). The early prognostic factors of glyphosate-surfactant intoxication. *The American Journal of Emergency Medicine*, *26*(3), 275–281.

Lee, J. H., Ju, J. E., Kim, B. I., Pak, P. J., Choi, E. K., Lee, H. S., et al. (2014). Rod-shaped iron oxide nanoparticles are more toxic than sphere-shaped nanoparticles to murine macrophage cells. *Environmental Toxicology and Chemistry*, *33*(12), 2759–2766.

Lehman, S. E., Morris, A. S., Mueller, P. S., Salem, A. K., Grassian, V. H., & Larsen, S. C. (2016). Silica nanoparticle-generated ROS as a predictor of cellular toxicity: mechanistic insights and safety by design. *Environmental Science Nano*, *3*(1), 56–66.

Li, J., Tang, M., & Xue, Y. (2019). Review of the effects of silver nanoparticle exposure on gut bacteria. *Journal of Applied Toxicology*, *39*(1), 27–37.

Linsinger, T., Chaudhry, Q., Dehalu, V., Delahaut, P., Dudkiewicz, A., Grombe, R., Larsen, E. H., & Löschner, K. et al. (2013). Validation of methods for the detection and quantification of engineered nanoparticles in food. *Food Chemistry*, *138*(2–3), 1959–1966.

Liu, Y., Hardie, J., Zhang, X., & Rotello, V. M. (2017). Effects of engineered nanoparticles on the innate immune system. *Seminars in Immunology*, *34*, 25–32.

Lomer, M. C., Hutchinson, C., Volkert, S., Greenfield, S. M., Catterall, A., Thompson, R. P., et al. (2004). Dietary sources of inorganic microparticles and their intake in healthy subjects and patients with Crohn's disease. *British Journal of Nutrition*, *92*(6), 947–955.

Magnuson, B. A., & Bouwmeester, H. (2011). Toxicology of nanomaterials in food. *Nanotechnology in the Agri-Food Sector*, 171–190.

Mahmoudi, M., Bertrand, N., Zope, H., & Farokhzad, O. C. (2016). Emerging understanding of the protein corona at the nano-bio interfaces. *Nano Today*, *11*(6), 817–832.

Maksimović, M., Omanović-Mikličanin, E., & Badnjević, A. (2019). Is nanofood safe? In M. Maksimović, E. Omanović-Mikličanin, & A. Badnjević (Eds.), *Nanofood and internet of nano things: for the next generation of agriculture and food sciences* (pp. 87–137). Cham: Springer International Publishing.

Mao, B. -H., Chen, Z. -Y., Wang, Y. -J., & Yan, S. -J. (2018). Silver nanoparticles have lethal and sublethal adverse effects on development and longevity by inducing ROS-mediated stress responses. *Scientific Reports*, *8*(1), 2445.

Martins, J. T., Ramos, Ó. L., Pinheiro, A. C., Bourbon, A. I., Silva, H. D., Rivera, M. C., Pastrana, L., & González-Fernández, Á. et al. (2015). Edible bio-based nanostructures: delivery, absorption and potential toxicity. *Food Engineering Reviews*, 7(4), 491–513.

Matin, A. H., Goddard, E., Vandermoere, F., Blanchemanche, S., Bieberstein, A., Marette, S., et al. (2012). Do environmental attitudes and food technology neophobia affect perceptions of the benefits of nanotechnology? *International Journal of Consumer Studies*, *36*(2), 149–157.

McCarron, E. (2016). *Nanotechnology and food: investigating consumers' acceptance of foods produced using nanotechnology*. Dublin Business School.

McClements, D. J. (2019). Nutraceuticals: superfoods or superfads? In *Future foods: how modern science is transforming the way we eat* (pp. 167–201). Cham: Springer International Publishing.

McClements, D. J., & Xiao, H. (2017). Is nano safe in foods? Establishing the factors impacting the gastrointestinal fate and toxicity of organic and inorganic food-grade nanoparticles. *NPJ Science of Food*, *1*(1), 6.

Merisko-Liversidge, E., & Liversidge, G. G. (2011). Nanosizing for oral and parenteral drug delivery: a perspective on formulating poorly-water soluble compounds using wet media milling technology. *Advanced Drug Delivery Reviews*, *63*(6), 427–440.

Miernicki, M., Hofmann, T., Eisenberger, I., von der Kammer, F., & Praetorius, A. (2019). Legal and practical challenges in classifying nanomaterials according to regulatory definitions. *Nature Nanotechnology*, *14*(3), 208–216.

Mills, N. L., Amin, N., Robinson, S. D., Anand, A., Davies, J., Patel, D., et al. (2006). Do inhaled carbon nanoparticles translocate directly into the circulation in humans? *American Journal of Respiratory and Critical Care Medicine*, *173*(4), 426–431.

Mintz, S. W., & Du Bois, C. M. (2002). The anthropology of food and eating. *Annual review of anthropology*, *31*(1), 99–119.

Morishige, T., Yoshioka, Y., Tanabe, A., Yao, X., Tsunoda, S. -i., Tsutsumi, Y., et al. (2010). Titanium dioxide induces different levels of IL-1β production dependent on its particle characteristics through caspase-1 activation mediated by reactive oxygen species and cathepsin B. *Biochemical and biophysical research communications*, *392*(2), 160–165.

Muhammad, Q., Jang, Y., Kang, S. H., Moon, J., Kim, W. J., & Park, H. (2020). Modulation of immune responses with nanoparticles and reduction of their immunotoxicity. *Biomaterials Science*, *8*(6), 1490–1501.

Müller, L., Riediker, M., Wick, P., Mohr, M., Gehr, P., & Rothen-Rutishauser, B. (2010). Oxidative stress and inflammation response after nanoparticle exposure: differences between human lung cell monocultures and an advanced three-dimensional model of the human epithelial airways. *Journal of the Royal Society Interface*, 7(suppl. 1), S27–S40.

Murugadoss, S., Lison, D., Godderis, L., Van Den Brule, S., Mast, J., Brassinne, F., et al. (2017). Toxicology of silica nanoparticles: an update. *Archives of Toxicology*, *91*(9), 2967–3010.

Oberdorster, G., Gelein, R., Johnston, C., Mercer, P., Corson, N., & Finkelstein, J. (1998). Ambient ultrafine particles: inducers of acute lung injury. In *Relationships between respiratory disease and exposure to air pollution* (pp. 216–229). ILSI Press.

OECD (2013). Regulatory Frameworks for Nanotechnology in Foods and Medical Products: Summary Results of a Survey Activity, OECD Science, Technology and Industry Policy Papers, No. 4, OECD Publishing, Paris.

Őrfi, E., & Szebeni, J. (2016). The immune system of the gut and potential adverse effects of oral nanocarriers on its function. *Advanced Drug Delivery Reviews*, *106*, 402–409.

Patel, P., Kansara, K., Senapati, V. A., Shanker, R., Dhawan, A., & Kumar, A. (2016). Cell cycle dependent cellular uptake of zinc oxide nanoparticles in human epidermal cells. *Mutagenesis*, *31*(4), 481–490.

Perillo, B., Di Donato, M., Pezone, A., Di Zazzo, E., Giovannelli, P., Galasso, G., et al. (2020). ROS in cancer therapy: the bright side of the moon. *Experimental & Molecular Medicine*, *52*(2), 192–203.

Peters, R., Brandhoff, P., Weigel, S., Marvin, H., Bouwmeester, H., Aschberger, K., et al. (2014). Inventory of nanotechnology applications in the agricultural, feed and food sector. *EFSA Supporting Publications*, *11*(7), 621E.

Powell, J., Ainley, C., Harvey, R., Mason, I., Kendall, M., Sankey, E., et al. (1996). Characterisation of inorganic microparticles in pigment cells of human gut associated lymphoid tissue. *Gut*, *38*(3), 390–395.

Powell, J. J., Faria, N., Thomas-McKay, E., & Pele, L. C. (2010). Origin and fate of dietary nanoparticles and microparticles in the gastrointestinal tract. *Journal of Autoimmunity*, *34*(3), J226–J233.

Ramasamy, M., Das, M., An, S. S. A., & Yi, D. K. (2014). Role of surface modification in zinc oxide nanoparticles and its toxicity assessment toward human dermal fibroblast cells. *International Journal of Nanomedicine*, *9*, 3707.

Rampado, R., Crotti, S., Caliceti, P., Pucciarelli, S., & Agostini, M. (2020). Recent advances in understanding the protein corona of nanoparticles and in the formulation of "stealthy" nanomaterials. *Frontiers in Bioengineering and Biotechnology*, *8*.

Renwick, L., Brown, D., Clouter, A., & Donaldson, K. (2004). Increased inflammation and altered macrophage chemotactic responses caused by two ultrafine particle types. *Occupational and Environmental Medicine*, *61*(5), 442–447.

Riasat, M. R., Guangjun, N., & Tariq, M. Z. (2016). Effects of nanoparticles on gastrointestinal disorders and therapy. *7JS z¢{x zxy~*, 216.

Rodriguez-Garraus, A., Azqueta, A., Vettorazzi, A., & López de Cerain, A. (2020). Genotoxicity of silver nanoparticles. *Nanomaterials*, *10*(2), 251.

Rollin, F., Kennedy, J., & Wills, J. (2011). Consumers and new food technologies. *Trends in Food Science & Technology*, *22*(2–3), 99–111.

Ronteltap, A., Van Trijp, J., Renes, R., & Frewer, L. (2007). Consumer acceptance of technology-based food innovations: lessons for the future of nutrigenomics. *Appetite*, *49*(1), 1–17.

Sanguansri, P., & Augustin, M. A. (2006). Nanoscale materials development–a food industry perspective. *Trends in Food Science & Technology*, *17*(10), 547–556.

Satterfield, T., Kandlikar, M., Beaudrie, C., Conti, J., & Harthorn, B. (2009). Anticipating the perceived risk of nanotechnologies. *Nature Nanotechnology*, *4*, 883.

Shreiner, A. B., Kao, J. Y., & Young, V. B. (2015). The gut microbiome in health and in disease. *Current Opinion in Gastroenterology*, *31*(1), 69–75.

Siegrist, M., & Cvetkovich, G. (2000). Perception of hazards: The role of social trust and knowledge. *Risk Analysis*, *20*(5), 713–720.

Siegrist, M., & Hartmann, C. (2020). Consumer acceptance of novel food technologies. *Nature Food*, *1*(6), 343–350.

Siemer, S., Hahlbrock, A., Vallet, C., McClements, D. J., Balszuweit, J., Voskuhl, J., et al. (2018). Nanosized food additives impact beneficial and pathogenic bacteria in the human gut: a simulated gastrointestinal study. *NPJ Science of Food*, *2*(1), 1–10.

Singh, R., Cheng, S., & Singh, S. (2020). Oxidative stress-mediated genotoxic effect of zinc oxide nanoparticles on *Deinococcus radiodurans*. *3 Biotech*, *10*(2), 66.

Sodano, V., Gorgitano Maria, T., Verneau, F., & Vitale Cosimo, D. (2016). Consumer acceptance of food nanotechnology in Italy. *British Food Journal*, *118*(3), 714–733.

Song, H. -Y., Kim, Y. -H., Seok, S. -J., Gil, H. -W., Yang, J. -O., Lee, E. -Y., et al. (2012). Cellular toxicity of surfactants used as herbicide additives. *Journal of Korean Medical Science*, *27*(1), 3–9.

Stampfli, N., Siegrist, M., & Kastenholz, H. (2010). Acceptance of nanotechnology in food and food packaging: a path model analysis. *Journal of Risk Research*, *13*(3), 353–365.

Stensberg, M. C., Wei, Q., McLamore, E. S., Porterfield, D. M., Wei, A., & Sepúlveda, M. S. (2011). Toxicological studies on silver nanoparticles: challenges and opportunities in assessment, monitoring and imaging. *Nanomedicine (Lond)*, *6*(5), 879–898.

Szakal, C., Roberts, S. M., Westerhoff, P., Bartholomaeus, A., Buck, N., Illuminato, I., et al. (2014). Measurement of nanomaterials in foods: integrative consideration of challenges and future prospects. *ACS Nano*, *8*(4), 3128–3135.

Szentkuti, L. (1997). Light microscopical observations on luminally administered dyes, dextrans, nanospheres and microspheres in the pre-epithelial mucus gel layer of the rat distal colon. *Journal of Controlled Release*, *46*(3), 233–242.

Tran, C. L., Buchanan, D., Cullen, R. T., Searl, A., Jones, A. D., & Donaldson, K. (2000). Inhalation of poorly soluble particles. II. Influence of particle surface area on inflammation and clearance. *Inhalation Toxicology*, *12*(12), 1113–1126.

Vidigal, M. C. T. R., Minim, V. P. R., Simiqueli, A. A., Souza, P. H. P., Balbino, D. F., & Minim, L. A. (2015). Food technology neophobia and consumer attitudes toward foods produced by new and conventional technologies: A case study in Brazil. *LWT - Food Science and Technology*, *60*(2 Part 1), 832–840.

Wardle, J., Parmenter, K., & Waller, J. (2000). Nutrition knowledge and food intake. *Appetite*, *34*(3), 269–275.

Warheit, D. B., Webb, T. R., Sayes, C. M., Colvin, V. L., & Reed, K. L. (2006). Pulmonary instillation studies with nanoscale TiO_2 rods and dots in rats: toxicity is not dependent upon particle size and surface area. *Toxicological Sciences*, *91*(1), 227–236.

World Health, A. (2000). The World Health Report 2000: health systems: improving performance. *Geneva, World Health Organization.*

Wu, J., Sun, J., & Xue, Y. (2010). Involvement of JNK and P53 activation in G2/M cell cycle arrest and apoptosis induced by titanium dioxide nanoparticles in neuron cells. *Toxicology Letters*, *199*(3), 269–276.

Xia, T., Korge, P., Weiss, J. N., Li, N., Venkatesen, M. I., Sioutas, C., et al. (2004). Quinones and aromatic chemical compounds in particulate matter induce mitochondrial dysfunction: implications for ultrafine particle toxicity. *Environmental Health Perspectives*, *112*(14), 1347–1358.

Xia, T., Kovochich, M., Liong, M., Madler, L., Gilbert, B., Shi, H., et al. (2008). Comparison of the mechanism of toxicity of zinc oxide and cerium oxide nanoparticles based on dissolution and oxidative stress properties. *ACS Nano*, *2*(10), 2121–2134.

Xu, L., Liu, Y., Bai, R., & Chen, C. (2010). Applications and toxicological issues surrounding nanotechnology in the food industry. *Pure and Applied Chemistry*, *82*(2), 349–372.

Yang, Y., & Hobbs, J. E. (2020). Food values and heterogeneous consumer responses to nanotechnology. *Canadian Journal of Agricultural Economics/Revue canadienne d'agroeconomie*, *68*(3), 289–313.

Yin, H., Chen, R., Casey, P. S., Ke, P. C., Davis, T. P., & Chen, C. (2015). Reducing the cytotoxicity of ZnO nanoparticles by a pre-formed protein corona in a supplemented cell culture medium. *RSC Advances*, *5*(90), 73963–73973.

Yue, C., Zhao, S., & Kuzma, J. (2015). Heterogeneous consumer preferences for nanotechnology and genetic-modification technology in food products. *Journal of Agricultural Economics*, *66*(2), 308–328.

Zhao, S., Yue, C., & Kuzma, J. (2020). 17- Consumer expectations and attitudes toward nanomaterials in foods. In S. M. Jafari (Ed.), *Handbook of food nanotechnology* (pp. 705–733). Academic Press.

Zolnik, B. S., González-Fernández, A., Sadrieh, N., & Dobrovolskaia, M. A. (2010). Nanoparticles and the immune system. *Endocrinology*, *151*(2), 458–465.

CHAPTER 2

Possible health risks associated with nanostructures in food

Atefe Rezaei[a], Marjan Daeihamed[b], Esra Capanoglu[c], Merve Tomas[d], Safoura Akbari-Alavijeh[e], Rezvan Shaddel[e], Sara Khoshnoudi-Nia[f], Sareh Boostani[g], Hadis Rostamabadi[h], Seid Reza Falsafi[h], Seid Mahdi Jafari[h]

[a]Department of Food Science and Technology, School of Nutrition and Food Science, Food Security Research Center, Isfahan University of Medical Sciences, Isfahan, Iran
[b]Department of Pharmaceutics, School of Pharmacy, Guilan University of Medical Sciences, Rasht, Iran
[c]Department of Food Engineering, Faculty of Chemical and Metallurgical Engineering, Istanbul Technical University, Maslak, Istanbul, Turkey
[d]Department of Food Engineering, Faculty of Engineering and Natural Sciences, Istanbul Sabahattin Zaim University, Halkali, Istanbul, Turkey.
[e]Department of Food Science and Technology, Faculty of Agriculture and Natural Resources, University of Mohaghegh Ardabili, Ardabil, Iran
[f]Seafood Processing Research Group, School of Agriculture, Shiraz University, Shiraz, Iran
[g]Department of Food Science and Technology, School of Agriculture, Shiraz University, Shiraz, Iran
[h]Department of Food Materials and Process Design Engineering, Gorgan University of Agricultural Sciences and Natural Resources, Gorgan, Iran

2.1 Introduction

According to the definition of the European Commission in the 2011, nanostructures have a size in the range of 1–100 nm and nanoparticles (NPs) are nanostructures that their three dimensions is lower than 100 nm (Smolkova, Dusinska, & Gabelova, 2017). However, the defined range of size for nanostructures varies across various governmental agencies and biomedical and engineering disciplines (Vita, Royse, & Pullen, 2019). Some sources have defined extended range of size for one or more dimensions of nanostructures including <200 nm (Lundy, Chen, Toh, & Hsieh, 2016), <500 nm (Rizvi & Saleh, 2018), as well as <1000 nm (Petros & DeSimone, 2010; Banerjee, Qi, Gogoi, Wong, & Mitragotri, 2016; Schutz, Juillerat-Jeanneret, Mueller, Lynch, & Riediker, 2013). The National Institute for Occupational Safety and Health (NIOSH) which is in agreement with the American Society for Testing and Materials (ASTM) and the International Organization for Standardization, ISO (Dananjaya et al., 2020) has confirmed the nanoscale range between 1 and 100 nm, but specifically added that the definition of nanoscale range is not fixed (Vita et al., 2019).

Safety and Regulatory Issues of Nanoencapsulated Food Ingredients
http://dx.doi.org/10.1016/B978-0-12-815725-1.00002-1

Nanotechnology has introduced many advantages in the food sector. For example, nanostructures can be used to enhance the quality, safety, nutritional values, and shelf life of food (Sozer & Kokini, 2009). Nanostructures that are found in food have different potential sources. Nanostructures can be used in food packaging, sensors, and materials that used to disinfect food-manufacturing plants and in this way can be entered indirectly into the foods. In other cases, nanostructures are directly entered to the food by incorporation as delivery systems for different food additives (McClements & Xiao, 2017a; Yada et al., 2014). In some cases, nanostructures are naturally presented in food such as casein micelles in milk with spherical clusters of 50–500 nm (Livney, 2010).

Nanostructures have unique properties that present new challenges for their application. The spherical shape of NPs makes a high surface for their interaction with other biomolecules in the cells. The NPs with smaller size have a larger surface-to-volume ratio. The little size provides the circumstances for easy intracellular uptake of the NPs (Buzea, Pacheco, & Robbie, 2007). The special physiological properties of food-based nanostructures including shape, particle size distribution, chemical composition, agglomeration state, crystal structure, surface characteristics (area, chemistry, charge), and porosity are important to be identified for understanding their possible mechanisms of injury and further harmful and toxic effects (Chau, Wu, & Yen, 2007; Jafari, Katouzian, & Akhavan, 2017a). However, it is still obscure to what extent various physicochemical properties of nanostructures may contribute to their toxicokinetics (Card, Jonaitis, Tafazoli, & Magnuson, 2011). On the other hand, various studies have also revealed that the methods used for the production of nanostructures may not necessarily generate products with adverse effects. For example, the toxicity of selenium NPs were significantly reduced compared to various selenium compounds (Wang, Zhang, & Yu, 2007).

Nanostructures can induce acute or chronic toxicity. As low levels of nanostructures may be consumed by a person in an extended time, chronic toxicity is more common. The toxicity of the nanostructures pertains to their ability to hurt cells and organs and therewith have influence on human health. After absorption of the nanostructures into the body, cellular or organ hurt may be arisen in diverse places within the gastrointestinal tract (GIT) (Abdelhamed et al., 2014; Buzea et al., 2007). Furthermore, the microbial cells that are normally presented in human GIT, may be damaged by nanostructures and indirectly affect on health (Fröhlich & Fröhlich, 2016).

After entrance of the NPs into the body, their surface will be modified physically or chemically by sorption of the proteins onto their surface and forming a "biocorona" (Monopoli, Åberg, Salvati, & Dawson, 2012). The biological fluids such as blood, intestinal fluid, and plasma have a high amount of proteins. These interactions cause in denaturing and unfolding the proteins (Shannahan et al., 2015). Also these modifications affect the activity, cellular uptake, pharmacokinetics, circulation time, and distribution of nanostructures in the body (Tenzer et al., 2011; Persaud et al., 2018). Nanostructures can pass through cell membrane, brain barrier and blood, and deliver encapsulated compounds in the target tissue (Smolkova et al., 2017). The size, shape, and surface chemistry of the nanostructures have effects on the cellular uptake (Nel et al., 2009, Treuel & Nienhaus, 2012). The modification of these factors can induce potential toxicity such as oxidative stress, which is related to immunotoxicity, genotoxicity, and inflammatory cell responses (Dusinska et al., 2015). Nanostructures can change the epigenetic regulatory mechanisms that induce some diseases such as cancer (Stoccoro, Karlsson, Coppede, & Migliore, 2013).

United States-based safety assessment agencies (such as EPA and FDA), AND European Food Safety Authority (EFSA), have published multiple guidelines about the possible health risks related to the application of nanomaterials in food and feed chain (Hardy et al., 2018). Risk assessment is an evaluation method of potential hazards for determination of the safety of substances to protect the environment and health (Tsuji et al., 2006).

The potential health risks of different nanostructures that are used in food science have not been reviewed comprehensively yet. In this paper, we have a review on different nanostructures that are mainly used in food sector. Different exposure ways and mechanisms related to adverse health effects of nanostructures are highlighted. Also, potential health risks associated to nanostructures and risk assessment of them for preparing future guidance on the safety of nanostructures are explained.

2.2 A brief classification of nanostructures

The composition of nanostructures has an important effect on their potential health risks and gastrointestinal fate in foods. The nanostructures, based on their compounds can be classified in two main groups: (1) inorganic and (2) organic (McClements & Xiao, 2017b). This classification is diagrammatically illustrated in Figs. 2.1 and 2.2.

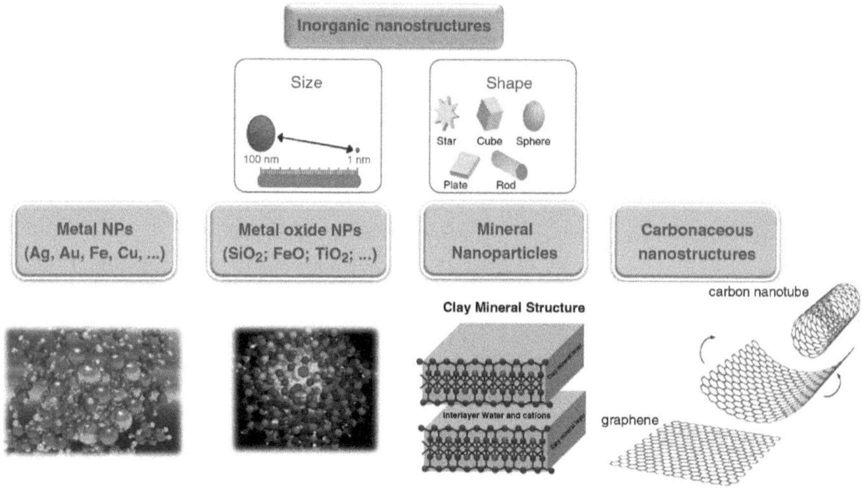

Figure 2.1 ***Schematic representation of inorganic nanostructures.***

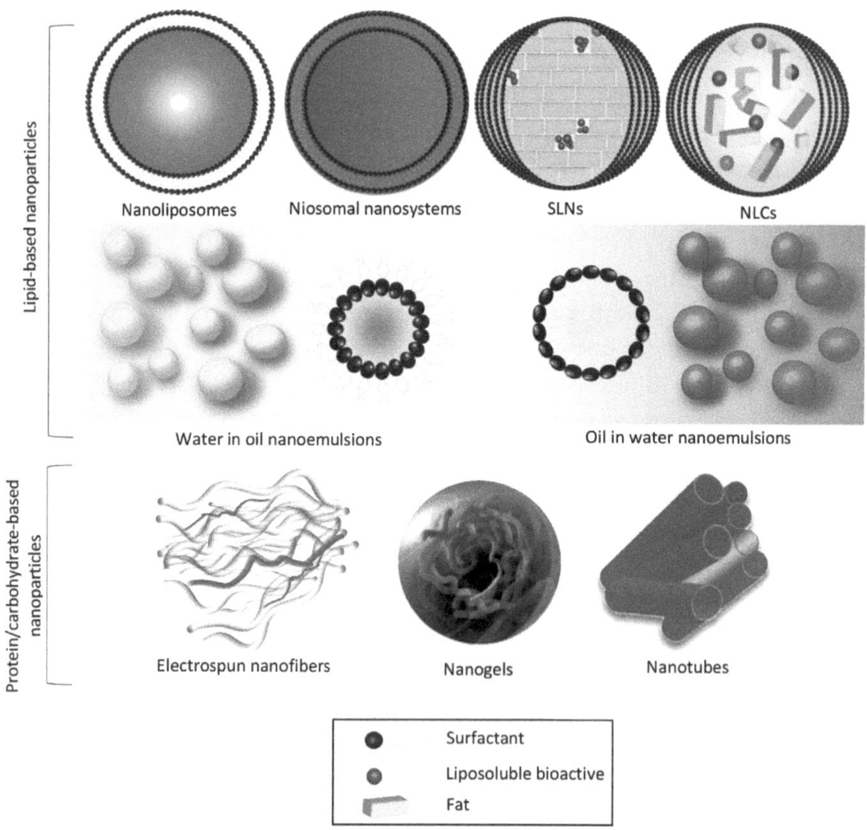

Figure 2.2 ***Schematic representation of organic nanostructures.***

2.2.1 Inorganic nanostructures

Inorganic NPs (INPs) or engineered NPs (ENPs) are a wide class of nanostructures produced by various synthesis methods and based on different core materials and coating (capping) agents (Tiede et al., 2008; Matsoukas, Desai, & Lee, 2015). INPs can be classified based on several factors such as, size, source, morphology, chemistry, etc. According to chemistry, INPs are classified into four major groups: (1) Metal NPs, (2) Metal oxide NPs, (3) Carbonaceous nanostructures, and (4) Mineral NPs, as shown in Fig. 2.1 (Wang, Feng, Tang, Li, & Tang, 2013a; Rizwan et al., 2017; Suárez-Iglesias, Collado, Oulego, & Díaz, 2017; Moini, Jahandideh, & Anderson, 2019). The main properties and applications of different groups of INPs are presented as follows.

2.2.1.1 Metal NPs

Rapid development of nanotechnology have enabled the large-scale production of metal NPs with unique physiochemical and antimicrobial characteristics. Metal-NPs have been adopted in a wide range of applications including healthcare, pharmaceutical, food packaging, agriculture, construction, defense, transportation, medicine, environmental, and other technological fields (Kessler, 2009). Antimicrobial (antifungal, antibacterial, and antiviral) activity of metallic NPs is one their most important properties from the food industry perspective. Antimicrobial action of metallic NPs might be due to damaging intercellular components and microbial cell wall, inhibition of the virus binding to the plasma membrane, interaction with double-stranded DNA, and interference with viral attachment (Baram-Pinto, Shukla, Perkas, Gedanken, & Sarid, 2009; Augustine & Hasan, 2020). Furthermore, studies have shown that several metallic NPs (such as silver, gold, and copper) may be applied in cancer diagnosis and cancer therapy, especially by the generation of reactive oxygen species (ROS) (Augustine & Hasan, 2020). Among the various metallic NPs, Ag and Au, the most important metal NPs, are briefly highlighted further. The application of several metal NPs is also presented in Table 2.1.

- **Silver (Ag) NPs**: Due to outstanding antimicrobial properties (antifungal and antibacterial activity) and high thermal resistance, Ag-NPs are mostly utilized in some commercial products (Table 2.1) (Ahamed, AlSalhi, & Siddiqui, 2010a; Asghari-Paskiabi, Jahanshiri, Shams-Ghahfarokhi, & Razzaghi-Abyaneh, 2020). Several companies like Duretham and Koda apply Ag-NPs in their packaging polymers for improving the shelf life of perishable food products (Hoseinnejad, Jafari, & Katouzian, 2018). The results of various studies showed that

Table 2.1 Classification and application of various inorganic nanostructures in the food industry.

Inorganic NPs		Application in food industry	References
Metal-NPs	Ag (Ag)	As antimicrobial agent in food packaging; Antioxidant agent; Bactericidal agents; Biosensors	(Sokolov et al., 2008, Ahamed et al., 2010a, Li & Chen, 2011, Bendale et al. 2017, Bai & Tang, 2020)
	Zinc (zn)	Antimicrobial agent, Biosensors (for detection food toxin)	(Kalita et al., 2012)
	Au (Au)	Antimicrobial agent; Biosensors; Catalytic agent	(Dizaj et al., 2015)
	Nickel (Ni)	Catalytic agent, water purification (as adsorbent); biosensor (for detection cholesterol and glucose)	(Singh, Kalita, Singh, & Malhotra, 2011, Chen et al., 2019a)
	Copper (Cu)	Antimicrobial agent; Nanodevices (e.g., nanoprobes, nano-sensors); Catalytic agents; Water purification (as adsorbent)	(Chandra, Kumar, & Tomar, 2014)
	Aluminum (Al)	Antimicrobial agent	(Singh et al., 2011)
	Palladium (PI)	Antioxidant agent; Catalytic agent; nonenzymatic glucose biosensors	(Lu et al., 2011, Kora & Rastogi, 2018)
	Platinum (Pt)	Antibacterial agent	(Tahir et al., 2017)
	Selenium (Se)	Antimicrobial agent; Antioxidant agent	(Khanna et al., 2019)

Metal oxide-NPs	Copper oxide (CuO, Cu_2O)	Agricultural biocides; Antimicrobial agent	(Duffy, Osmond-McLeod, Judy, & King, 2018)
	Nickel oxide (NiO)	Antimicrobial agent; Anti-biofilm agent	(Rakshit et al., 2013, Saleem, Ahmed, Khan, Al-Shaeri, & Musarrat, 2017)
	Titanium dioxide (TiO_2)	Antimicrobial agent, whitening agent	(McClements & Xiao, 2017a)
	Zinc oxide (ZnO)	Antimicrobial agent (in food packaging), fortification of food with zinc; ultraviolet light absorbers (protect foods against UV light).	(Wang et al., 2014)
	Iron oxide (Fe_2O_3 and Fe_3O_4)	Antimicrobial agent, fortification of food with iron; colorants	(Zimmermann & Hilty, 2011, McClements & Xiao, 2017b)
	Aluminum oxide (Al_2O_3)	Antimicrobial agent	(Şahin, Musevi, & Aslani, 2017)
	Magnesium oxide (MgO)	Antimicrobial agent; reinforced biofilm	(Jin & He, 2011, Sanuja, Agalya, & Umapathy, 2014)
	Silicon dioxide (SiO_2)	Anticaking agents (in salts, spices, dried milk, and etc.)	(Peters et al., 2012)
Carbonaceous NPs	Fullerenes	Antimicrobial agents; Antioxidant agent; Drug delivery; Organic solar cells (in food packaging)	(Hoheisel et al., 2010, Dizaj et al., 2015)
	Carbon Nanotubes	Antimicrobial agents; Biosensors; Reinforcement in composite;	(Dizaj et al., 2015)
	Graphene	Antimicrobial agents (especially graphene oxide); Drug delivery; Organic light emitting diodes (for food packaging); Renewable energy sources; biosensors; Ultra-sensitive devices (detection of biological molecules such as glucose, cholesterol, hemoglobin, and DNA)	(Siqueira & Oliveira, 2017, Pan et al., 2019)

(Continued)

Table 2.1 Classification and application of various inorganic nanostructures in the food industry (*Cont.*).

Inorganic NPs		Application in food industry	References
Clay minerals NPs	Kaolin	Drug Delivery; Enhancement of gel strength; Porous agent; Reactive clays; Superabsorbent (removal of brilliant green dye from water, removal of Cu(II), Cd, and Pb ions)	(Moini et al., 2019)
	Mica	Superabsorbent	
	Attapulgite	Superabsorbent (adsorption of Hg(II), La(III) and Ce(III) ions); Thermal stability	
	MMT	Antibacterial agent (Nontoxicity); Reinforcing materials (stretch ability and toughness); Self-healing; Superabsorbent; Thermoresponsive properties	
	Zeolite	Antibacterial; Removal of chemical oxygen demand (COD) wastewater; Controlled delivery of zinc micronutrient; Antimicrobial drug-delivery	
	Bentonite	Superabsorbent (removal and recovery of thorium (Siqueira Jr and Oliveira Jr))	
	Halloysite	Superabsorbent	
	Rectorite	Superabsorbent	
	Illite Smectite	Absorbent (absorption of methylene blue)	
	Sepiolite	Absorbent (Adsorption of cationic dye)	
	Tourmaline	Absorbent (Adsorption capacity for Pb^{2+} and Cu^{2+})	
	Hydrotalcite	Salt-resistance	
	Hydroxyapatite	Absorbent (e.g., Removal of Cu (II)); Reinforcing materials (highly extensible, tough and elastomeric nanocomposite)	
	Laponite	Biotechnological applications (such as injectable matrices, biomedical coatings, drug delivery, and regenerative medicine); Cell cultivation; Self-healable; Semiconductor NP-based Hydrogels; Swelling and deswelling agent	
	allophane	Carrier agent (Nontoxicity)	

the electromagnetic, antimicrobial activity, optical, catalytic, and other properties of Ag-NPs are basically influenced by shape, size, and concentration of particles, which is being dependent on the NP synthesis methods (Sokolov et al., 2008). Among various physical, chemical, and biological synthesized methods, biological ones have attracted the attention of scientists (Kalpana & Rajeswari, 2017).

- **Gold (Au) NPs:** Au-NPs exhibit a wide range of applications (e.g., pharmaceuticals, imaging, catalytic agents, biosensors, and catalyst agents) due to the unique size dependent properties (Table 2.1). Under the same experimental conditions, Au revealed very lower toxicity as compared to Ag (Bar-Ilan, Albrecht, Fako, & Furgeson, 2009). Moreover, lower concentration of Au-NPs than Ag-NPs are able to show apoptosis in malignant cells (Augustine & Hasan, 2020). Due to such interests, a number of recent studies have been focused on the biosynthesis of Au-NPs based on marine and fresh water alga (Ramakritinan, Kaarunya, Shankar, & Kumaraguru, 2013; Khanna, Kaur, & Goyal, 2019; Dananjaya et al., 2020). Conjugation of synthesized Ag, Au, and platinum NPs with antibiotics can improve the antibacterial potential of NPs (Dizaj, Mennati, Jafari, Khezri, & Adibkia, 2015). Ag-NPs and Au-NPs have good plasmonic properties which are greatly related to their physical attributes (e.g., size, shape and dielectric property of surrounding medium). Also, the sensing power of metal NPs is related to the adherence of biomolecules on the surface of metal NPs (Dizaj et al., 2015).

2.2.1.2 Metal oxide NPs

Metal oxides, as an interesting group of inorganic materials, have been extensively investigated in recent years because of their wide range of structures (more than metal NPs), morphologies, and physiochemical properties (Fawcett, Verduin, Shah, Sharma, & Poinern, 2017). In recent years, there have been many attempts to synthesize metal oxide NPs based on biological methods as a more green and cheaper method than chemical and physical methods. Biological sources such as bacteria and fungi, plants, algae, yeasts and actinomycetes can be considered as promising methods for the synthesis of metal and metal oxide NPs. Among the biological sources, plant parts (such as seed, leaf, bark, stem, fruit extracts, food, and agricultural wastes) show the best potential (regarding simplicity, safety, the rate of synthesis, and environmental considerations) to produce metal and metal oxide NPs (Kalpana & Rajeswari, 2017).

Metal oxide NPs have a variety of applications such as food packaging; fortification of food (e.g., iron oxide, zinc oxide); controlling powder flowability (e.g., silicon dioxide); whitening agent (e.g., titanium dioxide); antimicrobial agent (ZnO), antioxidant agent, biosensors, etc. in food field (Allahverdiyev, Abamor, Bagirova, & Rafailovich, 2011; McClements & Xiao, 2017a; Mishra et al. 2017). Due to the capability in acquiring sensitive data in a non-invasive way, metal oxide NPs exhibit great potential in bioimaging and biosensing technologies, especially in cell tracking and early-stage cancer diagnosis (Sokolov et al., 2008; Ahamed et al., 2010a; Li & Chen, 2011; Bendale et al. 2017; Bai & Tang, 2020). Table 2.1 briefly describes the various metal-oxide NPs and their applications.

2.2.1.3 Carbonaceous nanostructures

Carbon is one of the most abundant natural elements in the world (Siqueira & Oliveira, 2017). Due to its low specific weight besides allotropic characteristics and consequently chemical and thermal robustness of the various carbon forms, these materials are increasingly used in the manufacture of durable, high-performance, lightweight, and low-cost materials (Hoheisel, Schrettl, Szilluweit, & Frauenrath, 2010). Carbonaceous nanostructures are carbon materials in nanometer scale with controlled morphology and surface chemistry and topology. These nanostructures play a key role in various fields (Table 2.1) (Siqueira & Oliveira, 2017).

According to the dimension, carbonaceous nanostructures can be divided into four main groups: (1) zero-dimensional (e.g., fullerenes, diamond clusters), (2) one-dimensional (e.g., diamond nanorods and carbon nanotubes (CNTs)), (3) two-dimensional (e.g., diamond nanoplatelets and graphene sheets), and (4) three-dimensional structures (e.g., fullerite, nanocrystalline diamond films and CNT ropes) (Hu, Shenderova, & Brenner, 2007). The carbonaceous nanostructures that have the most application in food science are presented here.

1. Carbon nanotubes (CNTs)

CNTs are hollow and thin nano-sized tubes formed only by carbon. The two most important structures are single-walled carbon nanotubes (SWNT) and multiwalled carbon nanotubes (MWNT). There are many methods for synthesizing CNTs that produce various types with different properties. The three main methods are electric arc, pulsed laser, and chemical vapor deposition (CVD). They have different properties from other allotropes of carbon, with outstanding properties such as a high mechanical strength and elasticity, large surface area, lightweight, flexibility, optical transparency, chemical and thermal stability and high electrical conductivity. The importance of CNTs for nanotechnology is reflected in their numerous application possibilities, among these, the importance of the use of CNTs in sensors and biosensors (for detection of foodborne pathogens such as *staphylococcus aureus*) are worth to mention (Hoheisel et al., 2010; Pan et al., 2019). The other applications are summarized in Table 2.1.

2. Graphene and its derivatives

Graphene is a nanostructure in which carbon atoms form a hexagonal lattice in a two-dimensional layer. Graphene shows several attributes that are superior to any other material such as high electron mobility, mechanical resistance, chemical purity, large surface area, single atomic layer thickness, conductivity, absorption of a significant fraction of white light, biocompatible with different cell types both in vitro and in vivo, impermeability to any gas, and high electrical conductivity (>> Cu). On the other hand, graphene has a stable surface, strong *Van der Waals* forces, poor dispersibility and solubility. These properties can limit the use of this type of material. Therefore, various inorganic (e.g., SiO_2, Al_2O_3, Au; Pt) and organic materials (e.g., polychitosan and polyallylamine) have been applied to enhance the properties of graphene in the construction of food safety sensors.

Graphene and its derivatives have different applications in food industries. They can be applied in various types of sensors and biosensors for detection of food toxins (e.g., pesticides, mercury ion, aflatoxin, ochratoxins, citrinin, citreoviridin, fumonisins, patulin, trichothecenes, zearalenone, and sulfonamide residues in various foods); monitoring of glucose concentrations in fish (Takase et al., 2012, Singh et al., 2013); measuring H_2O_2 release from the living cells and foodborne pathogens) and also they can be used as antibacterial agents (such as functionalize polydopamine (PDA)-graphene nanosheets with silver NPs (AgNP-PDA-GNS)) (Singh et al., 2013; Yang, Chen, Ren, Zhang, & Yang, 2015a; Arduini, Cinti, Scognamiglio, & Moscone, 2016; Zeng, Zhu, Du, & Lin, 2016; Pan et al., 2019). Due to two-dimensional structure of graphene (larger surface area than CNTs) and easy functionalization, this nanostructure can be better as compared to CNTs, in biosensors (Hoheisel et al., 2010).

2.2.1.4 Clay mineral nanostructures

Clay minerals are usually characterized by a layered structure including tetrahedral silicon (Si) sheet(s) and octahedral aluminum (Al) sheet(s) (Barbas, 2013). Therefore, clay minerals can be also called phyllosilicate. Clay minerals often include variable amounts of $Fe^{2+,3+}$, Mg^{2+}, alkali metals, and other cations which can be present in the interlayer(s) space or in the lattice framework (Zhou & Keeling, 2013). Based on their layered structure, clay minerals can be classified into three main categories: (1) 1:1-type clay mineral is made up of one tetrahedral and one octahedral sheet per each clay layer (e.g., kaolinite, serpentine, and halloysite), (2) 2:1 clay mineral is composed of one octahedral sheet sandwiched between two tetrahedral sheets (e.g., montmorillonite, bentonite, laponite, micas (Mica is a generic term used for a group of complex aluminosilicate), talc, vermiculite, saponite, hectorite, sepiolite, and illite), and (3) 2:1:1 clays are consisted of a succession of the octahedral sheets close to a 2:1 layer (e.g., cloisite). In contrast to carbonaceous nanostructures, the clays exhibit hydrophilic nature. Therefore, it is common to add clay in hydrogel structure (Massaro, Colletti, Lazzara, & Riela, 2018).

The advantages of clay NPs such as chemical composition, high specific surface area, electrical charge, layered structure, low toxicity, high adsorption capacity, colloid and thixotropy, ion exchange capacity, chemical inertness, good rheological, and swelling property leads to a number of application possibilities (Massaro et al., 2018). These areas include pharmaceuticals, food packaging (as antimicrobial agent and absorbents), wastewater treatment, animal feeds, and coating paper (Zhou & Keeling, 2013). For a detailed listing of applications see Table 2.1.

2.2.2 Organic nanostructures

Nowadays, the researchers have considerably focused on the application of organic precursors for the development of NPs due to their high compatibility with the metabolic system and great capability for biological/biomimetic influences (Kavand, Anton, Vandamme, Serra, & Chan-Seng, 2020). In this context, lipids, proteins, and carbohydrates-based nano-formulations (Fig. 2.2) as three main classes of bio-nanostructures are described in following sections:

2.2.2.1 Lipid nanostructures

Lipid nanoparticulate systems, composed of biocompatible/biodegradable lipids, are among the most robust and flexible nanoencapsulation tools utilized for the protection of a rich variety of bioactive ingredients and enhancing their site-specific delivery and controlled release (Behbahani, Ghaedi, Abbaspour, Rostamizadeh, & Dashtian, 2019; Raikos, Hayward, Hayes, Meroni, & Ranawana, 2019). Irrespective of the nano-formulation strategies, diverse lipid NPs can be divided into two major classes: (1) vesicular arrangements, and (2) lipid-particulate systems (Dima, Assadpour, Dima, & Jafari, 2019; Jafari & McClements, 2018; Jain et al., 2017). More specifically, lipid-based nanocarriers can be categorized into nanoemulsion-based carriers (single/double), nano-phospholipid based carriers (e.g., liposomes or phytosomes), lipid nano-particulate systems comprised of solid lipid NPs (SLNs), or nanostructured lipid carriers (NLCs), as well as surfactant-based nano-vectors (e.g., niosomes, hexosomes, cubosomes, etc.) (Rostamabadi, Falsafi, & Jafari, 2019a). Among which, the leading lipid-based nano-systems engineered in diverse food formulations encompass (1) liposomal delivery systems, (2) nanoemulsions, (3) SLNs, and (4) NLCs (Akhavan, Assadpour, Katouzian, & Jafari, 2018; Arancibia, Miranda, Matiacevich, & Troncoso, 2017), as summarized in Table 2.2.

Compared to other organic NPs, namely, protein and carbohydrate-based NPs, lipid nano-formulations typically share a general architecture of lipid bilayers comprising hydrophilic-hydrophobic moieties and demonstrate various outstanding beneficial attributes, including (Brum et al., 2017; Rezaei, Fathi, & Jafari, 2019; Salvia-Trujillo & McClements, 2016; Sun & Xia, 2019):

- great potential in the encapsulation of bioactive components of diverse solubility (hydrophilic/hydrophobic),
- asynchronous/simultaneous delivery of lipophilic and hydrophilic nutraceuticals/drugs,
- possibility for large-scale production of nanodelivery systems via nature-inspired lipid ingredients,
- superior entrapment efficiency with lower toxicity, and
- potent targetability either in the food systems or in vivo.

Furthermore, the targetability of lipid nano-systems is practicably controllable via in vivo evaluations by active (e.g., in the presence of antibodies) or passive (e.g., relying on the size of the produced particles) strategies (Jafari, Katouzian, & Akhavan, 2017b; Maherani, Arab-Tehrany, R Mozafari, Gaiani, & Linder, 2011).

2.2.2.2 Protein nanostructures

Protein nanostructures are mainly grouped into four main categories; that is, nanofibers, NPs, nano-gels, and nanotubes (Table 2.1), which reckon as valuable candidates for the food/pharmaceutical applications over the last few decades owing to their facile nano-synthesis methods, being from abundant, cheap and renewable origins, powerful drug-binding capability, notable uptake into the targeted cells, excellent nutritional value, as well as

Table 2.2 An overview on organic nanostructures.

NP type	Constituent components	Advantage	Disadvantage	References
Lipid nanostructures				
Nano-liposomal carriers (100–200 nm)	One/more phospholipid bilayer(s) separated by internal aqueous compartments	• Safe, bioavailable, and easy to prepare • Low inherent toxicity owing to the compatibility with the body system. • Diversity of the composition to attain efficient formulations • Great capability in encapsulating hydrophilic/lipophilic/amphiphilic bioactives in food, pharmaceutical, agriculture, or cosmetic applications.	• High-price of food-grade lipid components and probable health issues for individuals. • Being not immunologically inert. • Poor physical/chemical resistibility in vivo or vs. co-ingredients present in food matrices. • Low capacity for bioactive loading.	(Davis, 2004; Koudelka et al., 2016; Rostamabadi, 2019a)
Niosomal carriers (10–3000 nm)	Non-ionic amphiphilic vesicles which create a closed bilayer in various unilamellar or multi-lamellar configurations	• Encapsulation both hydrophilic/lipophilic bioactives. • Possessing longer storage-time and stronger chemical stability in comparison with liposomes. • Compatible, available, low-toxic, and stabile versus acid/enzymatic hydrolysis. • Biodegradable with no immunogenic attributes.	• Poor physical resistance upon dispersion. • Risk of structural alteration (e.g., aggregation or fusion), cargo secretion, as well as the disintegration of the bioactive, in presence of liquid media.	(Paecharoenchai, 2013; Moghassemi & Hadjizadeh, 2014)

Nano-emulsions (typically <500 nm)	Mixtures of two incompatible phases either oil-in-water (O/W) or water-in-oil (W/O) stabilized via a proper surfactant	• Encapsulating both hydrophils and lipophils, simultaneously. • Possessing higher solubility as compared to micellar dispersions. • Applicable in food/pharmaceutical industry for targeted delivery of different active agents. • Possessing superior kinetic stability as compared to the coarse forms.	• Need for expensive and special apparatuses for preparation. • Possible risk of some unwanted phenomena, for example, Ostwald ripening, flocculation, membrane permeability, plus gelling, creaming, as well as particle agglomeration over storage. • Sensitivity to low pHs. • Difficulty of providing a well-ordered release mechanism resulted from their small size or the liquid state of the system.	(Jafari, Katouzian, Rajabi, & Ganje, 2017c; Mason, Wilking, Meleson, Chang, & Graves, 2006)
SLNs (50–1000 nm)	Lipid moieties that maintain their crystalline structure either in the body or at room temperature	• Ability to deliver both hydrophilic/lipophilic bioactives. • Providing a flexible, prolonged release manner owing to their solid architecture and slow disintegration. • No bio-toxicity. • Scalable, reproducible, and economical production processes without any need for organic solvents.	• The risk of particle assembly/expansion and bioactive expulsion caused by the unwanted transition in fat crystallinity over storage. • Low-level incorporation capacity for bioactive loading resulted from the stiff structure of fats. • Probable gelation-phenomenon.	Shi et al., 2014; Katouzian et al., 2017; Mehnert & Mäder, 2012; Parhi & Suresh, 2012

(*Continued*)

Table 2.2 An overview on organic nanostructures. (*Cont.*)

NP type	Constituent components	Advantage	Disadvantage	References
NLCs (50–1000 nm)	One liquid-lipid fraction within a basic solid-lipid lattice	• Excellent bioactive loading resulted from its imperfection crystallinity. • Proper for food applications and encapsulation purposes. • Promoting the bioactive stability/bioavailability/dispersibility.	• Difficulty in providing raw materials. • Environmental issues.	Obeidat et al., 2010; Montenegro et al., 2016
Carbohydrate nanostructures				
NPs	Nano-engineered carbohydrate particles based on chitosan, alginate, dextran, gum, etc.	• FDA-approved, biodegradable, and biocompatible • Non-toxic • Simple deletion from body through regular metabolic mechanisms. • Poorly immunogenic • Low cost	• Indigestible carbohydrate-based derivatives may have some safety challenges (especially in nano-sized scale) • Immune-related toxicity induced by nano-fabrication method (e. g., the presence of residual solvents following nano-formulation)	(Bae et al., 2019; Fathi, Martin, & McClements, 2014; Pérez-Álvarez et al., 2019)
Nano-gels (20–200 nm)	Nano-scaled hydrogels constructed by physically/covalently polymer chain cross-linking interactions	• Appropriate for biomedical applications • Low toxicity and high efficiency • High thermodynamic stability • Fast circulation through the body • Easy functionalization and high solubilization	• Low reproducibility • Difficulty in wide-scale industrialization • Weakness in delivering the payload to the site of action • Their arduous clearance following the accomplishment of their mission in vivo	(Pérez-Álvarez et al., 2019; Shitrit et al., 2019)

Nanofibers	Electrospun fibril nanostructures produced using electrospinning approach	• Proper for food applications and labile bioactive loading owing to the room temperature process • Improving the bioactive stability • Inherent high surface-area-to-volume ratio • Tunable morphologies and mechanical features, • Adjustable degradation mechanism of polymer matrix • Cost-effectiveness	• Difficulty in precise control of the solvent evaporation rate. • The risk of solvent residues on the superficial parts of the produced fibrous structures rendering the product cytotoxic and limiting their food/ biomedical applications. • Difficulty of handling a single fiber, owing to the direct collection of the produced fibers onto the plate/bath collector	(Afshar et al., 2019; Amini et al., 2018; Dev & Hemamalini, 2018; Hoseyni et al., 2020; Rostamabadi, Mahoonak, Allafchian, & Ghorbani, 2019AD)
Protein nanostructures				
Nanofibers	Electrospun fibril nanostructures produced using electrospinning method	• Cost-effectiveness • The room temperature process can diminish the risk of protein denaturation and improve the bioactive stability • High efficiency in bioactive loading • High surface-area-to-volume ratio • Tunable morphologies and mechanical features • Adjustable degradation mechanism of polymer matrix	• Difficulty in wide-scale industrialization, massively/ repeatedly. • The lack of extensive in vivo studies and systematic investigations regarding their environmental concerns for industrial up-scaling. • Limited long-term stability, especially about modified electrospun nanofibers.	(Alehosseini et al., 2019; Colin-Orozco et al., 2015; Deng et al., 2018; Gómez-Estaca et al., 2012)

(*Continued*)

Table 2.2 An overview on organic nanostructures. (*Cont.*)

NP type	Constituent components	Advantage	Disadvantage	References
NPs	Nano-dimensioned protein particles e. g., animal/plant proteins such as gelatin, collagen, milk proteins, silk proteins, soy proteins, zein, etc.	• Nature-inspired bio-NPs • Metabolizable and biodegradable • Safe and efficient in site-specific bioactive delivery • High loading capacity • High nutritional value • Good tolerance • Easy to process and high efficiency • Renewability and low toxicity/immunogenicity	• Stability and potential immunogenicity issues • High tendency to form aggregated structures • Immune-related toxicity induced by nano-fabrication method • Difficulty of industrial up scaling owing to their heterogeneous structure and excessive cost. • Limited controlled/sustained bioactive release resulted from the hydrophilic structure of animal protein nanocarrier and their fast solubilization in aqueous media.	(Mitragotri & Stayton, 2014; Owoseni-Fagbenro et al., 2019; Sadeghi, 2020a; Sadeghi, 2020b; Zou et al., 2016)
Nano-gels	Three-dimensional cross-linked hydrophilic networks	• High water content • Hydrophilic three-dimensional network • Low systemic toxicity and immunogenicity • Excellent biodegradability and colloid stability • High loading capacity and bioactivity	• Low scalability • Inconsistent batch-to batch reproducibility • Weakness in delivering the payload to the site of action • Their arduous clearance following the accomplishment of their mission in vivo	(Araiza-Calahorra & Sarkar, 2019; Chen et al., 2018; Pereira et al., 2015; Zhang, Zhhai, Wang, & Zhai, 2016)

Nanotubes	Cylindrical hollow structures fabricated through the self-assembly, pulsed electric field, layer-by-layer deposition methods, etc.	• Biocompatibility, biodegradability, and cost efficiency • Offering three-dimensional nanostructures with efficient, controlled bioactive delivery • The potential of acting as gelling agents • The functionalizing capability of internal and external layers • Improving the system viscosity by virtue of their low diameters (nano-dimension) and long length	• The risk of generation of problematic protein constructions, for example, prion protein and κ-casein amyloid fibrils (upon physiological conditions) that can lead to the mad cow disease and formation of cytotoxic products, respectively.	(Graveland-Bikker et al., 2006; Katouzian & Jafari, 2019; Nguyen et al., 2019; Uddin et al., 2018)

unique functional attributes (such as foaming, emulsification, gelation, and water-binding capacity) (Chen, Remondetto, & Subirade, 2006; Elzoghby, El-Fotoh, & Elgindy, 2011; Elzoghby, Samy, & Elgindy, 2012a).

A variety of food proteins are inherently occurring at the nano-scale dimensions (e.g., β-lactoglobulin, α-lactalbumin, casein micelles, ovalbumin, etc.) and tend to self-assemble into larger bio-constructions, which play vital roles in vivo (Katouzian & Jafari, 2019). In comparison with synthetic NPs, natural protein systems are much less expensive and bear inherent advantages such as superior biodegradability/biocompatibility/bioavailability and lower toxicity and therefore, have been introduced as effectual matrices for encapsulation of numerous food bioactive agents (Yin, Yuvienco, & Montclare, 2017). Protein nanostructures can also offer powerful platforms for delivering bioactives owing to their modifiable compositional sequence and two-dimensional/three-dimensional configuration (Elsadek & Kratz, 2012; Elzoghby, 2013; Nagarsekar & Ghandehari, 1999). The presence of diverse functional groups in the primary sequences of polypeptides opens a great possibility for protein-bioactive interactions (hydrophobic, electrostatic, or covalent bindings) and creation of a 3D protein network harboring a broad range of labile bioactive agents (Owoseni-Fagbenro et al., 2019). Furthermore, protein nanostructures have demonstrated valuable capability for surface modification by virtue of possessing various functional groups (carboxylic-amino units) on the NPs' surface, facilitating specific delivery of bioactives to their site of action (Elsabahy, Nazarali, & Foldvari, 2011).

Nano-scaled protein structures are considered as metabolizable naturally occurring biopolymers, and their hydrolysis induced by digestive enzymes may lead to the generation of non-toxic bioactive peptides, exerting various physiological influences in vivo (Elzoghby, Samy, & Elgindy, 2012b). Nevertheless, these structures still require surmounting hurdles to attain notable improvements in efficiency and de-risked immunogenicity/toxicity. It is worth mentioning that the main bottleneck, in this line, could be the risk of protein by-products and their likely adverse acute or chronic impacts, systemically/locally (Abolmaali, Tamaddon, & Dinarvand, 2013; Frokjaer & Otzen, 2005; Yin et al., 2017). Besides, some proteins may induce the creation of problematic nanostructures like prion protein structures (recognized as the mad cow disease), which can be generated through their misfolding and adapting a non-native infectious form (Dehle, Ecroyd, Musgrave, & Carver, 2010; Katouzian & Jafari, 2019). Another instance is the generation of κ-casein amyloid fibrils upon physiological environments, leading to a cytotoxic behavior (Thorn et al., 2005).

2.2.2.3 Carbohydrate nanostructures

Carbohydrates are a wide group of biomolecules comprised of a variety of functional groups, which can easily undergo (bio)chemical modifications, creating a range of carbohydrate based polymer derivatives (Meng & Edgar, 2016; Shishir, Xie, Sun, Zheng, & Chen, 2018; Rostamabadi, Falsafi, & Jafari, 2019b). These presence of hydroxyl, carboxyl, or amino groups in the structure of most carbohydrates (monosaccharides up to large polysaccharides) enables them a bioadhesive property through non-covalent interactions with biological tissues (e.g., epithelial or mucous surfaces), and therefore allow the substances to possess prolonged half-life in vivo (Rostamabadi, Falsafi, & Jafari, 2019c; Rostamabadi, Falsafi, & Jafari, 2019d). Their derivable functional groups lead to the creation of many diverse structures ranging in charge, configuration, and even molecular weight (Swierczewska, Han, Kim, Park, & Lee, 2016).

The prominent features of carbohydrates including availability, functionality, high stability, biodegradability/biocompatibility, hydrophilicity, as well as non-toxicity making them as attractive biomaterials for the production of engineering nanostructures in the food industry. The development of carbohydrate NPs has become a flourishing research area, consolidating the attributes of nano-dimensional structures with biofunctional characteristics. Nano-carbohydrates can also demonstrate potent multivalent effects, and dramatically promote the carbohydrate-mediated/bioactives interactions. Carbohydrate-based nano-systems in the form of nano-gels, NPs, or nano-fibers (Table 2.2) have proven to be advantageous platforms in bioactive delivery (Hadidi, Pouramin, Adinepour, Haghani, & Jafari, 2020; Hoseyni et al., 2020), biosensors, functional biomaterials, as well as tissue engineering.

2.3 Biological fate of nanostructures

The widespread applications of nanotechnology in the food industries has led to higher inevitable human exposures through consumption of these products or environmental contaminations, raising some safety concerns for human health (Martirosyan & Schneider, 2014; Bazila et al., 2018). Going through the nanoscale completely changes the toxicity profile and biological fate of materials compared to their conventional counterparts and understanding the fate of nanomaterials in human body and their biological interaction is fundamental in evaluations and management of their toxicity (Amini, Gilaki, & Karchani, 2014). The human body is exposed to food-related nanomaterials through direct or indirect exposures. Direct exposure usually occurs through addition of nanomaterials in food products like nanoscale food additives, while indirect exposure includes contaminations from other industries in food chain (i.e., agriculture and animal feed) or migrations from packaging and food contact materials.

Whether the exposure is direct or indirect, food-related nanomaterials may enter the human body via different exposure routs. Considering the potential applications of nanotechnology in food chain industries, there are three main possible routs of exposure for engineered nanomaterials: Oral and intestinal exposure sometimes referred to as "ingestion," dermal, and inhalation exposure (EFSA, 2018). Other exposure routs including intravenous or implantation to the body is unlikely to occur for food-related nanomaterials, although they are possible in some medical applications.

Oral and intestinal exposure are the major exposure routes in food-related nanomaterials, although dermal and inhalation exposures are also possible, generally through touching the products and surfaces containing nanomaterials or inhalation of the airborne particles in manufacturing processes or in some consumer products like pesticides. After entering the systemic circulation, nanomaterials or their degradation products are distributed throughout the body depending on their size and surface characteristics and may accumulate in some tissues. Finally, they may interact at cellular level or be subjected to elimination process (Bergin & Witzmann, 2013; Brandelli, 2020).

Although some studies have evaluated toxicokinetics of engineered nanomaterials (Chung et al., 2013; Park, 2013; Pujalté, Dieme, Haddad, Serventi, & Bouchard, 2017; Trbojevich & Torres, 2017), limited information is available on the ADME (Absorption, Distribution, Metabolism, and Elimination) aspects of food-related nanomaterials in human body due to some technical challenges and numerous influencing factors affecting this process (Cao et al., 2016). However, exposure route is a main determining factor in the assessment of toxicity, because in each route of exposure, nanomaterials face different physiological conditions (i.e., pH, enzymes, mechanical forces, and biological barriers) and hence possess different scenarios. The possible fate of food-related nanomaterials after exposure to human body is described in the following sections for each exposure route. Different physiological conditions and possible exposure scenarios in each route of exposure are summarized in Table 2.3 for food-related nanomaterials.

Table 2.3 Fate of food-related nanomaterials in different exposure routs and their possible toxicities.

Exposure route	Physiological conditions	Fate of nanomaterials	Possible toxicities
Ingestion	• pH changes (from 1.2 in the stomach to 7 in colon) • Mechanical forces • Changes in osmolarity and ion strength • Digestive enzymes • Bile salts and other surface active agents • Gut microbiota • Food components • Mucus layer • Epithelial layer	• Particle aggregation (mostly due to protein corona and pH variations) • Partial or complete degradation • Changes in particle characteristics like size and surface charge	• Local effects on endothelial cells (persorption) • Altering the normal function of gut microbiota • Systemic absorption of nanomaterials or their degradation products from different regions of gastrointestinal tract and accumulation in different organs
Inhalation	• Permeable and highly perfused epithelial layer • Mucociliary clearance • Macrophage uptake • Higher relative humidity and turbulence	• Aggregation (possibility of protein corona), high probability for hygroscopic particles • Changes in physichochemical properties	• Deposition in upper respiratory tract • Immediate inflammatory responses • Chronic lung disease • Systemic absorption from alveolar region and accumulation in different organs
Dermal	• The stratum corneum hinders the entrance of large particles • Skin appendages may act as a reservoir or an absorption pathway for nanomaterials • The barrier is more permeable in damaged skin	• Release of some nanomaterial components like metal ions	• Local effects including sensitivity and contact dermatitis • Systemic absorption of toxic metals from metal NPs • Systemic absorption of nanomaterials depending on their size

2.3.1 Oral and intestinal exposure

The oral and/or intestinal exposure is the main exposure route in food-related nanomaterials. In addition to nanomaterials that are introduced to human body through consumption of nanotechnology-based food and water products, particles captured by mucociliary clearance in the respiratory tract may also go through the GIT. However, in this exposure route, nanomaterials enter the complex gastrointestinal medium, which may alter the physicochemical characteristics of nanomaterials, hence altering their toxicity profile (Bergin & Witzmann, 2013; McClements & Xiao, 2017b).

The GIT is a complex environment, naturally designed for food digestion and nutrient uptake. Hence, orally entered NPs face different conditions including variable pH ranges and mechanical forces in each region, different electrolytes and digestive enzymes, some surface-active agents (i.e., bile salts and phospholipids), food components, and gut microbiota, which may alter the fate of nanomaterials (McClements & Xiao, 2017a; Berardi, Baldelli, & B, 2019). Food-related nanomaterials may be digested or remain intact throughout the GIT, while other scenarios are also possible including aggregation, release of core materials and alterations of physicochemical characteristics in different conditions of GIT or in the presence of food components (Cao et al., 2016; Jafari et al., 2017a). Although providing precise information regarding the fate of nanomaterials in GIT is not available, but this complex media will significantly influence the safety profile of products and considering the properties of nanomaterials in these conditions will help in safety evaluations.

If the nanomaterials are not degraded completely, they can diffuse through the mucus layer, pass the gastrointestinal epithelium, and translocate into the bloodstream, depending on their physicochemical properties (Bouwmeester et al., 2009). The epithelial layer separates GIT from blood capillaries and is covered by a protective mucus layer. The mucus layer which is composed of mucin proteins and carbohydrates, forms a hydrogel-like structure that can trap foreign pathogens and it is an acellular barrier hindering the interaction of NPs with epithelial cells. Nanomaterials with smaller size and positive charge are more likely to penetrate the mucus layer (Ensign, Cone, & Hanes, 2012; Brandelli, 2020). The epithelium which is composed of specific types of epithelial cells in each region, is the most resistant barrier against absorption of nanomaterials. The structural cells of epithelium vary in different regions of GIT. For example, the stomach and intestine are composed of columnar epithelium while squamous epithelium exists in oral cavity and the esophagus. However, some cell types are more permeable than others while the thickness of epithelial layer is also different in each region of GIT (Fröhlich & Roblegg, 2012).

In oral and intestinal exposure, the NPs will initially go through the mouth. The residence time in this area is usually very short due to continuous salivary secretions but it is generally dependent on the type of food product; for example, liquid products have almost no residence time in the mouth. However, the saliva is secreted in mouth which is a hypotonic solution with near neutral pH values (6.2–7.4). It is composed of mucin, some ions (i.e., calcium, phosphate, carbonate, etc.), and enzymes for initial degradation of fats and sugars (primarily amylase and lipase) (Fröhlich & Roblegg, 2012; Lundquist & Artursson, 2016; Teubl et al., 2018). If the nanomaterials are stable enough in this media, they can be absorbed from buccal and sublingual regions depending on their residence time (Hua, 2019). The surface area of the oral mucosa is relatively small (100–200 cm^2) and it is lined by an epithelial layer (with a thickness of 100–200 μm in the sublingual and 500–800 μm in the buccal region), which is covered by a 70–100 μm mucus layer. NPs with appropriate size (<200–300 nm) and surface characteristics (i.e., positive charge) in this medium, which can pass the mucosal and epithelial barriers, can be absorbed rapidly and directly into the systemic circulation via venous drainage to the superior vena cava (Fröhlich & Roblegg, 2012; Hua, 2019).

Once the nanomaterials in food matrix are swallowed, they will go through the esophagus and stomach. Although the residence time in esophagus is not very considerable but it contains some bicarbonate-based secretions and the epithelium consists of squamous cells with a thickness of 300–500 μm, which is the same as buccal region with significantly lower membrane thickness (Fröhlich & Roblegg, 2012). The NPs will then enter the stomach which has an acidic pH and contains some enzymes like pepsin. The gastric pH may vary from 1.2 to 3 depending on the fast and fed state of stomach (Bergin & Witzmann, 2013). However, reduction of pH values may alter the aggregation state and surface charge of nanomaterials or may dissolve the nanomaterials depending on its type and composition. For example, ZnO NPs showed considerable dissolution in gastric pH (Bian, Mudunkotuwa, Rupasinghe, & Grassian, 2011), while Ag NPs were shown to generate Ag^+ ions in gastric pH which may result in particle aggregation (Axson et al., 2015). Silica NPs were also shown to be aggregated in gastric fluids due to reduced solubility in acidic pH (Sakai-Kato, Hidaka, Un, Kawanishi, & Okuda, 2014). The residence time and surface area in the stomach is approximately 3 hours and 500 cm^2, respectively; and it has a relatively thick mucus layer with a 20–25 μm simple columnar epithelium (Fröhlich & Roblegg, 2012; Date, Hanes, & Ensign, 2016). However, the stomach is unlikely to be site of absorption for food-related nanomaterials.

The small intestine has a relatively large surface area (~ 30 m^2) due to the presence of villi and microvilli. It has an alkaline pH around 5–7 and contains bile salts, digestive enzymes (including lipase and protease), and some electrolytes. It is the major site for food digestion and absorption of nutrients. These conditions may change the properties and aggregation state of some nanomaterials. Furthermore, some NPs having starch, protein, or lipid-based structures can be digested by amylase, protease, or lipase enzymes (Fröhlich & Roblegg, 2016; McClements & Xiao, 2017a). For example, liposomes are subjected to degradation by intestinal lipase and bile salts (Daeihamed, Dadashzadeh, Haeri, & Akhlaghi, 2017). However, if the NPs are stable enough and can pass the mucus layer, they may be translocated to systemic circulation via endocytosis, transcytosis, or M-cell uptake, depending on their size or surface characteristics. Small NPs (<50 nm), like metal and metal oxide nanomaterials, are usually subjected to endocytosis, while M-cells can uptake larger particles as well. M-cells are specialized for translocation of foreign materials to lymphatic system and are located in some regions between the enterocytes in the epithelial layer (Fröhlich & Roblegg, 2012; Fröhlich & Roblegg, 2016). NPs can damage the intestinal cells via oxidative stress, inflammation and DNA damage, and some NPs can pass through these damaged areas. This mechanism is called persoption and happens to a lesser extent for food-related nanomaterials (Hillyer & Albrecht, 2001; Fröhlich & Roblegg, 2016).

After the earlier-mentioned steps, the existing nanomaterials will enter human colon, which has a thick mucus layer, a pH value around 6–7, and contains bacteria and undigested food components. The gut microbiota release some enzymes and biopolymers that may change the properties of nanomaterials (McClements & Xiao, 2017b). However, some nanomaterials like inorganic NPs with antibacterial effects may alter the normal bacterial functions in the GIT and higher exposures can cause clinical disorders including colitis and immunological dysfunctions (Fröhlich & Fröhlich, 2016). Studies have shown that most of engineered nanomaterials including silver, titanium dioxide, zinc oxide, and silica NPs can have harmful effects on gut microbiota (Pietroiusti, Magrini, & Campagnolo, 2016).

In conclusion, after oral exposure, food-related nanomaterials will encounter different conditions in the complex gastrointestinal media including pH variations, mechanical forces, different ionic strengths and osmolarities, numerous proteins, surface-active agents and enzymes, food components, and gut microbiota. This constantly changing media may alter the physicochemical properties and aggregation state of nanomaterials. Ultimately, the engineered nanomaterials may have local effects in the GIT or they can be absorbed and have systemic toxicities, which should be taken into considerations.

2.3.2 Inhalation exposure

Inhalation exposure is less common in food-related nanomaterials and it rarely happens through direct consumption of nanotechnology-based food products. However, applications of nanotechnology in the food industries may lead to respiratory uptake of these particles, especially fine powders, from environment or work places. Furthermore, some applications in food-related industries, like nano-pesticides in agriculture, may increase inhalation exposure to nanomaterials (Bierkandt, Leibrock, Wagener, Laux, & Luch, 2018).

The respiratory tract has a relatively large surface area (100–150 m^2), which is highly perfused and is separated from the systemic circulation by a thin monolayer of alveolar epithelium. The epithelial layer in the alveolar region is very thin (0.1 μm) and permeable to facilitate gas exchange to blood. Consequently, particles reaching the alveolar region may reach the systemic circulation to a great extent (Bierkandt et al., 2018; Brandelli, 2020). Particles entering the respiratory tract may be deposited in different regions of the respiratory tract according to their size and physicochemical properties including particle shape, surface charge, and density. Larger particles (>5 μm) usually deposit in upper respiratory tract and are usually captured and eliminated by mucociliary mechanism. Nanosized particles usually access the alveolar region of the lungs and can cross the alveolar epithelium and reach the systemic circulation (Bierkandt et al., 2018; Poh et al., 2018). However, the physicochemical properties and aggregation state of nanomaterials may change in respiratory tract. Hygroscopic particles may form large aggregates in the elevated relative humidity of respiratory tract or the surface charge of particles can change due to protein corona or in acidic environment of lysosomes in alveolar macrophages (Creutzenberg et al., 2012; Braakhuis et al., 2014a).

The inhaled nanomaterials may remain in the lungs and cause local respiratory toxicities, or pass the alveolar barrier and reach the bloodstream depending on their size and physicochemical characteristics. The biological responses to the inhaled NPs also range from immediate reactions like local inflammation or immunological responses to long-term chronic adverse effects due to accumulations in some organs including lung, liver, and brain (Bakand, Hayes, & Dechsakulthorn, 2012; Poh et al., 2018). Some nanomaterials like titanium dioxide and carbon nanotube have shown to induce chronic diseases such as pulmonary inflammation, pneumonia, airway hyperreactivity, and pulmonary granuloma due to accumulation in the lung (Ferreira, Cemlyn-Jones, Robalo, & C, 2013). In addition, studies have shown that nanomaterials can impair the phagocytic activity of macrophages and reduce natural respiratory protection mechanisms (Donaldson et al., 2002; Renwick, Brown, Clouter, & Donaldson, 2004).

2.3.3 Dermal exposure

Dermal exposure of food-related nanomaterials does not take place as frequently as oral exposure, but may happen during contacts of skin with the products or in manufacturing environments. The skin is composed of two main layers named the epidermis and the dermis layer. The epidermis consists of five layers of epithelial cells with the stratum corneum being the outer protective layer of epidermis, which is comprised of dead keratinocytes migrating from the basal layer towards the surface of skin. The dermis is composed of connective tissue and the blood vessels and skin appendages including hair follicles and sweat glands are located in this layer. The stratum corneum is a natural barrier against exogenous materials but some molecules can pass this barrier through cellular and/or intercellular pathways (Jatana & DeLouise, 2014).

Skin exposure of nanomaterials may cause local and systemic toxicities. Local effects including irritation and contact dermatitis has been observed for metal NPs or nanomaterials having metal catalytic residues. These nanomaterials can release some ions that induce local skin effects, like skin irritations and contact dermatitis, or may result in absorption of toxic metals (Larese Filon et al., 2016). However, other local effects are also possible, for examples TiO_2 NPs were shown to have skin-aging effects in chronic exposures (Wu et al., 2009).

In addition to local effects, nanomaterials can penetrate the skin and reach the systemic circulation via two major pathways: (1) The intercellular route, which is possible for very small nanomaterials (under 4 nm), (2) skin appendages (sweat glands and hair follicles) pathway for larger particles in the size range of 4–20 nm. It should be noted that particles >45 nm are unlikely to be absorbed and particles in the size range of 20–45 nm may be absorbed through damaged skin. Generally, particles <45 nm may be absorbed from skin and enter the systemic circulation although this size range may be wider for flexible nanomaterials including liposomes and micelles (Larese Filon et al., 2016). However, some larger particles may penetrate deeply into the hair follicles and stay there up to 10 days (Lademann et al., 2011).

2.4 Biological responses to the nanostructures entrance to the body

Application of nanostructure has many benefits in the food industry. However, in some cases, their application may have some adverse health effects. The adverse health effects of these nanostructures have been linked to higher possibility of direct or indirect interactions between these particles and biological systems. The size, chemical composition, surface characteristics, and shape of the nanostructures also affect these interactions. Recent advances indicated that NPs, especially metal-based NPs, greatly induce oxidative stress (Anjum, Gill, Duarte, & Pereira, 2019), inflammation (Luo, Chang, & Lin, 2015), genotoxicity (Magdolenova et al., 2014), epigenetic modification (Smolkova, El Yamani, Collins, Gutleb, & Dusinska, 2015), ER stress (Cao et al., 2017), and autophagy dysfunction (Stern, Adiseshaiah, & Crist, 2012), leading to adverse health effects. In the literature, these adverse health effects were mainly addressed and investigated for inorganic NPs (especially metal or metal oxides). However, studies on the negative health implications of organic nanomaterials including proteins, lipids (Doktorovova, Souto, & Silva, 2014), and carbohydrates (Ma & Yang, 2015) as well as their mechanisms are very limited. Therefore, in this section, we focus on different mechanisms related to the adverse health effects of inorganic nanostructures.

2.4.1 Oxidative stress

Effect of oxidative stress on the mechanisms of the nanostructures was studied extensively in the recent literatures (Manke, Wang, & Rojanasakul, 2013, Anjum et al., 2019, Evans, Jenkins, Doak, & Clift, 2019). Oxidative stress is the result of an imbalance between ROS and antioxidants. A number of nanostructures have been shown to generate oxidative stress, especially metal oxide NPs are known as stress agents that could release ions capable of catalyzing the formation of the highly reactive hydroxyl radical (•OH) through conversion of H_2O_2 by Fenton chemistry, which caused lipid peroxidation, DNA and protein damage (Evans et al., 2019). The cellular or organ/tissue levels of the ROS (O_2• −, OH•, and H_2O_2), reactive/protein carbonyls, thiobarbituric acid reactive substances/malondialdehyde, and the ROS-led protein and membrane lipid products are considered to be the important biomarkers of oxidative stress in plants (Anjum et al., 2019). On the other hand, following the induction of oxidative stress, the plants enhance the ability of antioxidant defense system as enzymatic (superoxide dismutase, glutathione sulfo-transferase, catalase, guaiacol peroxidase, ascorbate peroxidase, dehydroascorbate reductase, monodehydroascorbate reductase, glutathione reductase) and non-enzymatic (phenolics, carotenoids, ascorbate, glutathione, tocopherols) to overcome elevated ROS (Anjum et al., 2019). However, higher concentrations and prolonged exposure to NPs were observed to suppress the antioxidant defense system (Zoufan, Baroonian, & Zargar, 2020).

More recently, the toxic effects of Cu oxide NPs (12.5 μg/mL and 50 μg/mL), Cu-iron oxide nanopowders (50 μg/mL and 100 μg/mL) and carbon nanopowders (50 μg/mL and 100 μg/mL) in the human hepatoma HepG2 cells using oxidative stress (total ROS and superoxide anion production) assays for 3 or 24 h were examined by (Adeyemi et al., 2020). They showed that Cu oxide NPs and Cu-iron oxide nanopowder, increased the production of total ROS, whereas carbon nanopowder significantly reduced its levels. In addition, prolonged exposure to the NPs caused production of more superoxide anions. Another research on diet of standard cornmeal media mixed with Ag-NPs (50 and 100 μg/mL) fed third instar larvae of *D. melanogaster* for 24 and 48 h revealed that Ag-NPs induced oxidative stress (Ahamed et al., 2010b). It was observed that antioxidant glutathione content was significantly lower, whilst superoxide dismutase, catalase, and malondialdehyde levels, biomarkers of oxidative stress, were significantly higher in Ag-NPs exposed organisms. The authors concluded that Ag-NPs exposure induced the oxidative stress, which resulted in DNA damage and ultimately apoptosis.

A more recent study, (Ahamed, Akhtar, Khan, Alrokayan, & Alhadlaq, 2019) investigated the potential mechanisms of cytotoxicity of bismuth (III) oxide NPs (Bi_2O_3 NPs) (50–300 μg/mL) through oxidative stress in human breast cancer (MCF-7) cells. They reported that generation of ROS, higher lipid peroxidation, reduction of glutathione, and low superoxide dismutase enzyme activity were indicated as the oxidative stress responses of Bi_2O_3 NPs that induced cell death. Accordingly Ahamed et al. (2011) observed that nickel ferrite NPs ($NiFe_2O_4$) also induced oxidative stress in A549 cells by the induction of ROS and reduction of antioxidant glutathione, suggesting the potential mechanism of oxidative stress. Chen et al. (2019) showed that the glutathione level was lower, whereas ROS level was higher in the cells incubated with zinc oxide NPs (ZnO NPs). According to Choi et al. (2010), treatment with Ag-NPs induced malondialdehyde level by 1.5–1.7-folds in zebrafish compared to controls, which caused oxidative damage in the liver tissues. The catalase and glutathione peroxidase 1a, oxyradical-scavenging enzymes, were decreased in the tissues. On the other hand, the same researchers found that total glutathione level increased in the liver of zebrafish, coping with the increased level of oxyradicals caused by Ag-NPs.

More recently, Dąbrowska-Bouta, Sulkowski, Strużyński, and Strużyńska (2019) examined the oxidative stress-related biochemical parameters in myelin isolated from adult rat brain treated with Ag-NPs (2 weeks to 0.2 mg/kg b.w.). According to the results, it was observed that lipid peroxidation increased, the concentration of protein and non-protein SH groups on the myelin membrane decreased, and -SH-containing amino acids were converted to oxidized or S-nitro oxidized forms. These findings indicate that oxidative stress is an important mechanism for inducing Ag-NPs/Ag^+ neurotoxicity. By assessing the influence of Ag-NPs' intake on oxidative stress by measuring liver and heart superoxide anion production and liver malondialdehyde levels using Sprague–Dawley rats. Elle et al. (2013) reported that the production of superoxide anion was increased by 30% and 41% in the liver and cardiac after the rats received Ag-NPs compared with the control, respectively. Moreover, paraoxonase activity, marker of oxidative stress, was reduced by increased production of superoxide anion. However, malondialdehyde levels were unchanged. On the other hand, oxidative stress-responding transcription factors (i.e., nuclear factor κB (NF-κB) and nuclear factor-E2-related factor-2 (Nrf-2)) and the signal transduction pathway (i.e., mitogen-activated protein (MAP) kinase pathway) played key roles in oxidative stress. For example, Eom & Choi (2009a) showed that silica NPs (Si-NPs) led to the induction of heme oxygenase-1 via the Nrf-2–ERK MAP kinase signaling pathway. Similarly, the same researchers reported that cerium dioxide NPs (CeO_2NPs) led to significant increases in the cellular ROS concentrations, subsequently causing a great induction of heme oxygenase-1 via the p38-Nrf-2 signaling pathway (Eom & Choi, 2009b).

High concentration and prolonged exposure significantly exacerbated the oxidative stress. More recently, Zoufan et al. (2020) reported that the increasing concentration of ZnO-NPs enhanced the levels of ROS and malondialdehyde, and the proline content and antioxidant enzyme activities including catalase, guaiacol peroxidase, and superoxide dismutase showed a significant increase. According to the authors, all concentrations of ZnO-NPs used in the study greatly induced oxidative stress.

Several studies indicated that the size of NPs also affects their activity. Zhang et al. (2020) investigated the potential neurotoxicity of pegylated Au-NPs on the central nervous system, and found that the cytotoxicity in SHSY5Y cells of the different sized pegylated Au-NPs followed the order of 4.5 nm >13 nm >30 nm. In addition, 4.5 nm pegylated Au-NPs induced the overproduction of ROS and malondialdehyde, whereas superoxide dismutase, glutathione peroxidase, and catalase were decreased consistently with the oxidative stress. Similarly, Prasad et al. (2013) investigated the size effects (10 and 75 nm) and surface coating (citrate and polyvinylpyrrolidone) on the oxidative stress. The smaller (10-nm) Ag-NPs were more potent than the larger (75-nm) ones in each stress response pathway.

On the other hand, carbon nanotubes significantly affected the oxidative stress. For example, Manna et al. (2005) observed that single walled carbon nanotubes induced oxidative stress in human keratinocyte cell. In another study, Oberdörster (2004) showed that fullerenes increased the lipid peroxidation in brain of juvenile largemouth bass. It is also worth emphasizing that the potential to cause oxidative stress is characteristic of many pathogenic particles, and oxidative stress in cells exposed to the particles is considered to be an important pathogenic event that can lead to inflammation, genotoxicity, epigenetic modification, ER stress, and autophagy dysfunction (Manke et al., 2013). Table 2.4 shows some selected studies on the oxidative stress mechanism related to adverse health effects of inorganic NPs.

2.4.2 Inflammation

Inflammation is a natural part of our body's immune response to harmful conditions and can be classified as either acute or chronic. Acute inflammation is a short-term process occurring in response to tissue injury, usually appearing within minutes or hours, whereas chronic inflammation is a long-term inflammation, leading to severe chronic disease (Chen et al., 2018). Although inflammation can prevent infection, it is a complex and coordinated response group that is harmful in cases caused by chronic environmental factors such as exposure to NPs. Inflammatory effects, including fibrosis, oxidative stress, and DNA damage, predominantly dominate the pathogenic mechanisms triggered by NPs, making inflammation a target for the toxicology tests (Borm & Kreyling, 2004). The activation of oxidative stress-sensitive transcription factors stimulates the transcription of inflammatory factor genes and leads to the release of many inflammatory mediators (L'Azou & Marano, 2011). It is noted that cytokines/chemokines, acute phase proteins, immune-related effectors, reactive oxygen and

Table 2.4 Selected studies related to the oxidative stress caused by inorganic NPs

Nanomaterials	Model	Key outcomes	References
CuO NPs $CuFe_2O_3$ NPs	Human hepatoma HepG2 cell lines	• CuO NPs was highly toxic to human HepG2 cells • The excessive production of ROS, mainly superoxide anions, resulted in oxidative stress	(Adeyemi et al., 2020)
TiO_2 NPs	Human neuroblastoma (SH-SY5Y) cells	• Apoptosis increased by 4- to 10-fold compared to the control • ROS production values reached up to 10- fold of the control's	(Ferraro, Domingo, Etcheverrito, Olmedo, & Tasat, 2020)
MoS_2 NPs	Oecophylla smaragdina (Asian weaver ant)	• ROS, SOD and the lipid peroxidation (LPO) product were increased • CAT, GST, and GSH were reduced	(Sheeja, Anusri, Levna, Aneesh, & Lekha, 2020)
Au PEG NPs	C57B6/L mice	• ROS and MDA were increased • SOD, CAT and GSHPX were decreased • Smaller NPs were more toxic than the larger ones (4.5 nm >13 nm >30 nm)	(Zhang et al., 2020)
ZnO NPs	Chenopodium murale	• ROS, MDA, CAT, SOD, GPX, and proline content were incerased	(Zoufan et al., 2020)
Bi_2O_3 NPs	Human breast cancer (MCF-7) cells	• Higher ROS, higher lipid peroxidation, reduction of glutathione and low SOD enzyme activity was observed • Bi_2O_3 NPs induced oxidative stress and cell death	(Ahamed et al., 2019)
Ag-NPs	Wistar rats	• Higher lipid peroxidation • Lower concentrations of protein and non-protein –SH groups • Oxidative stress is a significant mechanism of AgNP/ Ag + -induced neurotoxicity	(Dąbrowska-Bouta et al., 2019)
	Sprague–Dawley rats	• Higher superoxide anion production • No change in the activity of SOD and MDA level • Lower plasma paraoxonase activity	(Elle et al., 2013)

(Continued)

Table 2.4 Selected studies related to the oxidative stress caused by inorganic NPs (*Cont.*)

Nanomaterials	Model	Key outcomes	References
	Drosophila melanogaster	• MDA level, an end product of lipid peroxidation, was significantly higher in Ag-NPs treated larvae • GSH was significantly lower in Ag-NPs exposed organisms • Antioxidant enzymes SOD and CAT were significantly higher in Ag-NPs treated larvae. • Ag-NPs increased the level of lipid peroxidation • DNA damage and ultimately apoptosis	(Ahamed et al., 2010b)
	Zebrafish	• Ag-NPs induce oxidative stress, DNA damage, and apoptosis • Higher MDA, glutathione • Lower oxyradical-scavenging enzymes such as CAT and GSHPX 1a	(Choi et al., 2010)
Fe_2O_3	Human Breast Cancer (MCF-7)Cell	• Higher ROS generation, higher lipid peroxidation and depletion of SOD, GSH, CAT activities • Oxidative stress caused cell death and DNA damage	(Alarifi, Ali, Alkahtani, & Alhader, 2014)
$NiFe_2O_4$	Human lung epithelial A549 cells	• Higher ROS and reduction of GSH	(Ahamed et al., 2011)
SiO_2 NPs	Human bronchial epithelial cell (Beas-2B)	• HO-1 induction via the Nrf-2–ERK MAP kinase signaling pathway	(Eom & Choi, 2009a)
CeO_2 NPs	Human bronchial epithelial cell (Beas-2B)	• HO-1 induction via the p38-Nrf-2 signaling pathway	(Eom & Choi, 2009b)
Carbon Nanotube	Human embryonic kidney	• IL-8, thiobarbituric acid reactive substances were increased. • Intracellular glutathione levels were decreased. • Physical properties of carbon NPs differently affected the oxidative stress.	(Rama Narsimha Reddy, Narsimha Reddy, Himabindu, Rama, & K, 2011)
	Human keratinocyte cell	• Single-walled carbon nanotubes increased oxidative stress.	(Manna et al., 2005)
Fullerenes, C_{60}	Brain of Juvenile Largemouth Bass	• Lipid peroxidation was elevated. • Total glutathione was depleted	(Oberdörster, 2004)

CAT, Catalase; GPX, Guaiacol peroxidase; GSH, Glutathione; GST, Glutathione S Transferase; GSHPX, Glutathione peroxidase; MAP, mitogen-activated protein; MDA, Malondialdehyde; Nrf-2, nuclear factor-E2-related factor-2; ROS, Reactive oxygen species; SOD, superoxide dismutase; HO-1, Heme oxygenase-1.

nitrogen species, prostaglandins and cyclooxygenase-related factors, and mediators such as transcription factors and growth factors are considered to be the important inflammation-related biomarkers (Brenner et al., 2014). The inflammatory response also coordinates the activation of three main pathways including NF-κB, mitogen-activated protein kinases (MAPK), and the Janus kinase/signal transducers and activators of transcription (JAK-STAT), playing major roles in inflammation, and causing inflammation-associated diseases (Chen et al., 2018).

Particle size, especially small size, is an important factor affecting the induction of inflammation. Braakhuis, Park, Gosens, De Jong, and Cassee (2014b) reported that exposure to 15 nm Ag-NPs enhanced the number of neutrophils and the level of monocyte chemoattractant protein-1 (MCP-1) and Interleukin-1 (IL-1b) by 175 and 5-folds, respectively, while 410 nm Ag particles had no effect. A possible explanation is that smaller particles dissolve faster compared to larger particles due to the high surface to volume ratio. In another study, larger mesoporous carbon nanomaterials induced more pronounced pro-inflammatory effects compared to their smaller counterparts (Chen et al., 2017).

More recently, Wang et al. (2020) investigated the effects of pulmonary exposure of ZnO-NPs and bulk-sized ZnO in mice. ZnO-NPs adversely affected the lung inflammation compared to the bulk-sized ZnO. Moreover, ZnO-NPs increased lung inflammation associated with lipopolysaccharides via increasing expression of IL-1β, IL-6, monocyte chemotactic protein-1α (MIP-1α), and granulocyte-macrophage colony stimulating factor (GM-CSF), while bulk-sized ZnO enhanced the lung inflammation via increasing expression of MCP-1 and MIP-1α.

Choi et al. (2010) investigated the inflammatory footprints of cerium oxide, nickel oxide, ZnO, and CuO NPs in the lung. Both acute (24 h) and chronic (4 weeks) inflammatory showed different inflammatory footprints. Chronic inflammatory responses differed after 4 weeks, with neutrophilic, neutrophilic/lymphocytic, eosinophilic/fibrotic/granulomatous, and fibrotic/granulomatous inflammation being caused respectively by CeO_{2-}NPs, NiO-NPs, ZnO-NPs, and CuO-NPs. On the other hand, patterns of neutrophil and eosinophil infiltrates differed after all these NPs treatments as the acute inflammatory response. In another study, IL-1, IL-6, IL-4, IL-10, IL-12, and transforming growth factor β (TGF-β) increased in a dose-dependent manner by the repeated administration of oral Ag-NPs (Park et al., 2010). More recently, Yu et al. (2019) evaluated the effects and related mechanisms of silica NPs on pulmonary inflammation through repeated intravenous administration in mice in a 45-day trial. In both serum and lung tissues, tumor-necrosis-factor-alpha (TNF-α), IL-1β, and IL-6 levels were increased in silica NPs-treated mice compared to the control group, causing pulmonary inflammation. Another study conducted by Cho et al. (2012) reported that both zeta potential under acidic conditions for low-solubility NPs and solubility to toxic species for high-solubility NPs induce the acute pulmonary inflammogenicity.

For a number of NPs, oxidative stress-related inflammatory reactions have been observed. In addition to the harmful effects on cellular proteins, lipids and DNA, the increase in ROS levels also causes the cells to activate the pro-inflammatory signaling cascade, leading to programmed cell death (Luo et al., 2015). Elle et al. (2013) examined the effects of Ag-NPs on markers of inflammation and oxidative stress in rats' liver and heart for 81 days. They found that liver tumor-necrosis-factor-alpha and interleukin-6 were increased by 9% and 12% in oral Ag NP supplementation, respectively, when compared with the control rat. This may also be explained with the fact that increases in the superoxide anion induced inflammation. Similarly, Guo et al. (2015) observed that silica NPs induced inflammatory cytokines, including interleukins 1 beta, 6, and 8, tumor-necrosis-factor-alpha, intercelluar adhesion molecule, vascular cell adhesion molecule-1, and monocyte chemoattractant protein-1. ROS generation elevated by silica NPs, induced proinflammatory responses. Besides, the authors concluded that silica NPs induced inflammation, oxidative stress, and nitric oxide (NO)/nitric oxide synthases (NOS) system imbalance, causing endothelial dysfunction via activation of the MAPK/Nrf2 (NF-E2 p45-related factor 2) pathway and nuclear factor-κB signaling. On the other hand, Lu et al. (2009) reported that the produced free radicals in vitro did not predict inflammation, while alumina 2 had no free radical activity but was inflammogenic.

Multi-walled carbon nanotubes led to severe inflammatory effects and DNA damage (Yamashita et al., 2010). On the other hand, graphene nanosheets and multiwalled carbon nanotubes induced site-specific Th2 inflammatory responses via the IL-33/ST2 axis (Wang et al., 2013b). In another study, Brown et al. (2007) investigated the in vitro effects of carbon nanotubes and nanofibers for their ability to stimulate the release of the pro-inflammatory cytokine TNF-α from human peripheral blood mononuclear cells. It was observed that TNF-α and ROS were produced by long, straight, well-dispersed nanofilaments compared with highly curved and entangled materials in monocytic cells. Some recent researches on the inflammation mechanisms related to adverse health effects of NPs are presented in Table 2.5.

Table 2.5 Some studies related to the inflammation caused by inorganic NPs.

Nanomaterials	Model	Key outcomes	References
ZnO NPs	Wild-type C57BL/6J mice	• ZnONPs increased lung inflammation • Expression of IL-1β, IL-6, MIP-1α and GM-CSF was increased	(Wang et al., 2020)
Si NPs	ICR mice	• TNF-α, IL-1β, and IL-6 levels were increased	(Yu et al., 2019)
	Human umbilical vein endothelial cells	• IL-1β, IL-6, IL-8, TNFα, VCAM-1, and MCP-1 were increased	(Guo et al., 2015)
	Endothelial cells	• TNF-α IL-1β, CRP, and IL-6 were increased	(Duan et al., 2014)
TiO_2 NPs	A/J Jms Slcmice	• Phospho-p38, NF-kB, and VCAM-1 were increased in a dose-dependent manner in the lung tissue of mice • Inflammation was observed in a TiO_2 dose-dependent manner after inhalation of TiO_2NPs	(Yu et al., 2015b)
Ag-NPs	Fischer rats	• Exposure to 15 nm, higher level of influx of neutrophils, pro-inflammatory cytokines, total glutathione were found • No effects were found after exposure to 410 nm.	(Braakhuis et al., 2014b)
Ag-NPs	ICR mice	• The pro-inflammatory cytokines including IL-1, IL-6, IL-4, IL-10, IL-12, and TGF-β were dose-dependently and increased.	(Park et al., 2010)
	RAW 264.7 cells	• IL1a, IL1b, IL6, IL10, MIP1a, MIP1b, MIP2, GCSF, and TNFα were increased	(Park et al., 2011)
CeO_2 NPs NiO NPs ZnO NPs CuO NPs	Wistar rats	• Neutrophilic, Neutrophilic/lymphocytic, Eosinophilic/fibrotic/granulomatous, Fibrotic/granulomatous inflammation were observed.	(Cho et al., 2010)
Carbon nanotubes	C57BL/6 mice	• Multi-walled carbon nanotubes led to cause severe inflammatory effects and DNA damage.	(Yamashita et al., 2010)

(Continued)

Table 2.5 Some studies related to the inflammation caused by inorganic NPs. (*Cont.*)

Nanomaterials	Model	Key outcomes	References
Graphene nanosheet Multiwalled carbon nanotubes	C57BL/6 mice	• These nanomaterials induce site-specific Th2 inflammatory responses via the IL-33/ST2 axis.	(Wang et al., 2013a)
Single carbon nanotubes and carbon nanofibers	C57BL/6 mice	• Pulmonary inflammatory and fibrogenic responses to single carbon nanotubes and carbon nanofibers varied depending upon the agglomeration state of the particles/fibers.	(Murray et al., 2012)
Mesoporous carbon nanomaterials	BALB/c female mice	• Larger mesoporous carbon nanomaterials induced more pronounced pro-inflammatory effects compared to the their smaller counterparts.	(Chen et al., 2017)
Carbon nanotubes and nanofibers	Human peripheral blood mononuclear cells	• The cellular response varied depending upon fiber morphology and state of aggregation. • TNF-α and ROS was significantly produced by long, straight, well-dispersed nanofilaments compared with highly curved and entangled materials in monocytic cells.	(Brown et al., 2007)

GMCSF, Granulocyte-macrophage colony stimulating factor; GCSF, granulocyte colony-stimulating factor; IL-1, interleukin-1; MCP-1, monocyte chemoattractant protein-1; MIP1a, macrophage protein 1-alpha; MIP-1α, monocyte chemotactic protein-1α; TGF-β, transforming growth factor β; TNFα, tumor-necrosis-factor-alpha; VCAM-1, vascular cell adhesion molecule-1.

2.4.3 Genotoxicity

Genotoxicity and genetic damages could result from the intake of any external sources including nanostructures. The genotoxicity mechanisms of nanostructures were classified into two groups including primary and secondary types. In primary genotoxicity, direct action of nanostructures on nucleic acids and induction of DNA damage could occur. In addition, formation of free radicals acting on the nucleic acid may led to indirect primary genotoxicity. On the other hand, the secondary genotoxicity can be a result of oxidative DNA attack by ROS via activated macrophages and neutrophils during NP-elicited inflammation (Magdolenova et al., 2014, Manickam,Velusamy, Lochana, Rajendran, & Tamizhselvi, 2017).

Ahamed et al. (2010b) investigated the effect of Cu oxide NPs induced genotoxic response through p53 pathway in human pulmonary epithelial cells. The upregulated expression pattern of Hsp70, p53 and DNA repair proteins by exposure to Cu oxide NPs indicated that the genotoxicity of NPs could be assessed with these proteins as molecular biomarkers. It was observed that Cu oxide NPs had a genotoxic potential that could be mediated through oxidative stress. More recently, Alaraby et al. (2020) indicated that 2 mM Cu oxide NPs greatly induced the DNA damage. The expression of the intestinal barrier genes (Duox, PPO2 and Peri A) was significantly down-regulated at the 0.4 and 2 mM doses of Co-NPs, whereas the repair gene (Ogg1) and the stress gene (Hsp70) were significantly down-regulated in a dose-dependent manner. The authors also showed that exposure to small sized (~ 50 nm) Cu oxide NPs was more genotoxic which allowed their pass into nucleus via transport receptors or during cell cycle to interact directly with DNA.

Similarly, Zhou et al. (2019) reported that the smallest silica NPs (10 nm) led to the largest effects of oxidative stress and genotoxicity to human umbilical vein endothelial cells (HUVECs) in culture. On the other hand, degree of agglomeration of NPs greatly affected genotoxicity. For example, Benameur et al. (2015) reported that exposure of cerium oxide NPs showed significantly higher genotoxic properties by a clastogenic mechanism. They also showed that elevated genotoxicity of cerium oxide NPs was related to the differences in their agglomeration process in dulbecco's modified Eagle's medium. It was also found that ROS and malondialdehyde level played an important role in inducing the genotoxic effects. Similarly, Titanium dioxide NPs (TiO_2-NPs) were found to be genotoxic at lower concentration (0.25 mM) followed by gradual decrease. It could be a possible reason for the agglomeration property of TiO_2-NPs (Ghosh, Bandyopadhyay, & Mukherjee, 2010).

In another study, AshaRani, Low Kah Mun, Hande, and Valiyaveettil (2009) showed that exposure of Ag-NPs led to the disruption of the mitochondrial respiratory chain, elevated ROS production and interruption of ATP synthesis, causing DNA damage. Ag-NPs induced the DNA breakage and micronucleus formation in a dose-dependent manner. Also, the genotoxicity of Ag-NPs was related with ROS production. Similarly, DNA damage and micronucleus formation was related to cell exposed to TiO_2-NPs with the induction of ROS and oxidative stress causing oxidative micronucleus formation and DNA damage, a possible mechanism of genotoxicity. In another study, (Wang et al., 2012) showed that exposure to 15 mg/L Cu oxide NPs increased the expression of p38 and p53, sensitive molecular biomarkers of genotoxicity, as early as after 4 h of the treatment. They concluded that intracellular Cu oxide NPs first produced ROS, which subsequently induced the expression of p38 and p53, finally causing irreversible DNA damage.

Lindberg et al. (2009) observed that carbon nanotubes induced a dose-dependent increase in DNA damage at all treatment times studied. The researchers also reported that graphite nanofibers caused DNA damage depending on the treatment time and dosage on human bronchial epithelial cells. On the other hand, the phosphorylated form of histone H2AX (γH2AX) foci formation indicated DNA damage. Guo, Zhang, Zheng, Yang, and Zhu (2011), for instance, evaluated the genotoxic effects of multi-wall carbon nanotubes on human umbilical vein endothelial cells. They showed that multi-wall carbon nanotubes induced the γH2AX foci formation, and reduced the viability of human umbilical vein endothelial cells. In another study, multi-wall carbon nanotubes induced a significant increase in percent tail DNA [2.96%–16.39%] compared to the control [2.02%] (Patlolla, Knighten, & Tchounwou, 2010). Graphene oxide nanosheets induced chromosomal aberrations in bone marrow and DNA damage in lung cells (El-Yamany et al., 2017). More studies about the genotoxicity mechanisms are presented in Table 2.6.

2.4.4 Epigenetic modification

Epigenetic modifications are considered as possible molecular targets after exposure to nanostructures (Sierra, Valdés, Fernández, Torrecillas, & Fraga, 2016). The possible mechanisms behind gene-environment interactions have been discussed in recent studies (Table 2.7) and include DNA methylation, histone modifications, and noncoding RNAs which are discussed further with details.

Table 2.6 Genotoxicity of inorganic NPs

Nanomaterials	Model	Key outcomes	References
Co NPs	Drosophila melanogaster	• The level of DNA strand breaks was significantly increased • Concentration of CoNPs affected the levels of DNA damage	(Alaraby et al., 2020)
Si NPs	Human umbilical vein endothelial cells	• The smallest silica NPs (10 nm) led to the highest effects of oxidative stress and genotoxicity	(Zhou et al., 2019)
CeO_2 NPs	Human dermal fibroblasts	• Clastogenic mechanism of chromosomal damage was found, causing structural chromosomal aberrations • Nano-CeO_2 of the material showed significantly higher genotoxic properties than micro-CeO_2	(Benameur et al., 2015)
CuO NPs and Ag-NPs	Mussels Mytilus galloprovincialis	• Both NPs possessed DNA damaging potential in hemocytes, showing genotoxicity	(Gomes et al., 2013)
CuO NPs	Human lung epithelial (A549) cells	• Oxidative stress was confirmed to be the first response of CuO NPs cytotoxicity, rather than the DNA damage • Expression of p38 and p53 was increased, leading to the irreversible DNA damage	(Wang et al., 2012)
TiO_2-NPs	Human epidermal cells	• DNA damage and micronucleus formation were observed, relating to oxidative stress, a possible mechanism	(Shukla et al., 2011)
	Human lymphocytes	• Lymphocytes showed significant genotoxic response at 0.25 mM, followed by gradual decrease	(Ghosh et al., 2010)
Ag-NPs	Murine peritoneal macrophage cell line	• Gene mutation was increased when treated with 113 nm • The concentrations of Ag-NPs induced an increase in gene mutation frequencies	(Park et al., 2011)

(*Continued*)

Table 2.6 Genotoxicity of inorganic NPs. (*Cont.*)

Nanomaterials	Model	Key outcomes	References
	Human lung cancer cell line, A549	• DNA damage induced by ROS, leading to ROS-induced genotoxicity	(Foldbjerg, Dang, & Autrup, 2011)
	Human bronchial epithelial (BEAS-2B) cells	• DNA breakage and micronucleus formation were induced in a dose-dependent manner	(Kim, Kim, Lee, Oh, & Chung, 2011)
	Human lung fibroblast cells (IMR-90) and human glioblastoma cells (U251)	• The disruption of the mitochondrial respiratory chain by Ag-NPs enhanced ROS generation and interruption of ATP synthesis, thus causing DNA damage • Extensive and dose-dependent damage to DNA was observed after exposure to Ag-NPs	(AshaRani et al., 2009)
	Sprague-Dawley Rats	• Ag-NPs was found in the brain, blood, liver, lungs, kidneys and testes. • Ag-NPs did not affect either the frequency of micronucleated polychromatic erythrocytes or the PCE/(PCE + NCE) ratio	(Kim et al., 2008)
CuO NPs	Human lung epithelial cells	• CuONPs induced genotoxicity through p53 pathway • The expression of Hsp70, Rad51 protein, and MSH2 protein was increased • DNA damage was induced in human lung cells • The lipid peroxidation and oxidative stress could be mediated DNA damage	(Ahamed et al., 2010b)

Carbon nanotubes and graphite nanofibers	Human bronchial epithelial cells	• Carbon nanotubes induced a dose-dependent increase in DNA damage at all treatment times (24 h, 48 h, or 72 h). • Graphite nanofibers caused DNA damage depending on the treatment time and dosage. • Carbon nanotubes and graphite nanofibers were genotoxic in human bronchial epithelial cells.	(Lindberg et al., 2009)
Multi-wall carbon nanotubes	Human umbilical vein endothelial cells	• Multi-wall carbon nanotubes induced the γH2AX foci (phosphorylated form of histone H2AX) formation. • Multi-wall carbon nanotubes reduced the viability of human umbilical vein endothelial cells.	(Guo et al., 2011)
	Human dermal fibroblast cells	• All the doses of multi-wall carbon nanotubes induced significantly increase in percent tail DNA [2.96%–16.39%] compared to control [2.02%].	(Patlolla et al., 2010)
Diamond NPs	Human peripheral lymphocytes	• Nanodiamond-mediated DNA oxidative damage (8-oxoG level) and changes in chromatin stability.	(Dworak, Wnuk, Zebrowski, Bartosz, & Lewinska, 2014)
Graphene oxide nanosheets	Albino mice	• Graphene oxide nanosheets induced chromosomal aberrations in bone marrow and DNA damage of lung cells • that were dose and time dependent.	(El-Yamany et al., 2017)

PCE + NCE, polychromatic erythrocyte/polychromatic erythrocyte + normochromatic erythrocyte.

Table 2.7 Epigenetic modifications caused by inorganic NPs

Nanomaterials	Model	Key outcomes	References
Epigenetic modifications (DNA methylation)			
CuO NPs	*Enchytraeus crypticus*	• Multigenerational exposure to Cu increased global DNA methylation and corresponded with phenotypic effects (reproduction)	(Bicho et al., 2020)
Ag NPs	Mouse hippocampal neuronal cells (HT22)	• The augmentation in 5-methylcytosine (5-mC) and DNMT1, DNMT2, DNMT3a, and DNMT3b levels were observed	(Mytych et al., 2017)
TiO_2-NPs and ZnO NPs	Lung fibroblast (MRC5) cell line	• A clear epigenetic alteration in DNA methylation in response to metal oxide NPs which was observed to be dose-dependent	(Patil et al., 2016)
Carbon based NS	Human lung adenocarcinoma cell (A549)	• Carbon based NPs exert significant effects on A549 cells and also contribute to increases in global DNA methylation	(Li et al., 2016)
TiO_2-NPs	Human epidermal keratinocytes cell (HaCaT)	• There is dysfunction of methylation cycle as a result of treatment with TiO_2-NP	(Tucci et al., 2013)
SiO_2 NPs	Human epidermal keratinocytes cells	• A dramatic decrease of PARP-1 expression on mRNA and protein level and an obvious increase in the level of PARP-1 methylation in nano-SiO_2-treated cells compared to the control group	(Gong et al., 2012)
Multi-walled and single walled carbon nanotubes	Human bronchial epithelial cells	• A higher number of single walled carbon nanotubes was deposited at both the nuclear and cellular level compared to multi walled carbon nanotubes after exposure. • The differences in the epigenetic response depend on carbon nanotubes-nucleus intearctions and the particle type.	(Öner et al., 2018)

	Human bronchial epithelial cell	• Only multi walled carbon nanotubes-exposed cells showed significant global DNA hypomethylation of cytosine.	(Emerce et al., 2019)
Epigenetic modifications (Histone modification)			
TiO_2 NPs	Human lung adenocarcinoma cell (A549)	• TiO_2NPs triggered the aberrant hypermethylation of the PARP-1 promoter and there was a cross talk between oxidative stress and PARP-1 methylation in the toxic effect of TiO_2NPs	(Bai et al., 2015)
Ag-NPs	Mouse erythroleukemia (MEL) cells	• Treatment with Ag-NPs has been observed to cause a significant reduction of global methylation level for histone 3 (H3) in erythroid MEL cells at sublethal concentrations, devoid of oxidative stress	(Qian et al., 2015)
Epigenetic modifications (Non-coding RNAs)			
Ag-NPs	Human T-cell leukemia	• Ag-NPs exposure resulted with alterations in expression of microRNAs	(Eom et al., 2014)
Au-NPs	Adult male and female Swiss mice	• Treatment with AuNP-100 nm, but not with AuNP-40 nm, produced significant changes in microRNA expression	(Balansky et al., 2013)

DNMT, DNA methyltransferases; PARP, Poly(ADP-ribose) polymerases-1.

2.4.4.1 DNA methylation

DNA methylation is defined as the covalent transfer of a methyl group to the C5 position of the cytosines catalyzed by DNA methyltransferases. This transformation regulates gene expression by altering DNA accessibility to the cellular transcriptional machinery (Lin & Chang, 2013). There are various recent studies focusing on epigenetic modification through DNA methylation.

TiO_2 is widely used in the confectionery and bakery products to provide an opaque or cloudy effect and white color. Exposure to TiO_2-NPs has been reported to result in several cellular responses. Tucci et al. (2013) investigated the effect of TiO_2 on the human keratinocyte cell line and reported that TiO_2 did not have a significant effect on cell cycle phase distribution or cell death. On the other hand, the authors reported dysfunction of DNA methylation cycle as a result of treatment with TiO_2-NP (Tucci et al., 2013). Patil, Gade, and Deobagkar (2016) investigated the DNA methylation-based epigenetic changes upon exposure to various concentrations of TiO_2 and ZnO NPs. Their results indicated that there is a significant correlation between NP exposure, endogenous DNA methyltransferase transcript levels, and global genomic DNA methylation together with DNA methyltransferase enzyme activities (Qian et al., 2015; Patil et al., 2016).

Mytych, Zebrowski, Lewinska, and Wnuk (2017) investigated the effect of Ag-NPs on cell proliferation, redox state, DNA damage response, and methylation in HT22 mouse hippocampal neuronal cells. The authors showed Ag-NPs-induced alterations in DNA methylation via increase in 5-methylcytosine and DNA methyltransferases levels. Furthermore, modifications in cell proliferation, DNA damage response, and epigenome acting were reported for the first time (Mytych et al., 2017). On the other hand, Gong et al. (2012) investigated the effect of epigenetic modification on nano silicon dioxide-induced expression of poly(ADP-ribose) polymerases-1(PARP-1), which is a pivotal repair gene. The authors observed a significant decrease of PARP-1 expression on mRNA and protein level. At the same time, the level of PARP-1 methylation in nano-silicon dioxide-treated cells was reported to increase compared to the control group (Gong et al., 2012).

Cu oxide-induced epigenetic modifications were investigated thoroughly. In a recent study, Bicho, Roelofs, Mariën, Scott-Fordsmand, and Amorim (2020) investigated the epigenetic effect of CuO nanostructures and Cu salt ($CuCl_2$) in *Enchytraeus crypticus*, for which global methylation was reported previously. It was reported that global DNA methylation increased due to multigenerational exposure to CuO nanostructures. The generation and Cu form were found to be effective in alterations in gene targets including epigenetic, stress and detoxification. The modifications were reported to occur in post-exposure generations as well, which was suggested to indicate transgenerational effects (Bicho et al., 2020).

In another study, changes that occur in A549 lung cells induced by carbon-based NPs (CNPs) were investigated using biospectroscopy techniques. DNA methylation levels in cells exposed to CNPs at 0.1 mg/L were also determined using HPLC-MS, and the genetic regulators for DNA methylation were checked by quantitative real-time RT-PCR. The results indicated that CNPs have significant effects on A549 lung cells and contribute to increases in global DNA methylation. The authors reported that real-world levels of NPs can change the methylome of exposed cells which should be considered for the regulatory assessment of these materials (Li et al., 2016). In another study, a higher number of single walled carbon nanotubes was deposited at both nuclear and cellular levels compared to multi walled carbon nanotubes after exposure to human bronchial epithelial cells (Öner et al., 2018). Moreover, the differences in the epigenetic response depend on carbon nanotubes-nucleus interactions and the particle type (Öner et al., 2018). More recently, only multi walled carbon nanotubes-exposed cells showed significant global DNA hypomethylation of cytosine (Emerce et al., 2019).

2.4.4.2 Histone modification

Histone modification is a post-translational modification that alters the terminal amino tails of histones and include acetylation, methylation, and phosphorylation. These modifications have been reported to affect the accessibility of the gene to the cellular transcription machinery. Recent research has suggested that histone post-translational modifications should also be considered as important molecular targets for nanostructures (Sierra et al., 2016).

Bai, Chen, and Gao (2015) investigated the interaction mechanisms between ROS and Poly(ADP-ribose) polymerase 1 (PARP-1) methylation triggered by TiO_2-NPs. The authors evaluated the toxic effect of TiO_2-NPs, PARP-1 methylation status, and oxidative damage using methyltransferase inhibitor 5-aza-2'-deoxycytidine and antioxidant α-lipoic acid. TiO_2-NPs' effect on cell viability was found to be dose-dependent. Generation of ROS was reported to increase following exposure to TiO_2-NPs. Use of α-lipoic acid was shown to be effective in decreasing TiO_2-NPs -induced ROS generation and reducing TiO_2-NPs -triggered PARP-1 methylation (Bai et al., 2015).

In another study, Qian et al. (2015) investigated potential effects of Ag-NPs on histone modifications and relevant gene expression in erythroid cells. The authors observed a significant decrease in global methylation level for histone 3(H3) due to exposure to Ag-NPs. The decrease in methylation was suggested to be associated with the decrease in histone methyltransferase levels and steric hindrance preventing methylation due to direct binding of Ag-NPs to histone. The authors indicated that Ag-NPs could modify the methylation status of histone (Qian et al., 2015).

2.4.4.3 Non-coding RNAs

The non-coding RNAs (ncRNAs) are functional RNA molecules that are not translated into proteins. They are abundant and classified according to their molecular properties and size. The ncRNAs play important roles in cellular processes and have been shown to alter gene expression by interacting with other epigenetic processes including expression of proteins in histone modifications, DNA methylation, and chromatin remodeling (Esteller, 2011).

(Balansky et al., 2013) evaluated the genotoxicity of Au-NPs with average size of 40-100 nm by investigating their ability to induce chromosome damage and effects on the post-transcriptional regulation of gene expression. Adult male and female Swiss mice (strain H) were treated with Au-NPs at a dose of 3.3 mg/kg body weight. The authors reported that treatment with 100 nm Au-NPs did not cause genotoxic damage in the bone marrow of adult mice. However, Au-NPs were observed to be genotoxic to both liver and peripheral blood of transplacentally exposed mice. Additionally, Au-NPs treatment was reported to result in significant changes in microRNA expression in the fetus lung and liver (Balansky et al., 2013).

In another study, Eom, Chatterjee, Lee, and Choi (2014) investigated the epigenetic mechanism behind the sensitivity of Jurkat T cells to Ag-NPs by conducting mRNA microarray and micro RNA microarray simultaneously. Different epigenetic mechanisms were reported to be involved in toxicity of Ag-NPs and Ag ions. The authors observed that Ag-NPs exposure resulted in alterations in expression of Micro RNAs. The expression of hsa-miR-219-5p was reported to be negatively correlated with the expression of metallothionein 1F (MT1F) and tribbles homolog 3 (TRIB3) in Jurkat T cells following Ag-NPs exposure (Eom et al., 2014).

2.4.5 Endoplasmic reticulum stress (ER stress)

ER is a crucial organelle involved in proper protein folding. However, high levels of misfolded proteins in the ER can result in ER stress, affecting the fate of cells and development of various diseases (Madkour, 2020). A diversity of NPs, particularly metal-based NPs, have been shown to be capable of activating ER stress pathway and morphological changes of ER both in vivo and in vitro as shown in Table 2.8. (Madkour, 2020). ER stress is initiated by three transmembrane proteins: (1) RNA-dependent protein kinase

Table 2.8 Selected studies related to ER stress of inorganic NPs.

Nanomaterials	Model	Key outcomes	References
Cu NPs	Zebrafish	• Cu NPs induced the generation of ER (indicated protein disulfide isomerase) and oxidative stress (indicated 4-hydroxy-2-nonenal) in the zebrafish cell • The ER and mitochondrial structure were damaged • Blocking the transportation of Cu to trans-Golgi network (*atp7a*$^{-/-}$) or mitochondria (*cox17*$^{-/-}$) could not mitigate Cu-induced intestinal developmental defects	(Zhao et al., 2020)
SiO_2 NPs	Human pulmonary alveolar epithelial cells	• The expression levels of BiP and CHOP were increased	(Zhang et al., 2019)
	Human hepatoma cells	• SiO_2NPs caused UPR induction and ER stress, activation of MAPK signaling, oxidative stress, and down-regulation of p53	(Christen, Camenzind, & Fent, 2014)
Si NPs	RAW 264.7 macrophage cells	• Upregulating of GRP78, CHOP, and ERO1α expression • Overexpression of CHOP significantly increased cell apoptosis	(Chen et al., 2019b)
	Human umbilical vein endothelial cells	• SiNPs induced ER stress • Up-regulated PERK/eIF2α/ATF4, IRE1/XBP1 and ATF6 pathways	Li, Ma et al. (2019)
Ag-NPs	Rat aortic endothelial cells	• Ag-NPs biocorona formation contributed to ER stress • Activation of the IREα and PERK pathways and downstream activation of both phosphorylated JNK (pJNK) and CHOP in RAEC cell exposed to 20 nm • Oxidative stress did not significantly affected the early ER stress response	(Persaud et al., 2019)

(Continued)

Table 2.8 Selected studies related to ER stress of inorganic NPs. (*Cont.*)

Nanomaterials	Model	Key outcomes	References
	The human Chang liver cells and Chinese hamster lung fibroblasts	• Ag-NPs induced apoptosis via the modulation of ER stress reaction • Mitochondrial Ca^{2+} overloading was observed • Ag-NPs induced GRP78/Bip release, PERK phosphorylation, IRE1 phosphorylation and ATF6 cleavage in a time-dependent manner • CHOP expression was induced	(Zhang et al., 2012)
ZnO-NPs	Male Kunming mice	• Up-regulated the IRE1α, XBP1s, BIP, and CHOP • The expression of caspase-3 was elevated • The number of sperms in the epididymis and the concentration of testosterone in serum were decreased in a dose dependent manner	(Tang et al., 2019)
	Murine liver	• Smaller ZnO-NPs (30 nm) induced significant ER stress • Higher expression of PERK, eIF2α, ATF4, Chop, JNK, caspase-12, caspase-9, GRP94, and Bax at the mRNA levels in 30 nm ZnO-NPs compared to 90 nm ZnO-NPs	(Kuang et al., 2016)
	Male C57BL/6 mice	• Ribosomal degranulation and ER swelling were observed in the liver tissues • Up-regulation of grp78, grp94, pdi-3, xbp-1 • PERK and eIF2α were enhanced • caspase-3, caspase-9, caspase-12, phosphorylation of JNK, and CHOP/GADD153, and up-regulation of pro-apoptotic genes (chop and bax)	(Yang et al., 2015b)

	Human umbilical vein endothelial cells	• Higher expression of spliced xbp-1, chop, and caspase-12 at the mRNA level • Activation of PERK-eIF2α-ATF4-CHOP/GADD34 and IRE/XBP-1 pathways was observed	(Chen et al., 2014)
TiO_2 NPs	Mouse hippocampal neuron HT22 cells	• The levels of ROS and Ca^{2+} were greatly enhanced • Upregulation of GRP78, IRE-1α, ATF6, CHOP and caspase-12 • TiO_2NPs induced apoptosis in HT22 cells through calcium imbalance-mediated ER stress and oxidative stress.	(He et al., 2018)
Au-NPs	Human neutrophils	• Au-NPs induced ER stress in human neutrophils • Activation of IRE1, PERK and ATF-6 was observed • Caspase-4 was activated	(Noël, Simard, & Girard, 2016)
	Human chronic myelogenous leukemia K562 cells	• Au-NPs led to cell death through induction of unmanageable ER stress • Prompt up-regulation of CHOP and cleaved caspase 3 were observed the Au-NPs-treated cells	(Tsai et al., 2011)
TiO_2 NPs	A/J Jms Slcmice	• The mitochondria and ER were disrupted • The expression levels of the Grp78/Bip, CHOP, and IRE1α were increased	(Yu et al., 2015b)
	Human bronchial epithelial cells	• The expression levels of Grp78/Bip, IRE-1α, phospho-IRE-1α, and CHOP were increased • TiO_2-NPs disrupted mitochondria by damaging ER membranes and cellular calcium homeostasis	(Yu et al., 2015a)

ATF6, Activating transcription factor 6; CHOP, C/EBP-homologous protein; CRP, C-reactive protein; ER, endoplasmic reticulum; IRE1, inositol-requiring enzyme 1; MAPK, mitogen-activated protein kinases; PERK, RNA-dependent protein kinase (PKR)-like ER kinase; UPR, unfolded protein response.

(PKR)-like ER kinase (Baram-Pinto et al., 2009), (2) the inositol-requiring enzyme 1 (IRE1), and (3) activating transcription factor 6 (ATF6) (Wu & Kaufman, 2006). In addition, induction of the glucose-regulated protein GRP78 (also referred as BiP) is a marker for ER stress. Furthermore, overexpression of C/EBP-homologous protein (CHOP) gene induced cell death, whereas deletion of the CHOP gene results in the alleviation of cell death triggered by ER stress.

(Zhang et al., 2012) reported that Ag-NPs induced the GRP78/Bip, marker of the unfolded protein response, release, PERK phosphorylation, IRE1 phosphorylation and ATF6 cleavage in a time-dependent manner. In another study, Cu-NPs were reported to induce generation of ER stress (indicated protein disulfide isomerase) and oxidative stress (indicated 4-hydroxy-2-nonenal) in the zebrafish cell. It was also observed that the Cu-NPs damaged the structure of ER and mitochondria (Zhao, Zhang, Sun, & Liu, 2020). ER stress also results in disruption of cellular Ca^{2+} homeostasis. (Yu et al., 2015a), for example, showed that TiO_2-NPs induced ER stress by increasing the levels of GRP78/Bip, IRE-1α, phospho-IRE-1α, and CHOP in human bronchial epithelial cells. In addition, the same researchers reported that TiO_2-NPs disrupted mitochondria by damaging ER membranes and cellular calcium homeostasis. Similarly, TiO_2-NPs were shown to induce apoptosis in mouse hippocampal neuron HT22 cells via calcium imbalance-mediated ER stress and oxidative stress (He et al., 2018).

In another study, ZnO-NPs were reported to induce ER stress via upregulation of the IRE1α, X-box-binding protein 1 splicing (XBP1s), BIP, and CHOP in male mice (Tang et al., 2019). Persaud et al. (2019) also reported that Ag-NPs biocorona formation contributed to ER stress. Activation of the PERK and IREα pathways and downstream activation of both phosphorylated JNK (pJNK) and CHOP were reported in the rat aortic endothelial cells (RAEC) exposed to 20 nm Ag-NPs. It was observed that oxidative stress did not significantly affect the early ER stress response in rat aortic endothelial cells after exposure to Ag-NPs. More recently, silica NPs were shown to induce ER stress via upregulation of PERK/eukaryotic translation initiation factor 2 α-subunit (eIF2α)/activating transcription factor 4 (ATF4), IRE1/XBP1 and ATF6 pathways in human umbilical vein endothelial cells (Li, Tang, & Xue, 2019a), while upregulation of GRP78, CHOP, and ER oxidoreduclin 1α (ERO1α) expression in RAW 264.7 macrophage cells (Chen et al., 2019a). These different observations could be attributed to the variations in cell type and culture conditions.

It is also worth emphasizing that the size of the NPs greatly affected the ER stress. For example, Kuang, Yang, Yang, Aguilar, and Xu (2016) investigated the toxicity of oral administration of ZnO-NPs with different sizes (90 nm and 30 nm) in mice. They reported that the 30 nm ZnO-NPs showed the highest increase in expression of eIF2α, ATF4, CHOP, caspase-7, caspase-3, caspase-12, caspase-9, apoptosis-inducing factor, and Bcl2-associated X protein, thereby causing the most severe liver damage. These results indicated that smaller sizes had higher tendencies to penetrate into the liver, thus, leading to more severe damage compared to the bigger NPs.

2.4.6 Autophagy dysfunction

Autophagy dysfunction is defined as excessive autophagy induction or blockade of autophagy flux. It also is recognized as a potential mechanism of cell death, resulting in either apoptosis or autophagic cell death. Nanostructures have been shown to elicit activation of the autophagy system via different pathways including: (1) induction of oxidative stress-dependent signaling (e.g., ER stress, mitochondrial damage), (2) suppression of Akt-mTOR (mammalian target of rapamycin) signaling, and (3) alteration of autophagy related gene/protein expression (Stern et al., 2012, Cohignac, Landry, Boczkowski, & Lanone, 2014, Popp & Segatori, 2015). It was noted that the light chain 3 (LC3) protein should be considered as an important biomarker for autophagy. LC3 protein also showed a molecular form conversion from cytosolic LC3-I into its enzymatic counterpart LC3-II (Yu et al., 2014).

Li, Hartono, Ong, Bay, and Yung (2010) reported in MRC-5 human lung fibroblasts exposed to Au-NPs, an increased expression of autophagy MAP-LC3 and Autophagy gene 7 (ATG 7) Beclin 1 (BECN1) proteins. It was also observed that Au-NPs induced formation of autophagosomes, protecting the cell from oxidative stress. Lee et al. (2014) evaluated the effect of Ag-NPs on mouse embryonic fibroblasts cell. They observed increased expression of LC3-II protein and accumulation of autophagosomes, suggesting that Ag-NPs could lead to blockage of autophagic flux. More recently, Pérez-Arizti et al. (2020) reported that TiO_2-NPs enhanced autophagy by 2.7-fold in rat astrocytes after 24 and 48 h by flow cytometry. According to Hussain et al. (2012), exposure to cerium dioxide NPs led to dose and time- dependent induction of autophagy.

In another study, Ma et al. (2011) showed that lysosomal pH increased by Au-NPs in rat kidney cell, leading to lysosomal dysfunction. In addition, oxidative stress conditions could be effective in activating autophagy. Yu et al. (2014) observed that exposure to silica NPs induced both autophagy and autophagic cell death, which were accompanied by ROS production in HepG2 cells. Moreover, the level of LC3-II protein was gradually enhanced in a dose-dependent way. Similarly, exposure to silica NPs disturbed the NO/NOS system, activated autophagy, induced the inflammatory response, and finally led to endothelial dysfunction via the PI3K/Akt/mTOR pathway. In another study, silica NPs were shown to damage lysosomes resulting in the fusion blockage of autophagosomes and lysosomes, in turn preventing autophagic cargo degradation (Wang et al., 2017). In another study, ZnO-NPs were reported to cause cell death through autophagic vacuole accumulation and mitochondrial damage in normal skin cells via generation of ROS. It was also observed that the autophagy-related proteins including ATG5, BECN1, and LC3-II, were increased (Yu et al., 2013).

Wan et al. (2013) showed that both single-walled carbon nanotubes and graphene oxide induced autophagosome accumulation. They found that the LC3-II/LC3-I ratio increased from 1.48 to 1.82 and 3.32 after exposure of 50 g/mL acid-functionalized single-walled carbon nanotubes and graphene oxides, respectively. It was also observed that graphene oxides were more potent than carbon nanotubes. According to Gao, Zhang, Yu, Ren, and Yang (2015), the ratio LC3 II /LC3 I and the expression of Beclin-1 were increased by multi-walled carbon nanotubes. Similarly, carbon black-metal ions co-exposure induced cell autophagy dysfunction via increasing the LC3-II/LC3-I (Kong et al., 2017). Some selected studies on autophagy dysfunction mechanism related to adverse health effects of NS are presented in Table 2.9.

2.5 Adverse health effect of nanostructures

Considerable attempts have been made to develop nano-sized food materials in the food industry. Despite the fact that nanotechnology opens up a new window of novel possibilities for the food industry, the entry of fabricated nanostructures into the food chain may lead to an accumulation of toxic and hazardous contaminants in food and unfavorably affect human health, as discussed in the previous sections. However, little is known about the bioavailability, uptake, distribution, potential accumulation and further toxicity of nanostructures in human body. So, "Precautionary Principle" makes necessity of more investigations on biological fate of food-based nanostructures in human body and their potential

Table 2.9 Autophagy dysfunction caused by inorganic NPs.

Nanomaterials	Model	Key Outcomes	References
TiO_2 NPs	Rat astrocytes	• TiO_2-NPs induced autophagy, mitochondrial damage and oxidative stress	(Pérez-Arizti et al., 2020)
	A/J Jms Slc mice	• LC3, p62, and Beclin 1 expression levels were increased • inhalation ofTiO2-NPs by mice leads to a dose-dependent accumulation of autophagy • The number of autophagosomes was greatly increased	(Yu et al., 2015b)
		• TiO_2-NP induced autophagy via increasing LC3 II/LC3 I ratio and expression of Sequestosome 1 (SQSTM1) /p62 and BECN1	(Yu et al., 2015b)
Ag-NPs	Mouse embryonic fibroblasts cell	• Increase of auto phagosomes formation • The expression levels of the LC3-II and p62 proteins was enhanced • Increase of abundant cytoplasmic acidic vesicular organelles formation	(Lee et al., 2014)
Si NPs	HepG2 cells	• Si NPs induces autophagy and autophagic cell death • Increasing ROS level correlated with the increasing of autophagy activation • The level of LC3-II protein gradually was enhanced with a dose-dependent way	(Yu et al., 2014)
	Endothelial cells	• Si NPs induced activated autophagy and inflammatory responses via the PI3-K/Akt/mTOR pathway	(Duan et al., 2014)

(Continued)

Table 2.9 Autophagy dysfunction caused by inorganic NPs. (*Cont.*)

Nanomaterials	Model	Key Outcomes	References
M-Fe NPs	RAW264.7 cells	• The mobility of cells was increased • Blocking of autophagy accelerated ER stress and apoptosis • Upregulation of BCL2 (B-cell lymphoma 2), LC3 (light chain 3), SQSTM1 (Sequestosome 1), ERN1 (Endoplasmic Reticulum To Nucleus Signaling 1), DDIT3 (DNA Damage Inducible Transcript 3), p-ERN1, and SOD2 (Superoxide dismutase 2) • Downregulation of BAX and BECN1	(Park et al., 2014)
ZnO NPs	Skin cells	• Upregulation of ATG5, BECN1, and LC3-II, ROS • Exposure ZnO-NPs caused cell death through mitochondria damage and autophagic vacuole and in normal skin cells via generation of ROS	(Yu et al., 2013)
CeO_2 NPs	Human Peripheral Blood Monocytes	• CeO_2-NPs exposure caused induction of autophagy in a dose- and time-dependent manner in human monocytes • p53 inhibition can cause an increase in autophagy • LysoTracker Red uptake and number of LC3b+ vesicles were increased	(Hussain et al., 2012)
Au-NPs	MRC-5human lung fibroblasts	• Increase of autophagosomes formation • Upregulation of autophagy proteins MAPLC3 and ATG7	(Li et al., 2010)
	Rat kidney cells	• Au-NPs exposure blocked the autophagic flux and degradation of an autophagy substrate • Lysosomal pH increased by Au-NPs in rat kidney cell, leading to lysosomal dysfunction	(Ma et al., 2011)

Single-walled carbon nanotubes and graphene oxides	Mouse peritoneal macrophages	• Both single-walled carbon nanotubes and graphene oxides induced autophagosome accumulation. • The LC3-II/LC3-I ratio increased from 1.48 to 1.82 and 3.32 after the exposure of 50 g/mL acid-functionalized single-walled carbon nanotubes and graphene oxides, respectively. • Graphene oxides were more potent than carbon nanotubes.	(Wan et al., 2013)
Carbon black-metal ions co-exposure	RAW264.7 macrophage-like cells C57BL6 mice	• Carbon black-metal ions co-exposure induced cell autophagy dysfunction. • The LC3-II/LC3-I was significantly increased.	(Kong et al., 2017)
Multi-walled carbon nanotubes	Wistar rats	• The LC3 II /LC3 I and the expression of Beclin-1 were increased. • Multi-walled carbon nanotubes could induce cognitive deficits in vivo via the autophagic pathway	(Gao et al., 2015)

ATG 7, Autophagy gene 7 protein; BECN1, Beclin 1; MAPLC3, microtubule-associated protein1light chain 3.

toxicity (Chau et al., 2007; Maksimović, Omanović-Mikličanin, & Badnjević, 2019). Nanostructures may have different possible health risks after entrance to the body such as disruption of normal GIT function, changing gut microbiota, increasing the unfavorable bioavailability (pesticide and..), accumulation within specific tissues and cellular malfunction that are presented in this section. A brief review of possible health risks associated with nanostructure on the most important organs that are affected by nanostructures is presented in Fig. 2.3.

It is worth mentioning that there have been few studies about the potential health risks associated with organic nanostructures in food. This may be related to the detection of organic nanostructures in foods or biological matrices that have similar components (such as polysaccharides, proteins

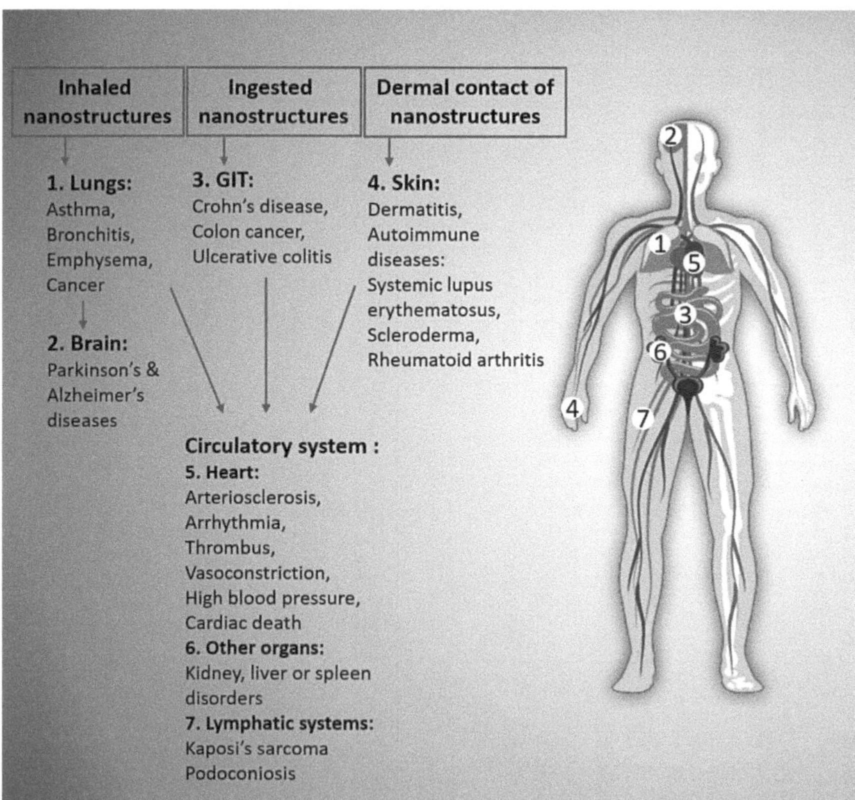

Figure 2.3 ***Possible health risks of inorganic nanostructures on different organs of the body.***

and lipids). However, advancement in analytical procedures and protocols of detection for these nanostructures would be helpful for the evaluation of their possible health risks.

2.5.1 Disruption of normal GIT functions

Over 100 years ago, the absorption of micro/nano materials through the GIT was reported for the first time (Florence, 1997). So, the study of toxicokinetics of nanostructures is not a new topic. Nevertheless, there is still a great need for exploring the absorption, metabolism, distribution, and excretion of orally administered nanostructures (Card et al., 2011).

After oral administration, nanostructures have to pass through the neutral pH and enzymatic environment of mouth, which contains saliva, electrolytes, and enzymes (e.g., amylase). Further, the nanostructures travel through the esophagus into the stomach with highly acidic condition (pH = 2-3), electrolytes, and digestive enzymes (e.g., lipase and pepsin). Moving through the pylorus sphincter, nanostructures enter the small bowel (pH = 5–7) with bile salts, electrolytes, phospholipids, and digestive enzymes such as amylase, proteases, and pancreatic lipase. The small-sized nanostructures with anionic or repulsive surface charge can diffuse through the negatively charged mucus layer and reach to the enterocytes. All nanostructures which are not absorbed in the previous sections may reach to the colon (pH = 6–7) and face with the colon microbiota and undigested food. GIT properties including different fluids and cellular structure and function may influence the characteristics of ingested NPs, which are briefly presented in Table 2.10 (Card et al., 2011; McClements & Xiao, 2017a, Vita et al., 2019).

Considering the whole information in Table 2.10, the properties of NPs may be altered passing through the GIT which may lead to their different fate and potential toxicity. GIT as the main port of entry for ingested nanostructures has three possible means including M-cells, normal enterocytes (intestinal epithelial cells), and paracellular or transcellular transport (Fig. 2.4). Adsorptive endocytosis may occur through M-cells involving phagocytosis, endocytosis, and clathrin-coated vesicles and pits. Another possible way of endocytosis may result in uptake of nanostructures through the apical side of normal enterocytes into the lymphatic system. The other possible route for uptake of nanostructures is the paracellular or transcellular pathway via passage between the intestinal cells. This pathway relies on the impaired and expanded tight junctions between the endothelial cells (Card et al., 2011; Vita et al., 2019).

Table 2.10 Some of the GIT characteristics which may change the properties of ingested NPs (Card et al., 2011; McClements & Xiao, 2017a; Vita et al., 2019).

Effective factors	Involved GIT fluids or actions	Influences on NPs
pH and ionic strength	Saliva, bile salts, electrolytes, and digestive enzymes	Aggregation of NPs and their interactions with other components
Surface-active components	Bile salts, surfactants, proteins, phospholipids, and free fatty acids	Adsorption to NP surfaces and alteration of their interfacial properties and biological fate
Enzyme activity	Digestive and metabolic enzymes: amylases, proteases, or lipases	Change of the properties of the NPs reaching specific regions of GIT
Biopolymers	Polysaccharides, proteins, and glycoproteins	Adsorbing biopolymers to NPs surfaces and alteration of their interfacial properties and further opposing or promoting flocculation
Mucus layer and GIT surfaces	Mucosae	Adhesion or permeation of NPs through the mucus layer and GIT cells depending on their properties
GIT microbiota	Microbial enzymes or biopolymers	Mutual effects of GIT bacteria and NPs: • Bacterial metabolites change the properties of NPs • NPs mainly inorganic ones alter the balance of GIT bacteria leading to adverse health effects
Mechanical forces	Chewing in the mouth, peristaltic movements of the esophagus, and agitating in the stomach and small bowel	Change of the aggregation state of the NPs by decomposition of flocculated systems

As mentioned earlier, the reactivity of nanostructures mainly depends on their chemical composition. Recently, several studies have reported that how different chemical composition of nanostructures affect their interaction with cellular or subcellular components. Moreover, the responses are both NP-specific and cell-type-specific (Vita et al., 2019). Nanostructures predominantly induce oxidative damage, disruption of the cellular redox

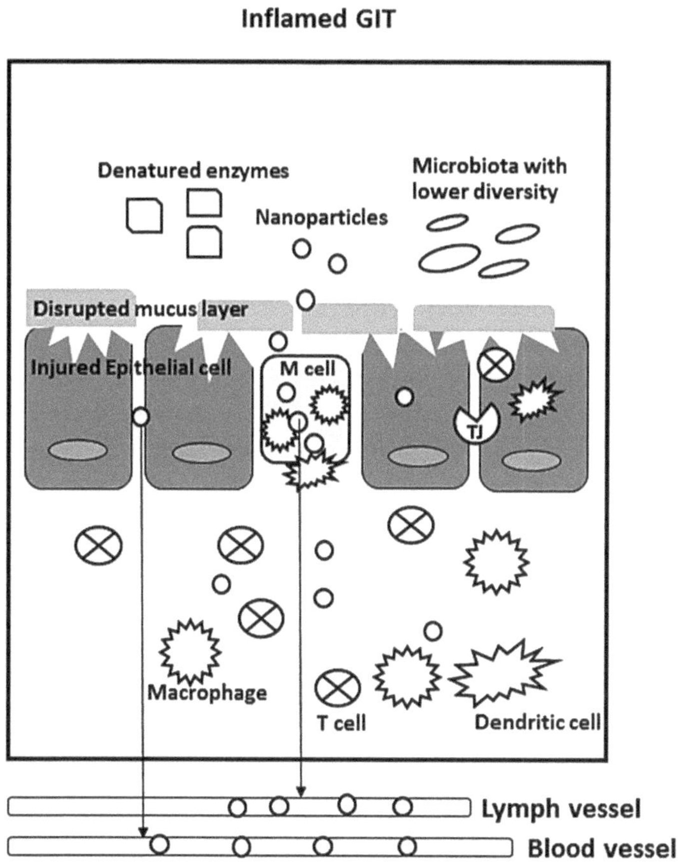

Figure 2.4 ***Typical disruption of normal GIT function caused by NPs' toxicity effects.*** Disrupted mucus layer and subsequent damages to epithelial cells and tight junctions facilitate the permeability of NPs leading to further accumulation and toxicity. Moreover, denatured enzymes as well as the lower diversity of gut microbiota trigger the immune and inflammatory responses. *TJ*, Tight junctions; *GIT*, Gastrointestinal tract.

homeostasis, and cause lysosomal rupture which may all lead to an inflammatory response and trigger a damage-associated molecular pattern in gut cells. Furthermore, NPs are capable of inducing an immune-stimulating molecular pattern (Khanna, Ong, Bay, & Baeg, 2015; Nakayama, 2018). This may cause health implications specifically for the patients with inflammatory bowel diseases (IBDs). Hereon, (Lomer, Thompson, & Powell, 2002) reported that the sites of TiO_2 accumulation in epithelial cells were responsible for inflammation in the cases with Crohn's disease. Colon cancer and Crohn's disease are two main GIT disorders with approved connections with the NPs entrance in the GIT (Maksimović et al., 2019).

Normal GIT functions may be disturbed by nanostructures specially with the small sized ones due to their large surface area for adsorption of GIT components such as metabolic or digestive enzymes (Fig. 2.4). Denaturation of enzyme structures after adsorption onto the NPs' surfaces may result in a lower catalytic activity and subsequently reduction of the rate or extent of digestion process in GIT. This is more likely about inorganic NPs or indigestible organic ones (McClements & Xiao, 2017b). McClements & Xiao (2017a) also reported that the mixture of lipid droplets with TiO_2-NPs led to a slightly reduction in lipid digestion.

Another adverse action of nanostructures is disruption of important GIT structures such as intercellular tight junction and microvilli of enterocytes, which may cause alteration of their protective function and abnormal nutrient absorption, as shown in Fig. 2.4 (Fröhlich & Fröhlich, 2016). For instance, Chalew and Schwab (2013) and Martirosyan, Bazes, and Schneider (2014) announced that Ag and TiO_2-NPs disrupted the tight junctions between enterocytes. Koeneman et al. (2010) also reported that the brush borders of Caco-2 cells collapsed via exposure to 10 μg/mL (and above) of TiO_2-NPs and an increase in intracellular-free calcium was induced. In another study, Faust, Doudrick, Yang, Westerhoff, and Capco (2014) found that TiO_2-NPs in the candy coating of chewing gum led to disruption of the microvilli in enterocytes at the concentration of 350 ng/mL.

Other investigations have also confirmed that some NPs such as Ag, SiO_2, ZnO, TiO_2, and Fe_3O_4 may accumulate at the cellular surface of different cell lines in vitro, as well as the cell surface within cytoplasm and subcellular surfaces of intestinal epithelial cells, M-cells, Peyer's patches, and macrophages inside the GIT (Vita et al., 2019). Brun et al. (2014) proved that TiO_2-NPs are able to cross the ileum epithelium (in vivo) and follicle-associated epithelium (ex vivo) and change their paracellular permeability by the induced remodeling of tight junctions. Moreover, in vitro studies indicated that Goblet cells and M-cells were influenced more than enterocytes through the endocytosis process. Another outcome was that TiO_2-NPs would persist in the gut epithelial cells leading to epithelium impairment. It should be also noted that this study was done in the worst situation regarding the applied dosage (Brun et al., 2014). However, not all NPs have adverse effects on the intestinal barrier function, as in vitro administration of Ag-NPs caused no loss of barrier integrity on monolayer Caco-2 cells (Vila, García-Rodríguez, Cortés, Marcos, & Hernández, 2018).

As depicted in Fig. 2.4, another adverse effect on human health caused by nanostructures in GIT is stimulation of immune response which is mostly associated with gut-associated lymphoid tissue (GALT) (Őrfi & Szebeni, 2016; McClements & Xiao, 2017a). Nanosilica particles are broadly utilized as a preservative in the food industry. Toda and Yoshino (2016) reported that these particles can stimulate lymphocytes such as T_H1, T_H2, and T_H17 via breakdown of ovalbumin-induced oral tolerance which is considered as a possible reason for food allergy. Shumakova et al. (2015) also found that oral exposure of silicon dioxide NPs to the gut microflora in rats may lead to a significant reduction in T_H cell counts and leukocyte and a corresponding elevation in plasmatumor necrosis factor alpha (TNF-alpha) and cytotoxic lymphocytes.

2.5.2 Changing gut microbiota

Unabsorbed nanostructures that can reach into the colon may have interference with colonic microflora and alter their diversity and viability. Despite the fact that gut microbiota have a major role in human wellbeing, any changes in their population may cause health implications (Fröhlich & Fröhlich, 2016; McClements & Xiao, 2017b). Furthermore, due to the progressive developments in sequencing methods and databases for identification of gut microflora, the type of gut microbiota can be considered as a diagnostic biomarker for most diseases (Fröhlich & Fröhlich, 2016). Taylor, Marcus, Guysi, and Walker (2015) reported the effect of TiO_2-NPs on gut microflora from healthy donors cultured in a colon reactor. They showed the alteration of hydrophobicity, short fatty acid profile, electrophoretic mobility, and sugar content of the extracellular matrix at the level of 0.01 g/L and of 3 mg/L of TiO_2-NPs. In another study, TiO_2-NPs which can be used as a food additive (E171 or INS171) in candy, gum, and puddings were evaluated concerning the potential toxic effects via model intestinal microbial community (Dudefoi, Moniz, Allen-vercoe, & Walker, 2017). TiO_2 additives at low concentrations showed only a minor effect on the profile of produced fatty acids by gut microbiota. Phylogenetic distributions and DNA profiles also confirmed the limited impacts on microbial population with a slight reduction in the abundance of *Bacteroides ovatus* in favor of *Clostridium cocleatum* (Dudefoi et al., 2017).

Hadrup et al. (2012) announced that Ag-NPs with average size of 14 nm did not change the composition of colon microbiota regarding the Bacteroides to Firmicutes ratio via oral ingestion in rats. Wilding et al. (2016) also reported no changes in the colon microbiota composition after exposure of mice to Ag-NPs with size range of 20–110 nm. However, Li et al. (2019a) reported that the toxic effects of Ag-NPs are dose-dependent, shape-dependent, and surface-coating dependent. Higher doses and larger particle sizes of Ag-NPs have less bactericidal effect due to their tendency to agglomerate, and only the Ag-NPs with the size range between 1 and 10 nm can influence directly on bacteria (Li et al., 2019b). Choi, Kim, Kim, and Lee (2018) declared that smaller Ag-NPs show higher bacterial toxic effects than the larger ones. In another study, ileal-mucosal microbiota changed after oral exposure of rats to 10 nm Ag-NPs with a higher proportion of Firmicutes phyla and the lower proportion of *Lactobacillus* genus (2015). Conversely, the proportion of lactic acid bacteria in the guts of quail increased using 25 mg/kg colloidal Ag-NPs in their drinking water (Sawosz et al., 2007).

A comparative study on Ag and Zn-NPs revealed that Zn-NPs are mostly toxic on enterocytes, while Ag-NPs potentially damage *E. coli* which confirmed that commercially available Ag-NPs may result in adverse effects on the GIT microbiota (Fröhlich & Fröhlich, 2016). All these contradictory findings are due to the differences in NPs sizes and concentrations as well as different used model systems for investigation. Furthermore, the location of specimen collection impacts on the obtained results because of different number and composition of microbiota along the GIT. Additionally, the analysis method of GIT microbiome may result in further bias (Fröhlich & Fröhlich, 2016).

2.5.3 Increasing the unfavorable bioavailability

NPs may have an increasing effect on the oral bioavailability of hydrophobic materials which could have adverse health impacts by the uptake of unfavorable non-polar materials in foods, such as pesticides and hormones. Zhang et al. (2019) investigated the effect of dietary fiber and emulsifier type in nanoemulsions on the bioavailability of chlorpyrifos on tomatoes. Chlorpyrifos is a hydrophobic organophosphorus pesticide commonly used for pest and insect control. Their results indicated that both chlorpyrifos bioavailability and lipid digestion depend on the type of dietary fibers and emulsifiers used to fabricate nanoemulsions regarding different electrical properties. These findings pointed out the use of lipid-based nanoemulsions in foods such as sauce, cream, dressing, and beverage may elevate the bioavailability of unfavorable hydrophobic pesticides on vegetables and fruits. This phenomenon may influence the level of absorbed pesticides in human body and further potential toxicity (McClements & Xiao, 2017a; Zhang et al., 2019).

Cao et al. (2019) investigated the effect of TiO_2 and SiO_2 NPs on the cytotoxicity and translocation and cellular uptake of the pesticide boscalid using an in vitro oral-GIT digestion system and quantified by LC/MS. Cytotoxicity assessments revealed that the cells exposed to the combined nanomaterials and boscalid showed greater toxic effect than the cells exposed to either nanomaterials or boscalid alone. On the other hand, the translocation of boscalid across the cellular layer was enhanced in the presence of TiO_2 and SiO_2 by 20% and 30%, respectively. These phenomena can be explained by the diminished health in epithelial cells and also the displacement of boscalid by the nanomaterials may also contribute to the enhanced translocation (Cao et al., 2019). In another study, Campos-Garcia, Martinez, Alves, Leonardo, and Barbieri (2015) reported that HNO_3-treated multi-walled carbon nanotubes triggered the acute toxicity of pesticide carbofuran to *Nile tilapia*, leading to five-fold increase in its LC50 values. Moreover, the exposure of *Nile tilapia* to nanotubes plus carbofuran resulted in decreased in both swimming capacity and oxygen consumption in comparison to the control group. These findings revealed that carbon nanotubes may act as pesticide carriers which can be effective on fish metabolism behavior, and survival. At present, this mechanism is not well understood and further investigations are still needed.

2.5.4 Accumulation within specific tissues

In vivo studies suggest that ingested nanostructures might be absorbed and accumulated in various tissues of human body. Nanostructures can transfer through the mucus layer and be absorbed via different passive or active transport mechanisms. Metabolization, transportation, and/or accumulation of nanostructures directly depend on the physicochemical characteristics of nanostructures such as dimension, morphology, composition, aggregation, and interfacial properties (McClements & Xiao, 2017b). Several studies have reported the transfer of TiO_2, Ag, and carbon NPs from GIT into the blood stream, and further to other organs like spleen, brain, liver, and reproductive system (Buzea et al., 2007; Maksimović et al., 2019). Long-term toxicity is expected due to the accumulation of NPs in specific tissues which is more important for the bio-persistent NPs not digested through the GIT (McClements & Xiao, 2017a).

One of the exposure routes that might cause toxic effects via accumulation is inhalation. Size and dose of nanostructures and the exposure time as well as the physique of the respiratory system and also the breathing pattern may influence the retention and deposition of nanostructures in various parts of lungs. Nanostructures <100 nm increase the risk of pulmonary diseases in comparison to the larger ones. On the other hand, the inhaled nanostructures can be absorbed through the blood circulation and lymph and reach to other organs or tissues (e.g., spleen, heart, bone marrow, lymph nodes, etc.) and bring about serious diseases (Maksimović et al., 2019). There is a direct connection between the inhalation of airborne nanostructures and prevalent pulmonary diseases such as bronchitis, emphysema, asthma, lung cancer, and some neurodegenerative disorders like Parkinson's and Alzheimer's diseases (Chau et al., 2007).

By the entrance of the nanostructures into the blood circulatory system, different upcoming disorders are possible such as heart diseases, arteriosclerosis, arrhythmia, autoimmune diseases, and a wide range of malfunctioning in other organs like liver, heart, brain, kidney, lung, etc. In this regard, certain NPs such as TiO_2, silicon dioxide, and Ag-NPs have been confirmed that are able to pass the physiological barriers to pose the toxicological effects. Moreover, ZnO-NPs can damage DNA and are considerably cytotoxic to liver and nephritic cells (Buzea et al., 2007; Maksimović et al., 2019). (De Jong et al., 2008) studied the size dependent biodistribution of Au-NPs and made a comparison between the 10-, 50-, and 250-nm sized NPs. Following an intravenous exposure to the rats, 10-nm Au-NPs accumulated in thymus, liver, kidney, spleen, testis, heart, brain, and lung while 50- and 250-nm Au-NPs were only observed in spleen and liver. A similar trend was observed following the oral administration of Au-NPs with size range of lower 50 nm and confirmed their higher biodistribution in spleen, kidney, brain, liver, or lungs while the larger NPs (58 nm) were detected almost inside the GIT (Card et al., 2011). Of note, direct comparison between these studies is not possible due to the differences between the experimental methods, route of exposure, and study design.

2.5.5 Cellular malfunction

NPs may cause cytotoxicity, cellular malfunction, and cell death through various mechanisms including apoptosis, oxidative stress, mitochondrial damage, and cell wall disruption (Dasgupta & Ranjan, 2018). These mechanisms are more likely to be triggered via inorganic NPs which can be absorbed in intestinal cells, since most of the organic NPs can be easily digested before absorption (McClements & Xiao, 2017b).

The generation of ROS such as superoxide, hydrogen peroxide, singlet oxygen, and hydroxyl radicals can promote the oxidative stress in cells. ROS are potent to interact with nucleic acid, protein, and lipid components of cell structure; further damage to nucleus, organelles, and cell membrane would be possible. Subsequently, many biochemical pathways such as DNA replication, gene expression, and ATP production might be adversely influenced. Silicon oxide NPs, Ag-NPs and ZnO-NPs have been identified as the susceptible inorganic NPs for interaction with the functioning cellular components leading to the cell malfunction (Athinarayanan, Periasamy, Alsaif, Al-Warthan, & Alshatwi, 2014; Wu, Yin, Wamer, Zeng, & Lo, 2014; Choi et al., 2015; Gaillet & Rouanet, 2015; Mc Clements, Li, & Xiao, 2015; Jafari et al., 2017a). Mitochondria as the main source of both normal cellular function and the regulation of programmed cell death (apoptosis) are the intracellular source of ROS production in cell. Apoptosis can be triggered via different signaling pathways. In this regard, caspases as a group of cytoplasmic proteases induce the production of apoptotic-related proteins. Caspases activated by cytochrome C are able to enter mitochondria to fracture the key substrates in electron transport chain, resulting in the elevated production of ROS (Ricci, Gottlieb, & Green, 2003). Wang et al. (2014) studied the effect of ZnO-NPs plus vitamin C on the activity of casapase-3 in GES-1 and NSCs intestinal cell lines. They observed that the combination of ZnO-NPs and vitamin C induced the higher activity of caspase-3 compared to the control group; while ZnO-NPs or vitamin C treated cells did not show any changes. This finding indicated the synergistic cytotoxic effect of the combination and the linked mitochondrial signal pathway (Wang et al., 2014).

The shape and size distribution of nanostructures are two main influencers of their cellular internalization. The optimum size is <50 nm and the most efficient shape is spherical (Jain, Ranjan, Dasgupta, & Ramalingam, 2018). On the other hand, the unique surface properties of nanostructures such as high conductivity, high tensile strength, electrical, physiochemical, and thermal characteristics and also the present chemical moieties may facilitate their interaction with the cells membrane. Ag and ZnO-NPs can electrostatically interact with the cell membrane of prokaryotic cells and cause toxicity, which may lead to the mitochondrial and morphological alterations and cytoplasmic accumulation within the cells as well. On the other hand, the semi-permeable membrane in eukaryotic cells can selectively provide the transport of NPs across the phospholipid membrane by specific protein channels or endocytosis (Jain et al., 2018). In this regard, cationic NPs are more potent to penetrate the lipid bilayer through clathrin-mediated pathway or via pore creating in the cell membrane that may trigger an imbalance in efflux and influx transport of ions, macromolecules, and proteins leading to cell death (Leroueil et al., 2008).

Anionic NPs internalization may happen through a nonspecific binding process on the cationic sites of cell membrane following by endocytosis (Shi, Thomas, Myc, Kotlyar, & Baker, 2007). Carbon nanotubes can interact with proteins and macromolecules upon arrival to the cells and cause damage to some functional proteins like RNase A via binding to the carboxylic groups (Zhao & Liu, 2012). Endocytosis, exocytosis, and phagocytosis of the cell membranes can also be affected by generation of protein corona as a mixture of the nanostructures bonded to the cellular proteins leading to the alteration of the action and stability of nanostructures (Jafari et al., 2017a).

Immunological responses are another type of cellular reaction to the absorbed nanostructures regarding the innate and adaptive immune systems. Since the nanostructures arrive the blood circulation, the complement system (a type of the innate humoral immune response) will be activated and then the nanostructures' surface will be marked via C3b and iC3b opsonic complement fragments which would be subsequently eliminated by phagocytic cells. Neutrophils, dendritic cells, monocytes, and macrophages are responsible for innate immune response while Type T and B lymphocytes are active in adaptive immune response (Jafari et al., 2017a). Moreover, some secreted agents such as cytokines (interleukin (IL)-1β, IL-6) and antibodies are responsible for inflammations. (Bettini et al., 2017) detected TiO_2-NPs in the immune cells of Peyer's patches (mainly T cells) in rats orally exposed for one week. In all treated rats with TiO_2-NPs a decreasing trend in Thelper (Th)-1 IFN-γ secretion was observed. Whereas, splenic Th1/Th17 secretion elevated considerably which caused inflammatory responses. Moreover, a 100-day TiO_2-NPs treatment promoted colon inflammation and triggered preneoplastic lesions (Bettini et al., 2017).

2.6 Risk assessment of NPs in food applications

Despite the many benefits of using nanostructures in the food industry, there are some concerns about the safety of these compounds. Risk assessment of the nanostructures can give us scientific information about their safety and toxicity and also some recommendations about the use of nanostructures. In 2011, the European Food Safety Authority (EFSA) published a guidance on the risk assessment of the application of nanoscience and nanotechnology in the food and feed chain (Committee, 2011) and recently published the revised guidance on human and animal health (Hardy et al., 2018). This guidance of EFSA scientific committee supplies an organized pathway for performing the safety assessment of nanostructure in food sector. The risk assessment of any compound have four steps; (1) hazard identification, (2) hazard characterization, (3) exposure assessment, and (4) risk characterization that are explained further briefly.

2.6.1 Hazard identification

Detailed identification of chemical composition, structure and physical properties, particle size, and coatings of nanostructures is a prerequisite for risk assessment of nanostructures (Tsuji et al., 2006). These parameters have influence on their interactions with tissues and identification of possible health risks associated with nanostructures.

2.6.2 Hazard characterization

In this step, the quantitative and/or qualitative assessment of the adverse effects related to nanostructures that may be presented in food is done. Indeed, a dose-response evaluation should be done. Determination of different factors such as uptake, distribution, metabolism, excretion, reactivity, and dosimetry are important for toxicity assessment of nanostructures (Tsuji et al., 2006). The hazard assessment of nanostructures is based on the standard tests introduced by Organization of Economic Co-operation and Development (OECD) guidelines. Now, there is a concern due to the lack of standardized assays and analytical procedures to test epigenetic effect of nanostructures (Smolkova et al., 2015). However, EFSA have presented guidance for in vitro or in vivo testing of nanostructures for identification and characterization of toxicological hazards of nanostructures (Hardy et al., 2018).

2.6.3 Exposure assessment

Evaluation of degree intake of nanostructures is performed in this step. The parameters that are important in this step are: nanostructures behavior, product uses, durability, receptors, and entry routes of nanostructures (Tsuji et al., 2006). The exposure of nanostructures depends on their characteristics (such as size, disperse, or aggregate form). For exposure assessment, we should consider estimation of exposure according to the consumption data and also provide average and high intake data for different population groups. The exposure to nanostructures may be directly (e.g., via food additives, novel foods, flavouring) or indirectly (e.g., migration from food packaging, transfer from animal feed to animals and from animals to food, carry over from pesticide to crop) (Hardy et al., 2018). According to the properties of nanostructures, there are different routes for internalization of the nanostructures into the body such as dermal, inhalation, ingestion, etc., that have been explained in previous sections. For example, the potential exposure routes for engineered nanostructures are skin penetration and ingestion while for airborne nanostructures, it is inhalation route (Maynard & Kuempel, 2005; Oberdörster et al., 2005).

2.6.4 Risk characterization

The information of the hazard identification, hazard characterization, and exposure assessment are combined to form risk characterization (Hardy et al., 2018). In this step, the likelihood of toxicity effects, nature of effects, and effectiveness of controls are discussed (Tsuji et al., 2006).

According to EU regulation (Reuland et al., 2013) No 2015/2283 on novel food, any food containing engineered nanostructures will be regarded as novel food and will need authorization. The risk assessment of the food containing nanostructures shall be done according to the EFSA guidance that is responsible for investigating up-to-date test methods for safety assessment (Hardy et al., 2018).

2.7 Conclusion

Nanotechnology has introduced many advantages in the food sector. However, there are some challenges about the safety of food containing NPs. We need to understand the biological responses after entrance of NPs into the body and also the biological fates of NPs to characterize their potential toxicity. Some NPs, especially inorganic ones, may have adverse health effects such as disruption of normal GIT functions, changing the gut microbiota, increasing the unfavorable bioavailability of some compounds in the body (such as pesticides and hormones), and accumulation of NPs within specific organs such as liver and cellular malfunction. Risk assessment can give us useful information about the safety of NPs before their application in a food product and commercialization. Although, available information on the toxicity of NPs are still limited and contrasting; and more study is needed.

References

Abdelhamed, S., Yokoyama, S., Refaat, A., Ogura, K., Yagita, H., Awale, S., et al. (2014). Piperine enhances the efficacy of TRAIL-based therapy for triple-negative breast cancer cells. *Anticancer Research*, *34*(4), 1893–1900.

Abolmaali, S. S, Tamaddon, A. M., & Dinarvand, R. (2013). A review of therapeutic challenges and achievements of methotrexate delivery systems for treatment of cancer and rheumatoid arthritis. *Cancer Chemotherapy and Arancibia Pharmacology*, *71*(5), 1115–1130.

Adeyemi, J. A., Machado, A. R. T., Ogunjimi, A. T., Alberici, L. C., Antunes, L. M. G., & Barbosa, F., Jr. (2020). Cytotoxicity, mutagenicity, oxidative stress and mitochondrial impairment in human hepatoma (HepG2) cells exposed to copper oxide, copper-iron oxide and carbon nanoparticles. *Ecotoxicology and Environmental Safety*, *189*, 109982.

Afshar, S., et al. (2019). Preparation and characterization of electrospun poly (lactic acid)-chitosan core-shell nanofibers with a new solvent system. *International journal of biological macromolecules*, *138*, 1130–1137.

Ahamed, M., AlSalhi, M. S., & Siddiqui, M. (2010a). Silver nanoparticle applications and human health. *Clinica chimica acta*, *411*(23–24), 1841–1848.

Ahamed, M., Posgai, R., Gorey, T. J., Nielsen, M., Hussain, S. M., & Rowe, J. J. (2010b). Silver nanoparticles induced heat shock protein 70, oxidative stress and apoptosis in *Drosophila melanogaster*. *Toxicology and Applied Pharmacology*, *242*(3), 263–269.

Ahamed, M., Akhtar, M. J., Siddiqui, M. A., Ahmad, J., Musarrat, J., Al-Khedhairy, A. A., et al. (2011). Oxidative stress mediated apoptosis induced by nickel ferrite nanoparticles in cultured A549 cells. *Toxicology*, *283*(2–3), 101–108.

Ahamed, M., Akhtar, M. J., Khan, M. A. M., Alrokayan, S. A., & Alhadlaq, H. A. (2019). Oxidative stress mediated cytotoxicity and apoptosis response of bismuth oxide (Bi2O3) nanoparticles in human breast cancer (MCF-7) cells. *Chemosphere*, *216*, 823–831.

Akhavan, S., Assadpour, E., Katouzian, I., & Jafari, S. M. (2018). Lipid nano scale cargos for the protection and delivery of food bioactive ingredients and nutraceuticals. *Trends in Food Science & Technology*.

Alaraby, M., Demir, E., Domenech, J., Velázquez, A., Hernández, A., & Marcos, R. (2020). In vivo evaluation of the toxic and genotoxic effects of exposure to cobalt nanoparticles using *Drosophila melanogaster*. *Environmental Science: Nano*, 7(2), 610–622.

Alarifi, S., Ali, D., Alkahtani, S., & Alhader, M. S. (2014). Iron oxide nanoparticles induce oxidative stress, DNA damage, and caspase activation in the human breast cancer cell line. *Biological Trace Element Research*, *159*(1–3), 416–424.

Alehosseini, A., et al. (2019). Electrospun curcumin-loaded protein nanofiber mats as active/bioactive coatings for food packaging applications. *Food Hydrocolloids*, *87*, 758–771.

Allahverdiyev, A. M., Abamor, E. S., Bagirova, M., & Rafailovich, M. (2011). Antimicrobial effects of TiO_2 and Ag_2O nanoparticles against drug-resistant bacteria and leishmania parasites. *Future Microbiology*, *6*(8), 933–940.

Amini, S. M., Gilaki, M., & Karchani, M. (2014). Safety of nanotechnology in food industries. *Electron Physician*, *6*(4), 962–968.

Amini, M., et al. (2018). Preparation and characterization of PVDF/Starch nanocomposite nanofibers using electrospinning method. *Materials Today: Proceedings*, *5*(7), 15613–15619.

Anjum, N. A., Gill, S. S., Duarte, A. C., & Pereira, E. (2019). Oxidative stress biomarkers and antioxidant defense in plants exposed to metallic nanoparticles. In *Nanomaterials and plant potential* (pp. 427–439). Springer.

Araiza-Calahorra, A., & Sarkar, A. (2019). Pickering emulsion stabilized by protein nanogel particles for delivery of curcumin: Effects of pH and ionic strength on curcumin retention. *Food Structure*, *21*, 100113.

Arancibia, C., Miranda, M., Matiacevich, S., & Troncoso, E. (2017). Physical properties and lipid bioavailability of nanoemulsion-based matrices with different thickening agents. *Food Hydrocolloids*, *73*, 243–254.

Arduini, F., Cinti, S., Scognamiglio, V., & Moscone, D. (2016). Nanomaterials in electrochemical biosensors for pesticide detection: advances and challenges in food analysis. *Microchimica Acta*, *183*(7), 2063–2083.

Asghari-Paskiabi, F., Jahanshiri, Z., Shams-Ghahfarokhi, M., & Razzaghi-Abyaneh, M. (2020). Antifungal nanotherapy: a novel approach to combat superficial fungal infections. In *Nanotechnology in skin, soft tissue, and bone infections* (pp. 93–107). Springer.

AshaRani, P.V., Low Kah Mun, G., Hande, M. P., & Valiyaveettil, S. (2009). Cytotoxicity and genotoxicity of silver nanoparticles in human cells. *ACS Nano*, *3*(2), 279–290.

Athinarayanan, J., Periasamy, V. S., Alsaif, M. A., Al-Warthan, A. A., & Alshatwi, A. A. (2014). Presence of nanosilica (E551) in commercial food products: TNF-mediated oxidative stress and altered cell cycle progression in human lung fibroblast cells. *Cell Biology and Toxicology*, *30*(2), 89–100.

Augustine, R., & Hasan, A. (2020). Emerging applications of biocompatible phytosynthesized metal/metal oxide nanoparticles in healthcare. *Journal of Drug Delivery Science and Technology*, *56*, 101516.

Axson, J. L., Stark, D. I., Bondy, A. L., Capracotta, S. S., Maynard, A. D., Philbert, M. A., et al. (2015). Rapid kinetics of size and pH-dependent dissolution and aggregation of silver nanoparticles in simulated gastric fluid. *Journal of Physical Chemistry. C, Nanomaterials and Interfaces*, *119*(35), 20632–20641.

Bae, W., et al. (2019). DNA nanostructures in vitro, in vivo and on membranes. *Nano Today*, *26*, 98–107.

Bai, C., & Tang, M. (2020). Toxicological study of metal and metal oxide nanoparticles in zebrafish. *Journal of Applied Toxicology*, *40*(1), 37–63.

Bai, W., Chen, Y., & Gao, A. (2015). Cross talk between poly (ADP-ribose) polymerase 1 methylation and oxidative stress involved in the toxic effect of anatase titanium dioxide nanoparticles. *International Journal of Nanomedicine*, *10*, 5561.

Bakand, S., Hayes, A., & Dechsakulthorn, F. (2012). Nanoparticles: a review of particle toxicology following inhalation exposure. *Inhalation Toxicology*, *24*(2), 125–135.

Balansky, R., Longobardi, M., Ganchev, G., Iltcheva, M., Nedyalkov, N., Atanasov, P., et al. (2013). Transplacental clastogenic and epigenetic effects of gold nanoparticles in mice. *Mutation Research/Fundamental and Molecular Mechanisms of Mutagenesis*, *751*, 42–48.

Banerjee, A., Qi, J., Gogoi, R., Wong, J., & Mitragotri, S. (2016). Role of nanoparticle size, shape and surface chemistry in oral drug delivery. *Journal of Controlled Release*, *238*, 176–185.

Bar-Ilan, O., Albrecht, R. M., Fako, V. E., & Furgeson, D.Y. (2009). Toxicity assessments of multisized gold and silver nanoparticles in zebrafish embryos. *Small*, *5*(16), 1897–1910.

Baram-Pinto, D., Shukla, S., Perkas, N., Gedanken, A., & Sarid, R. (2009). Inhibition of herpes simplex virus type 1 infection by silver nanoparticles capped with mercaptoethane sulfonate. *Bioconjugate Chemistry*, *20*(8), 1497–1502.

Barbas, J. (2013). *The use of on-line characterization technologies during the manufacture of nanostructured materials.*

Bazila, N., Gaurav, S., Ovais Shafiq, Q., Soban Ahmad, F., Rayees Ul, I., & Kaiser, Y. (2018). Importance and health hazards of nanoparticles used in the food industry. *Nanotechnology Reviews*, 7(6), 623–641.

Behbahani, E. S., Ghaedi, M., Abbaspour, M., Rostamizadeh, K., & Dashtian, K. (2019). Curcumin loaded nanostructured lipid carriers: In vitro digestion and release studies. *Polyhedron*, *164*, 113–122.

Benameur, L., Auffan, M., Cassien, M., Liu, W., Culcasi, M., Rahmouni, H., et al. (2015). DNA damage and oxidative stress induced by CeO_2 nanoparticles in human dermal fibroblasts: Evidence of a clastogenic effect as a mechanism of genotoxicity. *Nanotoxicology*, *9*(6), 696–705.

Bendale, Y., Bendale, V., & Paul, S. (2017). Evaluation of cytotoxic activity of platinum nanoparticles against normal and cancer cells and its anticancer potential through induction of apoptosis. *Integrative Medicine Research*, *6*(2), 141–148.

Berardi, A., Baldelli, F., & Bombelli (2019). Oral delivery of nanoparticles - let's not forget about the protein corona. *Expert Opinion on Drug Delivery*, *16*(6), 563–566.

Bergin, I. L., & Witzmann, F. A. (2013). Nanoparticle toxicity by the gastrointestinal route: evidence and knowledge gaps. *International Journal of Biomedical Nanoscience and Nanotechnology*, *3*(1–2).

Bettini, S., Boutet-Robinet, E., Cartier, C., Coméra, C., Gaultier, E., Dupuy, J., et al. (2017). Food-grade TiO_2 impairs intestinal and systemic immune homeostasis, initiates preneoplastic lesions and promotes aberrant crypt development in the rat colon. *Scientific Reports*, 7, 40373–140373.

Bian, S. W., Mudunkotuwa, I. A., Rupasinghe, T., & Grassian, V. H. (2011). Aggregation and dissolution of 4 nm ZnO nanoparticles in aqueous environments: influence of pH, ionic strength, size, and adsorption of humic acid. *Langmuir*, *27*(10), 6059–6068.

Bicho, R. C., Roelofs, D., Mariën, J., Scott-Fordsmand, J. J., & Amorim, M. J. B. (2020). Epigenetic effects of (nano) materials in environmental species–Cu case study in Enchytraeus crypticus. *Environment International*, *136*, 105447.

Bierkandt, F. S., Leibrock, L., Wagener, S., Laux, P., & Luch, A. (2018). The impact of nanomaterial characteristics on inhalation toxicity. *Toxicology Research (Camb)*, 7(3), 321–346.

Borm, P. J. A., & Kreyling, W. (2004). Toxicological hazards of inhaled nanoparticles—potential implications for drug delivery. *Journal of Nanoscience and Nanotechnology*, *4*(5), 521–531.

Bouwmeester, H., Dekkers, S., Noordam, M.Y., Hagens, W. I., Bulder, A. S., de Heer, C., et al. (2009). Review of health safety aspects of nanotechnologies in food production. *Regulatory Toxicology and Pharmacology*, *53*(1), 52–62.

Braakhuis, H. M., Gosens, I., Krystek, P., Boere, J. A. F., Cassee, F. R., Fokkens, P. H. B., et al. (2014a). Particle size dependent deposition and pulmonary inflammation after short-term inhalation of silver nanoparticles. *Particle and Fibre Toxicology*, *11*(1), 1–16.

Braakhuis, H. M., Park, M.V. D. Z., Gosens, I., De Jong, W. H., & Cassee, F. R. (2014b). Physicochemical characteristics of nanomaterials that affect pulmonary inflammation. *Particle and Fibre Toxicology*, *11*(1), 18.

Brandelli, A. (2020). The interaction of nanostructured antimicrobials with biological systems: Cellular uptake, trafficking and potential toxicity. *Food Science and Human Wellness*, *9*(1), 8–20.

Brenner, D. R., Scherer, D., Muir, K., Schildkraut, J., Boffetta, P., Spitz, M. R., et al. (2014). A review of the application of inflammatory biomarkers in epidemiologic cancer research. *Cancer Epidemiology and Prevention Biomarkers*, *23*(9), 1729–1751.

Brown, D. M., Kinloch, I. A., Bangert, U., Windle, A. H., Walter, D. M., Walker, G. S., et al. (2007). An in vitro study of the potential of carbon nanotubes and nanofibres to induce inflammatory mediators and frustrated phagocytosis. *Carbon*, *45*(9), 1743–1756.

Brum, A. A., dos Santos, P. P., da Silva, M. M., Paese, K., Guterres, S. S., Costa, T. M. H., et al. (2017). Lutein-loaded lipid-core nanocapsules: Physicochemical characterization and stability evaluation. *Colloids and Surfaces A: Physicochemical and Engineering Aspects*, *522*, 477–484.

Brun, E., Barreau, F., Veronesi, G., Fayard, B., Sorieul, S., Chanéac, C., et al. (2014). Titanium dioxide nanoparticle impact and translocation through ex vivo, in vivo and in vitro gut epithelia. *Particle and Fibre Toxicology*, *11*(1), 13–113.

Buzea, C., Pacheco, I. I., & Robbie, K. (2007). Nanomaterials and nanoparticles: sources and toxicity. *Biointerphases*, *2*(4), MR17–MR71.

Campos-Garcia, J., Martinez, D. S. T., Alves, O. L., Leonardo, A. F. G., & Barbieri, E. (2015). Ecotoxicological effects of carbofuran and oxidised multiwalled carbon nanotubes on the freshwater fish Nile tilapia: Nanotubes enhance pesticide ecotoxicity. *Ecotoxicology and Environmental Safety*, *111*, 131–137.

Cao, Y., Li, J., Liu, F., Li, X., Jiang, Q., Cheng, S., et al. (2016). Consideration of interaction between nanoparticles and food components for the safety assessment of nanoparticles following oral exposure: A review. *Environmental Toxicology and Pharmacology i*, *46*, 206–210.

Cao, Y., Long, J., Liu, L., He, T., Jiang, L., Zhao, C., et al. (2017). A review of endoplasmic reticulum (ER) stress and nanoparticle (NP) exposure. *Life Sciences*, *186*, 33–42.

Cao, X., DeLoid, G. M., Bitounis, D., De La Torre-Roche, R., White, J. C., Zhang, Z., et al. (2019). Co-exposure to the food additives SiO_2 (E551) or TiO_2 (E171) and the pesticide

boscalid increases cytotoxicity and bioavailability of the pesticide in a tri-culture small intestinal epithelium model: potential health implications. *Environmental Science: Nano, 6*(9), 2786–2800.
Card, J. W., Jonaitis, T. S., Tafazoli, S., & Magnuson, B. A. (2011). An appraisal of the published literature on the safety and toxicity of food-related nanomaterials. *Critical Reviews in Toxicology, 41*(1), 20–49.
Chalew, T. E. A., & Schwab, K. J. (2013). Toxicity of commercially available engineered nanoparticles to Caco-2 and SW480 human intestinal epithelial cells. *Cell Biology and Toxicology, 29*(2), 101–116.
Chandra, S., Kumar, A., & Tomar, P. K. (2014). Synthesis and characterization of copper nanoparticles by reducing agent. *Journal of Saudi Chemical Society, 18*(2), 149–153.
Chau, C. F., Wu, S. H., & Yen, G. C. (2007). The development of regulations for food nanotechnology. *Trends in Food Science and Technology, 18*(5), 269–280.
Chen, L., Remondetto, G. E., & Subirade, M. (2006). Food protein-based materials as nutraceutical delivery systems. *Trends in Food Science & Technology, 17*(5), 272–283.
Chen, R., Huo, L., Shi, X., Bai, R., Zhang, Z., Zhao, Y., et al. (2014). Endoplasmic reticulum stress induced by zinc oxide nanoparticles is an earlier biomarker for nanotoxicological evaluation. *ACS Nano, 8*(3), 2562–2574.
Chen, Y., Yang, Y., Xu, B., Wang, S., Li, B., Ma, J., et al. (2017). Mesoporous carbon nanomaterials induced pulmonary surfactant inhibition, cytotoxicity, inflammation and lung fibrosis. *Journal of Environmental Sciences, 62*, 100–114.
Chen, L., Deng, H., Cui, H., Fang, J., Zuo, Z., Deng, J., et al. (2018). Inflammatory responses and inflammation-associated diseases in organs. *Oncotarget, 9*(6), 7204.
Chen, F., Jin, J., Hu, J., Wang, Y., Ma, Z., & Zhang, J. (2019a). Endoplasmic reticulum stress cooperates in silica nanoparticles-induced macrophage apoptosis via activation of CHOP-mediated apoptotic signaling pathway. *International Journal of Molecular Sciences, 20*(23), 5846.
Chen, P., Wang, H., He, M., Chen, B., Yang, B., & Hu, B. (2019b). Size-dependent cytotoxicity study of ZnO nanoparticles in HepG2 cells. *Ecotoxicology and Environmental Safety, 171*, 337–346.
Cho, W. -S., Duffin, R., Poland, C. A., Howie, S. E. M., MacNee, W., Bradley, M., et al. (2010). Metal oxide nanoparticles induce unique inflammatory footprints in the lung: important implications for nanoparticle testing. *Environmental Health Perspectives, 118*(12), 1699–1706.
Cho, W. -S., Duffin, R., Thielbeer, F., Bradley, M., Megson, I. L., MacNee, W., et al. (2012). Zeta potential and solubility to toxic ions as mechanisms of lung inflammation caused by metal/metal oxide nanoparticles. *Toxicological Sciences, 126*(2), 469–477.
Choi, J. E., Kim, S., Ahn, J. H., Youn, P., Kang, J. S., Park, K., et al. (2010). Induction of oxidative stress and apoptosis by silver nanoparticles in the liver of adult zebrafish. *Aquatic Toxicology, 100*(2), 151–159.
Choi, J., Kim, H., Kim, P., Jo, E., Kim, H. -M., Lee, M. -Y., et al. (2015). Toxicity of zinc oxide nanoparticles in rats treated by two different routes: single intravenous injection and single oral administration. *Journal of Toxicology and Environmental Health, Part A, 78*(4), 226–243.
Choi, Y., Kim, H. -A., Kim, K. -W., & Lee, B. -T. (2018). Comparative toxicity of silver nanoparticles and silver ions to *Escherichia coli*. *Journal of Environmental Sciences, 66*, 50–60.
Christen, V., Camenzind, M., & Fent, K. (2014). Silica nanoparticles induce endoplasmic reticulum stress response, oxidative stress and activate the mitogen-activated protein kinase (MAPK) signaling pathway. *Toxicology Reports, 1*, 1143–1151.
Chung, H. E., Yu, J., Baek, M., Lee, J. A., Kim, M. S., Kim, S. H., et al. (2013). Toxicokinetics of zinc oxide nanoparticles in rats. *Journal of Physics: Conference Series, 429*, 012037.

Cohignac, V., Landry, M. J., Boczkowski, J., & Lanone, S. (2014). Autophagy as a possible underlying mechanism of nanomaterial toxicity. *Nanomaterials*, *4*(3), 548–582.
Colin-Orozco, J., et al. (2015). Properties of poly (ethylene oxide)/whey protein isolate nanofibers prepared by electrospinning. *Food Biophysics*, *10*(2), 134–144.
Committee, E. S. (2011). Guidance on the risk assessment of the application of nanoscience and nanotechnologies in the food and feed chain. *EFSA Journal*, *9*(5), 2140.
Creutzenberg, O., Bellmann, B., Korolewitz, R., Koch, W., Mangelsdorf, I., Tillmann, T., et al. (2012). Change in agglomeration status and toxicokinetic fate of various nanoparticles in vivo following lung exposure in rats. *Inhalation Toxicology*, *24*(12), 821–830.
Dąbrowska-Bouta, B., Sulkowski, G., Strużyński, W., & Strużyńska, L. (2019). Prolonged exposure to silver nanoparticles results in oxidative stress in cerebral myelin. *Neurotoxicity Research*, *35*(3), 495–504.
Daeihamed, M., Dadashzadeh, S., Haeri, A., & Akhlaghi, M. F. (2017). Potential of liposomes for enhancement of oral drug absorption. *Current Drug Delivery*, *14*(2), 289–303.
Dananjaya, S., Thao, N. T., Wijerathna, H., Lee, J., Edussuriya, M., Choi, D., et al. (2020). In vitro and in vivo anticandidal efficacy of green synthesized gold nanoparticles using Spirulina maxima polysaccharide. *Process Biochemistry*, *92*, 138–148.
Dasgupta, N., & Ranjan, S. (2018). In *Nano-food toxicity and regulations* (pp. 151–179). Springer.
Date, A. A., Hanes, J., & Ensign, L. M. (2016). Nanoparticles for oral delivery: Design, evaluation and state-of-the-art. *Journal of Controlled Release*, *240*, 504–526.
Davis, S. S. (2004). Coming of age of lipid-based drug delivery systems. *Advanced Drug Delivery Reviews*, *56*(9), 1241–1242.
De Jong, W. H., Hagens, W. I., Krystek, P., Burger, M. C., Sips, A. J. A. M., & Geertsma, R. E. (2008). Particle size-dependent organ distribution of gold nanoparticles after intravenous administration. *Biomaterials*, *29*(12), 1912–1919.
Dehle, F. C., Ecroyd, H., Musgrave, I. F., & Carver, J. A. (2010). AB-Crystallin inhibits the cell toxicity associated with amyloid fibril formation by κ-casein and the amyloid-β peptide. *Cell Stress and Chaperones*, *15*(6), 1013–1026.
Deng, L., et al. (2018). Characterization of gelatin/zein nanofibers by hybrid electrospinning. *Food Hydrocolloids*, *75*, 72–80.
Dev, V. G., & Hemamalini, T. (2018). Porous electrospun starch rich polycaprolactone blend nanofibers for severe hemorrhage. *International Journal of Biological Macromolecules*.
Dima, C., Assadpour, E., Dima, S., & Jafari, S. M. (2019). Bioactive-loaded nanocarriers for functional foods: from designing to bioavailability. *Current Opinion in Food Science*.
Dizaj, S. M., Mennati, A., Jafari, S., Khezri, K., & Adibkia, K. (2015). Antimicrobial activity of carbon-based nanoparticles. *Advanced Pharmaceutical Bulletin*, *5*(1), 19.
Doktorovova, S., Souto, E. B., & Silva, A. M. (2014). Nanotoxicology applied to solid lipid nanoparticles and nanostructured lipid carriers–a systematic review of in vitro data. *European Journal of Pharmaceutics and Biopharmaceutics*, *87*(1), 1–18.
Donaldson, K., Brown, D., Clouter, A., Duffin, R., MacNee, W., Renwick, L., et al. (2002). The pulmonary toxicology of ultrafine particles. *Journal of Aerosol Medicine*, *15*(2), 213–220.
Duan, J., Yu, Y., Yu, Y., Li, Y., Wang, J., Geng, W., et al. (2014). Silica nanoparticles induce autophagy and endothelial dysfunction via the PI3K/Akt/mTOR signaling pathway. *International Journal of Nanomedicine*, *9*, 5131.
Dudefoi, W., Moniz, K., Allen-vercoe, E., & Walker, V. K. (2017). Impact of food grade and nano-TiO_2 particles on a human intestinal community. *Food and Chemical Toxicology*, *106*, 242–249.
Duffy, L. L., Osmond-McLeod, M. J., Judy, J., & King, T. (2018). Investigation into the antibacterial activity of silver, zinc oxide and copper oxide nanoparticles against poultry-relevant isolates of *Salmonella* and *Campylobacter*. *Food Control*, *92*, 293–300.
Dusinska, M., Boland, S., Saunders, M., Juillerat-Jeanneret, L., Tran, L., Pojana, G., et al. (2015). Towards an alternative testing strategy for nanomaterials used in nanomedicine: lessons from NanoTEST. *Nanotoxicology*, *9*(Suppl. 1), 118–132.

Dworak, N., Wnuk, M., Zebrowski, J., Bartosz, G., & Lewinska, A. (2014). Genotoxic and mutagenic activity of diamond nanoparticles in human peripheral lymphocytes in vitro. *Carbon, 68*, 763–776.

EFSA (2018). Annual report of the EFSA Scientific Network of Risk Assessment of Nanotechnologies in Food and Feed for 2017. *EFSA Supporting Publications, 15*(3), 1393E.

El-Yamany, N. A., Mohamed, F. F., Salaheldin, T. A., Tohamy, A. A., Abd El-Mohsen, W. N., & Amin, A. S. (2017). Graphene oxide nanosheets induced genotoxicity and pulmonary injury in mice. *Experimental and Toxicologic Pathology, 69*(6), 383–392.

Elle, R. E., Gaillet, S., Vide, J., Romain, C., Lauret, C., Rugani, N., et al. (2013). Dietary exposure to silver nanoparticles in Sprague–Dawley rats: effects on oxidative stress and inflammation. *Food and Chemical Toxicology, 60*, 297–301.

Elsabahy, M., Nazarali, A., & Foldvari, M. (2011). Non-viral nucleic acid delivery: Key challenges and future directions. *Current Drug Delivery, 8*(3), 235–244.

Elsadek, B., & Kratz, F. (2012). Impact of albumin on drug delivery-new applications on the horizon. *Journal of Controlled Release, 157*, 4–28.

Elzoghby, A. O., El-Fotoh, W. S. A., & Elgindy, N. A. (2011). Casein-based formulations as promising controlled release drug delivery systems. *Journal of Controlled Release, 153*(3), 206–216.

Elzoghby, A. O., Samy, W. M., & Elgindy, N. A. (2012a). Albumin-based nanoparticles as potential controlled release drug delivery systems. *Journal of Controlled Release, 157*(2), 168–182.

Elzoghby, A. O., Samy, W. M., & Elgindy, N. A. (2012b). Protein-based nanocarriers as promising drug and gene delivery systems. *Journal of Controlled Release, 161*(1), 38–49. doi: 10.1016/j.jconrel.2012.04.036.

Elzoghby, A. O. (2013). Gelatin-based nanoparticles as drug and gene delivery systems: Reviewing three decades of research. *Journal of Controlled Release, 172*(3), 1075–1091.

Emerce, E., Ghosh, M., Öner, D., Duca, R. -C., Vanoirbeek, J., Bekaert, B., et al. (2019). Carbon nanotube-and asbestos-induced DNA and RNA methylation changes in bronchial epithelial cells. *Chemical Research in Toxicology, 32*(5), 850–860.

Ensign, L. M., Cone, R., & Hanes, J. (2012). Oral drug delivery with polymeric nanoparticles: the gastrointestinal mucus barriers. *Adv Drug Deliv Rev, 64*(6), 557–570.

Eom, H. -J., & Choi, J. (2009a). Oxidative stress of CeO_2 nanoparticles via p38-Nrf-2 signaling pathway in human bronchial epithelial cell, Beas-2B. *Toxicology Letters, 187*(2), 77–83.

Eom, H. -J., & Choi, J. (2009b). Oxidative stress of silica nanoparticles in human bronchial epithelial cell, Beas-2B. *Toxicology in Vitro, 23*(7), 1326–1332.

Eom, H. -J., Chatterjee, N., Lee, J., & Choi, J. (2014). Integrated mRNA and micro RNA profiling reveals epigenetic mechanism of differential sensitivity of Jurkat T cells to AgNPs and Ag ions. *Toxicology Letters, 229*(1), 311–318.

Esteller, M. (2011). Non-coding RNAs in human disease. *Nature Reviews Genetics, 12*(12), 861–874.

Evans, S. J., Jenkins, G. J., Doak, S. H., & Clift, M. J. D. (2019). Cellular defense mechanisms following nanomaterial exposure: a focus on oxidative stress and cytotoxicity. In *Biological Responses to Nanoscale Particles* (pp. 243–254). Springer.

Fathi, M., Martin, A., & McClements, D. J. (2014). Nanoencapsulation of food ingredients using carbohydrate based delivery systems. *Trends in Food Science & Technology, 39*(1), 18–39.

Faust, J. J., Doudrick, K., Yang, Y., Westerhoff, P., & Capco, D. G. (2014). Food grade titanium dioxide disrupts intestinal brush border microvilli in vitro independent of sedimentation. *Cell Biology and Toxicology, 30*(3), 169–188.

Fawcett, D., Verduin, J. J., Shah, M., Sharma, S. B., & Poinern, G. E. J. (2017). A review of current research into the biogenic synthesis of metal and metal oxide nanoparticles via marine algae and seagrasses. *Journal of Nanoscience, 2017*.

Ferraro, S. A., Domingo, M. G., Etcheverrito, A., Olmedo, D. G., & Tasat, D. R. (2020). Neurotoxicity mediated by oxidative stress caused by titanium dioxide nanoparticles in human neuroblastoma (SH-SY5Y) cells. *Journal of Trace Elements in Medicine and Biology*, *57*, 126413.

Ferreira, A. J., Cemlyn-Jones, J., Robalo, C., & Cordeiro (2013). Nanoparticles, nanotechnology and pulmonary nanotoxicology. *Revista Portuguesa de Pneumologia*, *19*(1), 28–37.

Florence, A. T. (1997). The oral absorption of micro-and nanoparticulates: neither exceptional nor unusual. *Pharmaceutical Research*, *14*(3), 259–266.

Foldbjerg, R., Dang, D. A., & Autrup, H. (2011). Cytotoxicity and genotoxicity of silver nanoparticles in the human lung cancer cell line, A549. *Archives of Toxicology*, *85*(7), 743–750.

Fröhlich, E. E., & Fröhlich, E. (2016). Cytotoxicity of nanoparticles contained in food on intestinal cells and the gut microbiota. *International Journal of Molecular Sciences*, *17*(4), 509.

Fröhlich, E., & Roblegg, E. (2012). Models for oral uptake of nanoparticles in consumer products. *Toxicology*, *291*(1–3), 10–17.

Fröhlich, E., & Roblegg, E. (2016). Oral uptake of nanoparticles: human relevance and the role of in vitro systems. *Archives of Toxicology*, *90*(10), 2297–2314.

Frokjaer, S., & Otzen, D. E. (2005). Protein drug stability: A formulation challenge. *Nature Reviews Drug Discovery*, *4*(4), 298–306.

Gómez-Estaca, J., et al. (2012). Formation of zein nanoparticles by electrohydrodynamic atomization: Effect of the main processing variables and suitability for encapsulating the food coloring and active ingredient curcumin. *Food Hydrocolloids*, *28*(1), 82–91.

Gaillet, S., & Rouanet, J. -M. (2015). Silver nanoparticles: their potential toxic effects after oral exposure and underlying mechanisms–a review. *Food and Chemical Toxicology*, *77*, 58–63.

Gao, J., Zhang, X., Yu, M., Ren, G., & Yang, Z. (2015). Cognitive deficits induced by multi-walled carbon nanotubes via the autophagic pathway. *Toxicology*, *337*, 21–29.

Ghosh, M., Bandyopadhyay, M., & Mukherjee, A. (2010). Genotoxicity of titanium dioxide (TiO_2) nanoparticles at two trophic levels: plant and human lymphocytes. *Chemosphere*, *81*(10), 1253–1262.

Gomes, T., Araújo, O., Pereira, R., Almeida, A. C., Cravo, A., & Bebianno, M. J. (2013). Genotoxicity of copper oxide and silver nanoparticles in the mussel *Mytilus galloprovincialis*. *Marine Environmental Research*, *84*, 51–59.

Gong, C., Tao, G., Yang, L., Liu, J., Liu, Q., Li, W., et al. (2012). Methylation of PARP-1 promoter involved in the regulation of nano-SiO_2-induced decrease of PARP-1 mRNA expression. *Toxicology Letters*, *209*(3), 264–269.

Graveland-Bikker, J., et al. (2006). Structural and mechanical study of a self-assembling protein nanotube. *Nano letters*, *6*(4), 616–621.

Guo, Y. -Y., Zhang, J., Zheng, Y. -F., Yang, J., & Zhu, X. -Q. (2011). Cytotoxic and genotoxic effects of multi-wall carbon nanotubes on human umbilical vein endothelial cells in vitro. *Mutation Research/Genetic Toxicology and Environmental Mutagenesis*, *721*(2), 184–191.

Guo, C., Xia, Y., Niu, P., Jiang, L., Duan, J., Yu, Y., et al. (2015). Silica nanoparticles induce oxidative stress, inflammation, and endothelial dysfunction in vitro via activation of the MAPK/Nrf2 pathway and nuclear factor-κB signaling. *International Journal of Nanomedicine*, *10*, 1463.

Hadidi, M., Pouramin, S., Adinepour, F., Haghani, S., & Jafari, S. M. (2020). Chitosan nanoparticles loaded with clove essential oil: Characterization, antioxidant and antibacterial activities. *Carbohydrate Polymers*, 116075.

Hadrup, N., Loeschner, K., Bergström, A., Wilcks, A., Gao, X., Vogel, U., et al. (2012). Subacute oral toxicity investigation of nanoparticulate and ionic silver in rats. *Archives of Toxicology*, *86*(4), 543–551.

Hardy, A., Benford, D., Halldorsson, T., Jeger, M. J., Knutsen, H. K., More, S., et al. (2018). Guidance on risk assessment of the application of nanoscience and nanotechnologies in the food and feed chain: Part 1, human and animal health. *EFSA Journal*, *16*(7.), .

He, Q., Zhou, X., Liu, Y., Gou, W., Cui, J., Li, Z., et al. (2018). Titanium dioxide nanoparticles induce mouse hippocampal neuron apoptosis via oxidative stress-and calcium imbalance-mediated endoplasmic reticulum stress. *Environmental Toxicology and Pharmacology*, *63*, 6–15.

Hillyer, J. F., & Albrecht, R. M. (2001). Gastrointestinal persorption and tissue distribution of differently sized colloidal gold nanoparticles. *Journal of Pharmaceutical Sciences*, *90*(12), 1927–1936.

Hoheisel, T. N., Schrettl, S., Szilluweit, R., & Frauenrath, H. (2010). Nanostructured carbonaceous materials from molecular precursors. *Angewandte Chemie International Edition*, *49*(37), 6496–6515.

Hoseinnejad, M., Jafari, S. M., & Katouzian, I. (2018). Inorganic and metal nanoparticles and their antimicrobial activity in food packaging applications. *Critical Reviews in Microbiology*, *44*(2), 161–181.

Hoseyni, S. Z., Jafari, S. M., Tabarestani, H. S., Ghorbani, M., Assadpour, E., & Sabaghi, M. (2020). Production and characterization of catechin-loaded electrospun nanofibers from Azivash gum-polyvinyl alcohol. *Carbohydrate Polymers*, 115979.

Hu, Y., Shenderova, O. A., & Brenner, D. W. (2007). Carbon nanostructures: morphologies and properties. *Journal of Computational and Theoretical Nanoscience*, *4*(2), 199–221.

Hua, S. (2019). Advances in nanoparticulate drug delivery approaches for sublingual and buccal administration. *Frontiers in Pharmacology*, *10*, 1328.

Hussain, S., Al-Nsour, F., Rice, A. B., Marshburn, J., Yingling, B., Ji, Z., et al. (2012). Cerium dioxide nanoparticles induce apoptosis and autophagy in human peripheral blood monocytes. *ACS Nano*, *6*(7), 5820–5829.

Jafari, S. M., & McClements, D. J. (2018). *Nanoemulsions: Formulation, Applications, and Characterization*. Academic Press.

Jafari, S. M., Katouzian, I., & Akhavan, S. (2017a). 15 - Safety and regulatory issues of nanocapsules. In S. M. Jafari (Ed.), *Nanoencapsulation technologies for the food and nutraceutical industries* (pp. 545–590). Academic Press.

Jafari, S. M., Katouzian, I., & Akhavan, S. (2017b). In *Safety and regulatory issues of nanocapsules* (pp. 545–590). Elsevier.

Jafari, S. M., Katouzian, I., Rajabi, H., & Ganje, M. (2017c). In *Bioavailability and release of bioactive components from nanocapsules* (pp. 494–523). Elsevier Inc..

Jain, A., Sharma, G., Kushwah, V., Thakur, K., Ghoshal, G., Singh, B., et al. (2017). Fabrication and functional attributes of lipidic nanoconstructs of lycopene: An innovative endeavour for enhanced cytotoxicity in MCF-7 breast cancer cells. *Colloids and Surfaces B: Biointerfaces*, *152*, 482–491.

Jain, A., Ranjan, S., Dasgupta, N., & Ramalingam, C. (2018). Nanomaterials in food and agriculture: an overview on their safety concerns and regulatory issues. *Critical Reviews in Food Science and Nutrition*, *58*(2), 297–317.

Jatana, S., & DeLouise, L. A. (2014). Understanding engineered nanomaterial skin interactions and the modulatory effects of ultraviolet radiation skin exposure. *Wiley Interdisciplinary Reviews. Nanomedicine and Nanobiotechnology*, *6*(1), 61–79.

Jin, T., & He, Y. (2011). Antibacterial activities of magnesium oxide (MgO) nanoparticles against foodborne pathogens. *Journal of Nanoparticle Research*, *13*(12), 6877–6885.

Kalita, P., Singh, J., Kumar Singh, M., Solanki, P. R., Sumana, G., & Malhotra, B. (2012). Ring like self assembled Ni nanoparticles based biosensor for food toxin detection. *Applied Physics Letters*, *100*(9), 093702.

Kalpana, V., & Rajeswari, V. D. (2017). Biosynthesis of metal and metal oxide nanoparticles for food packaging and preservation: a green expertise. In *Food biosynthesis* (pp. 293–316). Elsevier.

Katouzian, I., & Jafari, S. M. (2019). Protein nanotubes as state-of-the-art nanocarriers: Synthesis methods, simulation and applications. *Journal of Controlled Release*, *303*, 302–318. doi: 10.1016/j.jconrel.2019.04.026.

Katouzian, I., et al. (2017). Formulation and application of a new generation of lipid nanocarriers for the food bioactive ingredients. *Trends in Food Science and Technology, 68*, 14–25.

Kavand, A., Anton, N., Vandamme, T., Serra, C. A., & Chan-Seng, D. (2020). Synthesis and functionalization of hyperbranched polymers for targeted drug delivery. *Journal of Controlled Release.*

Kessler, V. G. (2009). The chemistry behind the sol–gel synthesis of complex oxide nanoparticles for bio-imaging applications. *Journal of Sol-Gel Science and Technology, 51*(3), 264.

Khanna, P., Ong, C., Bay, B. H., & Baeg, G. H. (2015). Nanotoxicity: an interplay of oxidative stress, inflammation and cell death. *Nanomaterials, 5*(3), 1163–1180.

Khanna, P., Kaur, A., & Goyal, D. (2019). Algae-based metallic nanoparticles: synthesis, characterization and applications. *Journal of Microbiological Methods, 163*, 105656.

Kim, Y. S., Kim, J. S., Cho, H. S., Rha, D. S., Kim, J. M., Park, J. D., et al. (2008). Twenty-eight-day oral toxicity, genotoxicity, and gender-related tissue distribution of silver nanoparticles in Sprague-Dawley rats. *Inhalation Toxicology, 20*(6), 575–583.

Kim, H. R., Kim, M. J., Lee, S. Y., Oh, S. M., & Chung, K. H. (2011). Genotoxic effects of silver nanoparticles stimulated by oxidative stress in human normal bronchial epithelial (BEAS-2B) cells. *Mutation Research/Genetic Toxicology and Environmental Mutagenesis, 726*(2), 129–135.

Koeneman, B. A., Zhang, Y., Westerhoff, P., Chen, Y., Crittenden, J. C., & Capco, D. G. (2010). Toxicity and cellular responses of intestinal cells exposed to titanium dioxide. *Cell Biology and Toxicology, 26*(3), 225–238.

Kong, H., Xia, K., Pan, L., Zhang, J., Luo, Y., Zhang, Y., et al. (2017). Autophagy and lysosomal dysfunction: a new insight into mechanism of synergistic pulmonary toxicity of carbon black-metal ions co-exposure. *Carbon, 111*, 322–333.

Kora, A. J., & Rastogi, L. (2018). Green synthesis of palladium nanoparticles using gum ghatti (*Anogeissus latifolia*) and its application as an antioxidant and catalyst. *Arabian Journal of Chemistry, 11*(7), 1097–1106.

Koudelka, S., Mikulik, R., Mašek, J., Raška, M., Knotigová, P. T., Miller, A. D., et al. (2016). Liposomal nanocarriers for plasminogen activators. *Journal of Controlled Release, 227*, 45–57.

Kuang, H., Yang, P., Yang, L., Aguilar, Z. P., & Xu, H. (2016). Size dependent effect of ZnO nanoparticles on endoplasmic reticulum stress signaling pathway in murine liver. *Journal of Hazardous Materials, 317*, 119–126.

L'Azou, B., & Marano, F. (2011). Nanoparticle toxicity mechanisms: oxidative stress and inflammation. In *Nanoethics and Nanotoxicology* (pp. 87–109). Springer.

Lademann, J., Richter, H., Schanzer, S., Knorr, F., Meinke, M., Sterry, W., et al. (2011). Penetration and storage of particles in human skin: perspectives and safety aspects. *European Journal of Pharmaceutics and Biopharmaceutics*, 77(3), 465–468.

Larese Filon, F., Bello, D., Cherrie, J. W., Sleeuwenhoek, A., Spaan, S., & Brouwer, D. H. (2016). Occupational dermal exposure to nanoparticles and nano-enabled products: Part I-Factors affecting skin absorption. *International Journal of Hygiene and Environmental Health, 219*(6), 536–544.

Lee, Y. -H., Cheng, F. -Y., Chiu, H. -W., Tsai, J. -C., Fang, C. -Y., Chen, C. -W., et al. (2014). Cytotoxicity, oxidative stress, apoptosis and the autophagic effects of silver nanoparticles in mouse embryonic fibroblasts. *Biomaterials, 35*(16), 4706–4715.

Leroueil, P. R., Berry, S. A., Duthie, K., Han, G., Rotello, V. M., McNerny, D. Q., et al. (2008). Wide varieties of cationic nanoparticles induce defects in supported lipid bilayers. *Nano Letters, 8*(2), 420–424.

Li, Y. F., & Chen, C. (2011). Fate and toxicity of metallic and metal-containing nanoparticles for biomedical applications. *Small*, 7(21), 2965–2980.

Li, J. J., Hartono, D., Ong, C. -N., Bay, B. -H., & Yung, L. -Y. L. (2010). Autophagy and oxidative stress associated with gold nanoparticles. *Biomaterials, 31*(23), 5996–6003.

Li, J., Tian, M., Cui, L., Dwyer, J., Fullwood, N. J., Shen, H., et al. (2016). Low-dose carbon-based nanoparticle-induced effects in A549 lung cells determined by biospectroscopy are associated with increases in genomic methylation. *Scientific Reports, 6*, 20207.

Li, J., Tang, M., & Xue, Y. (2019a). Review of the effects of silver nanoparticle exposure on gut bacteria. *Journal of Applied Toxicology, 39*, 27–37.

Li, Y., Ma, R., Liu, X., Qi, Y., Abulikemu, A., Zhao, X., et al. (2019b). Endoplasmic reticulum stress-dependent oxidative stress mediated vascular injury induced by silica nanoparticles in vivo and in vitro. *NanoImpact, 14*, 100169.

Lin, Y. -Z., & Chang, P. -L. (2013). Colorimetric determination of DNA methylation based on the strength of the hydrophobic interactions between DNA and gold nanoparticles. *ACS Applied Materials & Interfaces, 5*(22), 12045–12051.

Lindberg, H. K., Falck, G. C. M., Suhonen, S., Vippola, M., Vanhala, E., Catalán, J., et al. (2009). Genotoxicity of nanomaterials: DNA damage and micronuclei induced by carbon nanotubes and graphite nanofibres in human bronchial epithelial cells in vitro. *Toxicology Letters, 186*(3), 166–173.

Livney, Y. D. (2010). Milk proteins as vehicles for bioactives. *Current Opinion in Colloid & Interface Science, 15*(1–2), 73–83.

Lomer, M. C. E., Thompson, R. P. H., & Powell, J. J. (2002). Fine and ultrafine particles of the diet: influence on the mucosal immune response and association with Crohn's disease. *Proceedings of the Nutrition Society, 61*(1), 123–130.

Lu, S., Duffin, R., Poland, C., Daly, P., Murphy, F., Drost, E., et al. (2009). Efficacy of simple short-term in vitro assays for predicting the potential of metal oxide nanoparticles to cause pulmonary inflammation. *Environmental Health Perspectives, 117*(2), 241–247.

Lu, L. -M., Li, H. -B., Qu, F., Zhang, X. -B., Shen, G. -L., & Yu, R. -Q. (2011). In situ synthesis of palladium nanoparticle–graphene nanohybrids and their application in nonenzymatic glucose biosensors. *Biosensors and Bioelectronics, 26*(8), 3500–3504.

Lundquist, P., & Artursson, P. (2016). Oral absorption of peptides and nanoparticles across the human intestine: Opportunities, limitations and studies in human tissues. *Advanced Drug Delivery Reviews, 106*, 256–276.

Lundy, D. J., Chen, K. -H., Toh, E. K. W., & Hsieh, P. C. H. (2016). Distribution of systemically administered nanoparticles reveals a size-dependent effect immediately following cardiac ischaemia-reperfusion injury. *Scientific Reports, 6*(1), 1–10.

Luo, Y. -H., Chang, L. W., & Lin, P. (2015). Metal-based nanoparticles and the immune system: activation, inflammation, and potential applications. *BioMed Research International, 2015*.

Ma, Y., & Yang, Q. (2015). Nanotoxicology aspects of carbohydrate nanostructures. *Carbohydrate Nanotechnology*, 423–452.

Ma, X., Wu, Y., Jin, S., Tian, Y., Zhang, X., Zhao, Y., et al. (2011). Gold nanoparticles induce autophagosome accumulation through size-dependent nanoparticle uptake and lysosome impairment. *ACS Nano, 5*(11), 8629–8639.

Madkour, L. H. (2020). Endoplasmic reticulum (ER) stress as a mechanism for NP-induced toxicity. In *Nanoparticles induce oxidative and endoplasmic reticulum stresses* (pp. 403–450). Springer.

Magdolenova, Z., Collins, A., Kumar, A., Dhawan, A., Stone, V., & Dusinska, M. (2014). Mechanisms of genotoxicity. A review of in vitro and in vivo studies with engineered nanoparticles. *Nanotoxicology, 8*(3), 233–278.

Maherani, B., Arab-Tehrany, E., R Mozafari, M., Gaiani, C., & Linder, M. (2011). Liposomes: A review of manufacturing techniques and targeting strategies. *Current Nanoscience*, 7(3), 436–452.

Maksimović, M., Omanović-Mikličanin, E., & Badnjević, A. (2019). In *Is nanofood safe?* (pp. 87–137). Springer.

Manickam, V., Velusamy, R. K., Lochana, R., Rajendran, B., & Tamizhselvi, R. (2017). Applications and genotoxicity of nanomaterials in the food industry. *Environmental Chemistry Letters, 15*(3), 399–412.

Manke, A., Wang, L., & Rojanasakul, Y. (2013). Mechanisms of nanoparticle-induced oxidative stress and toxicity. *BioMed Research International, 2013*.
Manna, S. K., Sarkar, S., Barr, J., Wise, K., Barrera, E. V., Jejelowo, O., et al. (2005). Single-walled carbon nanotube induces oxidative stress and activates nuclear transcription factor-κB in human keratinocytes. *Nano Letters, 5*(9), 1676–1684.
Martirosyan, A., & Schneider, Y. J. (2014). Engineered nanomaterials in food: implications for food safety and consumer health. *International Journal of Environmental Research and Public Health, 11*(6), 5720–5750.
Martirosyan, A., Bazes, A., & Schneider, Y. -J. (2014). In vitro toxicity assessment of silver nanoparticles in the presence of phenolic compounds–preventive agents against the harmful effect?". *Nanotoxicology, 8*(5), 573–582.
Mason, T. G., et al. (2006). Nanoemulsions: formation, structure, and physical properties. *Journal of Physics: Condensed Matter, 18*(41), R635.
Massaro, M., Colletti, C. G., Lazzara, G., & Riela, S. (2018). The use of some clay minerals as natural resources for drug carrier applications. *Journal of Functional Biomaterials, 9*(4), 58.
Matsoukas, T., Desai, T., & Lee, K. (2015). Engineered nanoparticles and their applications. *Journal of Nanomaterials, 2015*.
Maynard, A. D., & Kuempel, E. D. (2005). Airborne nanostructured particles and occupational health. *Journal of Nanoparticle Research, 7*(6), 587–614.
McClements, D. J., & Xiao, H. (2017a). Is nano safe in foods? Establishing the factors impacting the gastrointestinal fate and toxicity of organic and inorganic food-grade nanoparticles. *NPJ Science of Food, 1*, 6.
McClements, D. J., & Xiao, H. (2017b). Is nano safe in foods? Establishing the factors impacting the gastrointestinal fate and toxicity of organic and inorganic food-grade nanoparticles. *NPJ Science of Food, 1*(1), 1–13.
Mc Clements, D. J., Li, F., & Xiao, H. (2015). The nutraceutical bioavailability classification scheme: classifying nutraceuticals according to factors limiting their oral bioavailability. *Annual Review of Food Science and Technology, 6*, 299–327.
Mehnert, W., & Mäder, K. (2012). Solid lipid nanoparticles: production, characterization and applications. *Advanced Drug Delivery Reviews, 64*, 83–101.
Meng, X., & Edgar, K. J. (2016). Click" reactions in polysaccharide modification. *Progress in Polymer Science, 53*, 52–85.
Mishra, P. K., Mishra, H., Ekielski, A., Talegaonkar, S., & Vaidya, B. (2017). Zinc oxide nanoparticles: a promising nanomaterial for biomedical applications. *Drug Discovery Today, 22*(12), 1825–1834.
Mitragotri, S., & Stayton, P. (2014). Organic nanoparticles for drug delivery and imaging. *Mrs Bulletin, 39*(3), 219–223.
Moghassemi, S., & Hadjizadeh, A. (2014). Nano-niosomes as nanoscale drug delivery systems: an illustrated review. *Journal of Controlled Release, 185*, 22–36.
Moini, N., Jahandideh, A., & Anderson, G. (2019). Inorganic nanocomposite hydrogels: present knowledge and future challenge. In *Sustainable polymer composites and nanocomposites* (pp. 805–853). Springer.
Monopoli, M. P., Åberg, C., Salvati, A., & Dawson, K. A. (2012). Biomolecular coronas provide the biological identity of nanosized materials. *Nature Nanotechnology, 7*, 779.
Montenegro, L., et al. (2016). From nanoemulsions to nanostructured lipid carriers: A relevant development in dermal delivery of drugs and cosmetics. *Journal of Drug Delivery Science and Technology, 32*, 100–112.
Murray, A. R., Kisin, E. R., Tkach, A.V., Yanamala, N., Mercer, R., Young, S. -H., et al. (2012). Factoring-in agglomeration of carbon nanotubes and nanofibers for better prediction of their toxicity versus asbestos. *Particle and Fibre Toxicology, 9*(1), 1–19.
Mytych, J., Zebrowski, J., Lewinska, A., & Wnuk, M. (2017). Prolonged effects of silver nanoparticles on p53/p21 pathway-mediated proliferation, DNA damage response, and

methylation parameters in HT22 hippocampal neuronal cells. *Molecular Neurobiology*, *54*(2), 1285–1300.

Nagarsekar, A., & Ghandehari, H. (1999). Genetically engineered polymers for drug delivery. *Journal of Drug Targeting*, 7(1), 11–32.

Nakayama, M. (2018). Macrophage recognition of crystals and nanoparticles. *Frontiers in Immunology*, *9*, 103–1103.

Nel, A. E., Mädler, L., Velegol, D., Xia, T., Hoek, E. M.V., Somasundaran, P., et al. (2009). Understanding biophysicochemical interactions at the nano–bio interface. *Nature Materials*, *8*, 543.

Nguyen, P. D., et al. (2019). Nonlinear vibration of carbon nanotube reinforced composite truncated conical shells in thermal environment. *International Journal of Structural Stability and Dynamics*, *19*(12), 1950158.

Noël, C., Simard, J. -C., & Girard, D. (2016). Gold nanoparticles induce apoptosis, endoplasmic reticulum stress events and cleavage of cytoskeletal proteins in human neutrophils. *Toxicology in Vitro*, *31*, 12–22.

Obeidat, W. M., et al. (2010). Preservation of nanostructured lipid carriers (NLC). *European Journal of Pharmaceutics and Biopharmaceutics*, *76*(1), 56–67.

Oberdörster, G., Oberdörster, E., & Oberdörster, J. (2005). Nanotoxicology: an emerging discipline evolving from studies of ultrafine particles. *Environmental Health Perspectives*, *113*(7), 823–839.

Oberdörster, E. (2004). Manufactured nanomaterials (fullerenes, C60) induce oxidative stress in the brain of juvenile largemouth bass. *Environmental Health Perspectives*, *112*(10), 1058–1062.

Öner, D., Ghosh, M., Bové, H., Moisse, M., Boeckx, B., Duca, R. C., et al. (2018). Differences in MWCNT-and SWCNT-induced DNA methylation alterations in association with the nuclear deposition. *Particle and Fibre Toxicology*, *15*(1), 1–19.

Őrfi, E., & Szebeni, J. (2016). The immune systemof the gut and potential adverse effects of oral nanocarriers on its function. *Advanced Drug Delivery Reviews Journal*, *15*(106), 402–409.

Owoseni-Fagbenro, K. A., Saifullah, S., Imran, M., Perveen, S., Rao, K., & Fasina, T. M. (2019). Egg proteins stabilized green silver nanoparticles as delivery system for hesperidin enhanced bactericidal potential against resistant *S. aureus*. *Journal of Drug Delivery Science and Technology*, *50*, 347–354.

Pérez-Álvarez, L., et al. (2019). Chitosan nanogels as nanocarriers of polyoxometalates for breast cancer therapies. *Carbohydrate Polymers*, *213*, 159–167.

Pérez-Arizti, J. A., Ventura-Gallegos, J. L., Juárez, R. E. G., del Pilar Ramos-Godinez, M., Colín-Val, Z., & López-Marure, R. (2020). Titanium dioxide nanoparticles promote oxidative stress, autophagy and reduce NLRP3 in primary rat astrocytes. *Chemico-Biological Interactions*, *317*, 108966.

Paecharoenchai, O. (2013). Development of cationic niosomes containing novel synthesized cationic lipids for nucleic acid delivery, A Thesis Submitted in Partial Fulfillment of the Requirements for the Degree Doctor of Philosophy Program in Pharmaceutical Technology Graduate School, Silpakorn University.

Pan, M., Yin, Z., Liu, K., Du, X., Liu, H., & Wang, S. (2019). Carbon-based nanomaterials in sensors for food safety. *Nanomaterials*, *9*(9), 1330.

Parhi, R., & Suresh, P. (2012). Preparation and characterization of solid lipid nanoparticles-a review. *Current Drug Discovery Technologies*, *9*(1), 2–16.

Park, E. -J., Bae, E., Yi, J., Kim, Y., Choi, K., Lee, S. H., et al. (2010). Repeated-dose toxicity and inflammatory responses in mice by oral administration of silver nanoparticles. *Environmental Toxicology and Pharmacology*, *30*(2), 162–168.

Park, M.V. D. Z., Neigh, A. M., Vermeulen, J. P., de la Fonteyne, L. J. J., Verharen, H.W., Briedé, J. J., et al. (2011). The effect of particle size on the cytotoxicity, inflammation, developmental toxicity and genotoxicity of silver nanoparticles. *Biomaterials*, *32*(36), 9810–9817.

Park, E. -J., Choi, D. -H., Kim, Y., Lee, E. -W., Song, J., Cho, M. -H., et al. (2014). Magnetic iron oxide nanoparticles induce autophagy preceding apoptosis through mitochondrial damage and ER stress in RAW264. 7 cells. *Toxicology in Vitro*, *28*(8), 1402–1412.

Park, K. (2013). Toxicokinetic differences and toxicities of silver nanoparticles and silver ions in rats after single oral administration. *Journal of Toxicology and Environmental Health, Part A*, *76*(22), 1246–1260.

Patil, N. A., Gade, W. N., & Deobagkar, D. D. (2016). Epigenetic modulation upon exposure of lung fibroblasts to TiO2 and ZnO nanoparticles: alterations in DNA methylation. *International Journal of Nanomedicine*, *11*, 4509.

Patlolla, A., Knighten, B., & Tchounwou, P. (2010). Multi-walled carbon nanotubes induce cytotoxicity, genotoxicity and apoptosis in normal human dermal fibroblast cells. *Ethnicity & Disease*, *20*(1 Suppl 1), S1.

Pereira, P., et al. (2015). Biocompatibility of a self-assembled glycol chitosan nanogel. *Toxicology in Vitro*, *29*(3), 638–646.

Persaud, I., Shannahan, J. H., Raghavendra, A. J., Alsaleh, N. B., Podila, R., & Brown, J. M. (2018). Biocorona formation contributes to silver nanoparticle induced endoplasmic reticulum stress. *Ecotoxicology and Environmental Safety*, *170*, 77–86.

Persaud, I., Shannahan, J. H., Raghavendra, A. J., Alsaleh, N. B., Podila, R., & Brown, J. M. (2019). Biocorona formation contributes to silver nanoparticle induced endoplasmic reticulum stress. *Ecotoxicology and Environmental Safety*, *170*, 77–86.

Peters, R., Kramer, E., Oomen, A. G., Herrera Rivera, Z. E., Oegema, G., Tromp, P. C., et al. (2012). Presence of nano-sized silica during in vitro digestion of foods containing silica as a food additive. *ACS Nano*, *6*(3), 2441–2451.

Petros, R. A., & DeSimone, J. M. (2010). Strategies in the design of nanoparticles for therapeutic applications. *Nature Reviews Drug Discovery*, *9*(8), 615–627.

Pietroiusti, A., Magrini, A., & Campagnolo, L. (2016). New frontiers in nanotoxicology: Gut microbiota/microbiome-mediated effects of engineered nanomaterials. *Toxicology and Applied Pharmacology*, *299*, 90–95.

Poh, T. Y., Ali, N. A. t. B. M., Mac Aogáin, M., Kathawala, M. H., Setyawati, M. I., Ng, K. W., et al. (2018). Inhaled nanomaterials and the respiratory microbiome: clinical, immunological and toxicological perspectives. *Particle and Fibre Toxicology*, *15*(1), 46.

Popp, L., & Segatori, L. (2015). Differential autophagic responses to nano-sized materials. *Current Opinion in Biotechnology*, *36*, 129–136.

Prasad, R. Y., McGee, J. K., Killius, M. G., Suarez, D. A., Blackman, C. F., DeMarini, D. M., et al. (2013). Investigating oxidative stress and inflammatory responses elicited by silver nanoparticles using high-throughput reporter genes in HepG2 cells: effect of size, surface coating, and intracellular uptake. *Toxicology in Vitro*, *27*(6.), .

Pujalté, I., Dieme, D., Haddad, S., Serventi, A. M., & Bouchard, M. (2017). Toxicokinetics of titanium dioxide (TiO(2)) nanoparticles after inhalation in rats. *Toxicology Letters*, *265*, 77–85.

Qian, Y., Zhang, J., Hu, Q., Xu, M., Chen, Y., Hu, G., et al. (2015). Silver nanoparticle-induced hemoglobin decrease involves alteration of histone 3 methylation status. *Biomaterials*, *70*, 12–22.

Raikos, V., Hayward, N., Hayes, H., Meroni, E., & Ranawana, V. (2019). Optimising the ratio of long-to short-chain triglycerides of the lipid phase to enhance physical stability and bioaccessibility of lycopene-loaded beverage emulsions. *International Journal of Food Science & Technology,*, *54*(4), 1355–1362.

Rakshit, S., Ghosh, S., Chall, S., Mati, S. S., Moulik, S., & Bhattacharya, S. C. (2013). Controlled synthesis of spin glass nickel oxide nanoparticles and evaluation of their potential antimicrobial activity: a cost effective and eco friendly approach. *RSC Advances*, *3*(42), 19348–19356.

Rama Narsimha Reddy, A., Narsimha Reddy, Y., Himabindu, V., Rama, D., & Krishna (2011). Induction of oxidative stress and cytotoxicity by carbon nanomaterials is dependent on physical properties. *Toxicology and Industrial Health*, *27*(1), 3–10.

Ramakritinan, C., Kaarunya, E., Shankar, S., & Kumaraguru, A. (2013). In *Antibacterial effects of Ag, Au and bimetallic (Ag-Au) nanoparticles synthesized from red algae*Solid State Phenomena, Trans Tech Publ.

Renwick, L. C., Brown, D., Clouter, A., & Donaldson, K. (2004). Increased inflammation and altered macrophage chemotactic responses caused by two ultrafine particle types. *Occupational and Environmental Medicine, 61*(5), 442–447.

Reuland, D. J., Khademi, S., Castle, C. J., Irwin, D. C., McCord, J. M., Miller, B. F., et al. (2013). Upregulation of phase II enzymes through phytochemical activation of Nrf2 protects cardiomyocytes against oxidant stress. *Free Radical Biology and Medicine, 56*, 102–111.

Rezaei, A., Fathi, M., & Jafari, S. M. (2019). Nanoencapsulation of hydrophobic and low-soluble food bioactive compounds within different nanocarriers. *Food Hydrocolloids, 88*, 146–162.

Ricci, J. -E., Gottlieb, R. A., & Green, D. R. (2003). Caspase-mediated loss of mitochondrial function and generation of reactive oxygen species during apoptosis. *The Journal of Cell Biology, 160*(1), 65–75.

Rizvi, S. A. A., & Saleh, A. M. (2018). Applications of nanoparticle systems in drug delivery technology. *Saudi Pharmaceutical Journal, 26*(1), 64–70.

Rizwan, M., Ali, S., Qayyum, M. F., Ok, Y. S., Adrees, M., Ibrahim, M., et al. (2017). Effect of metal and metal oxide nanoparticles on growth and physiology of globally important food crops: A critical review. *Journal of Hazardous Materials, 322*, 2–16.

Rostamabadi, H., Falsafi, S. R., & Jafari, S. M. (2019a). Nanoencapsulation of carotenoids within lipid-based nanocarriers. *Journal of Controlled Release.*

Rostamabadi, H., Mahoonak, A. S., Allafchian, A., & Ghorbani, M. (2019). Fabrication of β-carotene loaded glucuronoxylan-based nanostructures through electrohydrodynamic processing. *International Journal of Biological Macromolecules.*

Rostamabadi, H., Falsafi, S. R., & Jafari, S. M. (2019b). Nano-helices of amylose for encapsulation of food ingredients. *Biopolymer Nanostructures for Food Encapsulation Purposes.* Elsevier.

Rostamabadi, H., Falsafi, S. R., & Jafari, S. M. (2019c). Nanostructures of starch for encapsulation of food ingredients. *Biopolymer nanostructures for food encapsulation purposes.* Elsevier.

Rostamabadi, H., Falsafi, S. R., & Jafari, S. M. (2019d). Starch-based nanocarriers as cutting-edge natural cargos for nutraceutical delivery. *Trends in Food Science & Technology.*

Sadeghi, Z., et al. (2020a). Mannosylated chitosan nanoparticles loaded with FliC antigen as a novel vaccine candidate against *Brucella melitensis* and *Brucella abortus* infection. *Journal of biotechnology, 310*, 89–96.

Sadeghi, Z. (2020b). Nanoparticle-based vaccines for brucellosis: calcium phosphate nanoparticles-adsorbed antigens induce cross protective response in mice. *International Journal of Nanomedicine, 15*, 3877.

Şahin, E., Musevi, S. J., & Aslani, A. (2017). Antibacterial activity against *Escherichia coli* and characterization of ZnO and ZnO–Al_2O_3 mixed oxide nanoparticles. *Arabian Journal of Chemistry, 10*, S230–S235.

Sakai-Kato, K., Hidaka, M., Un, K., Kawanishi, T., & Okuda, H. (2014). Physicochemical properties and in vitro intestinal permeability properties and intestinal cell toxicity of silica particles, performed in simulated gastrointestinal fluids. *Biochimica et Biophysica Acta (BBA) - General Subjects, 1840*(3), 1171–1180.

Saleem, S., Ahmed, B., Khan, M. S., Al-Shaeri, M., & Musarrat, J. (2017). Inhibition of growth and biofilm formation of clinical bacterial isolates by NiO nanoparticles synthesized from Eucalyptus globulus plants. *Microbial Pathogenesis, 111*, 375–387.

Salvia-Trujillo, L., & McClements, D. J. (2016). Enhancement of lycopene bioaccessibility from tomato juice using excipient emulsions: Influence of lipid droplet size. *Food Chemistry, 210*, 295–304.

Sanuja, S., Agalya, A., & Umapathy, M. (2014). Studies on magnesium oxide reinforced chitosan bionanocomposite incorporated with clove oil for active food packaging application. *International Journal of Polymeric Materials and Polymeric Biomaterials, 63*(14), 733–740.

Sawosz, E., Binek, M., Grodzik, M., Zielińska, M., Sysa, P., Szmidt, M., et al. (2007). Influence of hydrocolloidal silver nanoparticles on gastrointestinal microflora and morphology of enterocytes of quails. *Archives of Animal Nutrition*, *61*(6), 444–451.

Schutz, C. A., Juillerat-Jeanneret, L., Mueller, H., Lynch, I., & Riediker, M. (2013). Therapeutic nanoparticles in clinics and under clinical evaluation. *Nanomedicine (Lond)*, *8*(3), 449–467.

Shannahan, J. H., Podila, R., Aldossari, A. A., Emerson, H., Powell, B. A., Ke, P. C., et al. (2015). Formation of a protein corona on silver nanoparticles mediates cellular toxicity via scavenger receptors. *Toxicological Sciences*, *143*(1), 136–146.

Sheeja, C. C., Anusri, A., Levna, C., Aneesh, P. M., & Lekha, D. (2020). MoS2 nanoparticles induce behavioral alteration and oxidative stress mediated cellular toxicity in the social insect Oecophylla smaragdina (Asian weaver ant). *Journal of Hazardous Materials*, *385*, 121624.

Shi, X., Thomas, T. P., Myc, L. A., Kotlyar, A., & Baker, J. R., Jr. (2007). Synthesis, characterization, and intracellular uptake of carboxyl-terminated poly (amidoamine) dendrimer-stabilized iron oxide nanoparticles. *Physical Chemistry Chemical Physics*, *9*(42), 5712–5720.

Shi, S., et al. (2014). Dual drugs (microRNA-34a and paclitaxel)-loaded functional solid lipid nanoparticles for synergistic cancer cell suppression. *Journal of Controlled Release*, *194*, 228–237.

Shishir, M. R. I, Xie, L., Sun, C., Zheng, X., & Chen, W. (2018). Advances in micro and nano-encapsulation of bioactive compounds using biopolymer and lipid-based transporters. *Trends in Food Science & Technology*, *78*, 34–60.

Shitrit, Y., et al. (2019). Shear thinning pectin hydrogels physically cross-linked with chitosan nanogels. *Carbohydrate Polymers*, *225*, 115249.

Shukla, R. K., Sharma, V., Pandey, A. K., Singh, S., Sultana, S., & Dhawan, A. (2011). ROS-mediated genotoxicity induced by titanium dioxide nanoparticles in human epidermal cells. *Toxicology in Vitro*, *25*(1), 231–241.

Shumakova, A. A., Efimochkina, N. R., Minaeva, L. P., Bykova, I. B., Batishchava, S. Y., Markova, Y. M., et al. (2015). Toxicological assessment of nanostructured silica. III. Microecological, hematological indices, state of cellular immunity. *Voprosy Pitaniia*, *84*(4), 55–65.

Sierra, M. I., Valdés, A., Fernández, A. F., Torrecillas, R., & Fraga, M. F. (2016). The effect of exposure to nanoparticles and nanomaterials on the mammalian epigenome. *International Journal of Nanomedicine*, *11*, 6297.

Singh, J., Kalita, P., Singh, M. K., & Malhotra, B. (2011). Nanostructured nickel oxide-chitosan film for application to cholesterol sensor. *Applied Physics Letters*, *98*(12), 123702.

Singh, C., Srivastava, S., Ali, M. A., Gupta, T. K., Sumana, G., Srivastava, A., et al. (2013). Carboxylated multiwalled carbon nanotubes based biosensor for aflatoxin detection. *Sensors and Actuators B: Chemical*, *185*, 258–264.

Siqueira, J., Jr., & Oliveira, O. N., Jr. (2017). Carbon-based nanomaterials. In *Nanostructures* (pp. 233–249). Elsevier.

Smolkova, B., El Yamani, N., Collins, A. R., Gutleb, A. C., & Dusinska, M. (2015). Nanoparticles in food. Epigenetic changes induced by nanomaterials and possible impact on health. *Food and Chemical Toxicology*, *77*, 64–73.

Smolkova, B., Dusinska, M., & Gabelova, A. (2017). Nanomedicine and epigenome. Possible health risks. *Food and Chemical Toxicology*, *109*(Pt 1), 780–796.

Sokolov, V., Chernov, M., Shlykov, V., Razgulina, O., Yurkovets, D., & Krupskaya, V. (2008). Mineral nanoparticles in dispersed soils. *Journal of Surface Investigation. X-ray, Synchrotron and Neutron Techniques*, *2*(5), 759–763.

Sozer, N., & Kokini, J. L. (2009). Nanotechnology and its applications in the food sector. *Trends in Biotechnology*, *27*(2), 82–89.

Stern, S. T., Adiseshaiah, P. P., & Crist, R. M. (2012). Autophagy and lysosomal dysfunction as emerging mechanisms of nanomaterial toxicity. *Particle and Fibre Toxicology*, *9*(1), 20.

Stoccoro, A., Karlsson, H. L., Coppede, F., & Migliore, L. (2013). Epigenetic effects of nano-sized materials. *Toxicology*, *313*(1), 3–14.
Suárez-Iglesias, O., Collado, S., Oulego, P., & Díaz, M. (2017). Graphene-family nanomaterials in wastewater treatment plants. *Chemical Engineering Journal*, *313*, 121–135.
Sun, R., & Xia, Q. (2019). Nanostructured lipid carriers incorporated in alginate hydrogel: Enhanced stability and modified behavior in gastrointestinal tract. *olloids and Surfaces A: Physicochemical and Engineering Aspects*, *574*, 197–206.
Swierczewska, M., Han, H. S., Kim, K., Park, J. H., & Lee, S. (2016). Polysaccharide-based nanoparticles for theranostic nanomedicine. *Advanced Drug Delivery Reviews*, *99*, 70–84.
Tahir, K., Nazir, S., Ahmad, A., Li, B., Khan, A. U., Khan, Z. U. H., et al. (2017). Facile and green synthesis of phytochemicals capped platinum nanoparticles and in vitro their superior antibacterial activity. *Journal of Photochemistry and Photobiology B: Biology*, *166*, 246–251.
Takase, M., Yoneyama, Y., Murata, M., Hibi, K., Ren, H., & Endo, H. (2012). Carbon nanotube enhanced mediator-type biosensor for real-time monitoring of glucose concentrations in fish. *Analytical and Bioanalytical Chemistry*, *403*(4), 1187–1190.
Tang, Y., Chen, B., Hong, W., Chen, L., Yao, L., Zhao, Y., et al. (2019). ZnO nanoparticles induced male reproductive toxicity based on the effects on the endoplasmic reticulum stress signaling pathway. *International Journal of Nanomedicine*, *14*, 9563.
Taylor, A. A., Marcus, I. M., Guysi, R. L., & Walker, S. L. (2015). Metal oxide nanoparticles induce minimal phenotypic changes in a model colon gut microbiota. *Environmental Engineering Science*, *32*(7), 602–612.
Tenzer, S., Docter, D., Rosfa, S., Wlodarski, A., Kuharev, J., Rekik, A., et al. (2011). Nanoparticle size is a critical physicochemical determinant of the human blood plasma corona: a comprehensive quantitative proteomic analysis. *ACS Nano*, *5*(9), 7155–7167.
Teubl, B. J., Stojkovic, B., Docter, D., Pritz, E., Leitinger, G., Poberaj, I., et al. (2018). The effect of saliva on the fate of nanoparticles. *Clinical Oral Investigations*, *22*(2), 929–940.
Thorn, D. C., Meehan, S., Sunde, M., Rekas, A., Gras, S. L., MacPhee, C. E., et al. (2005). Amyloid fibril formation by bovine milk κ-casein and its inhibition by the molecular chaperones αS-and β-casein. *Biochemistry*, *44*(51), 17027–17036.
Tiede, K., Boxall, A. B., Tear, S. P., Lewis, J., David, H., & Hassellöv, M. (2008). Detection and characterization of engineered nanoparticles in food and the environment. *Food Additives and Contaminants*, *25*(7), 795–821.
Toda, T., & Yoshino, S. (2016). Amorphous nanosilica particles block induction of oral tolerance in mice. *Journal of Immunotoxicology*, *13*(5), 723–728.
Trbojevich, R. A., & Torres, A. M. (2017). Biological synthesis, pharmacokinetics, and toxicity of different metal nanoparticles. In P. D. M. Rai, & P. D. R. Shegokar (Eds.), *Metal nanoparticles in pharma* (pp. 451–468). Cham, Springer International Publishing.
Treuel, L., & Nienhaus, G. U. (2012). Toward a molecular understanding of nanoparticle–protein interactions. *Biophysical Reviews*, *4*(2), 137–147.
Tsai, Y. -Y., Huang, Y. -H., Chao, Y. -L., Hu, K. -Y., Chin, L. -T., Chou, S. -H., et al. (2011). Identification of the nanogold particle-induced endoplasmic reticulum stress by omic techniques and systems biology analysis. *ACS Nano*, *5*(12), 9354–9369.
Tsuji, J. S., Maynard, A. D., Howard, P. C., James, J. T., Lam, C. -w., Warheit, D. B., et al. (2006). Research strategies for safety evaluation of nanomaterials, part IV: risk assessment of nanoparticles. *Toxicological Sciences*, *89*(1), 42–50.
Tucci, P., Porta, G., Agostini, M., Dinsdale, D., Iavicoli, I., Cain, K., et al. (2013). Metabolic effects of TiO_2 nanoparticles, a common component of sunscreens and cosmetics, on human keratinocytes. *Cell Death & Disease*, *4*(3), e549–e1549.
Uddin, I., et al. (2018). A generic self-assembly process in microcompartments and synthetic protein nanotubes. *Small*, *14*(19), 1704020.
Vila, L., García-Rodríguez, A., Cortés, C., Marcos, R., & Hernández, A. (2018). Assessing the effects of silver nanoparticles on monolayers of differentiated Caco-2 cells, as a model of intestinal barrier. *Food and Chemical Toxicology*, *116*, 1–10.

Vita, A. A., Royse, E. A., & Pullen, N. A. (2019). Nanoparticles and danger signals: Oral delivery vehicles as potential disruptors of intestinal barrier homeostasis. *Journal of Leukocyte Biology*, *106*(1), 95–103.

Wan, B., Wang, Z. -X., Lv, Q. -Y., Dong, P. -X., Zhao, L. -X., Yang, Y., et al. (2013). Single-walled carbon nanotubes and graphene oxides induce autophagosome accumulation and lysosome impairment in primarily cultured murine peritoneal macrophages. *Toxicology Letters*, *221*(2), 118–127.

Wang, H., Zhang, J., & Yu, H. (2007). Elemental selenium at nano size possesses lower toxicity without compromising the fundamental effect on selenoenzymes: comparison with selenomethionine in mice. *Free Radical Biology and Medicine*, *42*(10), 1524–1533.

Wang, Z., Li, N., Zhao, J., White, J. C., Qu, P., & Xing, B. (2012). CuO nanoparticle interaction with human epithelial cells: cellular uptake, location, export, and genotoxicity. *Chemical Research in Toxicology*, *25*(7), 1512–1521.

Wang, F., Feng, L., Tang, M. R., Li, J. Y., & Tang, Q. G. (2013a). In *Advances in water treatment application of sepiolite mineral materials*Advanced Materials Research, Trans Tech Publ.

Wang, X., Podila, R., Shannahan, J. H., Rao, A. M., & Brown, J. M. (2013b). Intravenously delivered graphene nanosheets and multiwalled carbon nanotubes induce site-specific Th2 inflammatory responses via the IL-33/ST2 axis. *International Journal of Nanomedicine*, *8*, 1733.

Wang, Y., Yuan, L., Yao, C., Ding, L., Li, C., Fang, J., et al. (2014). A combined toxicity study of zinc oxide nanoparticles and vitamin C in food additives. *Nanoscale*, *6*(24), 15333–15342.

Wang, J., Yu, Y., Lu, K., Yang, M., Li, Y., Zhou, X., et al. (2017). Silica nanoparticles induce autophagy dysfunction via lysosomal impairment and inhibition of autophagosome degradation in hepatocytes. *International Journal of Nanomedicine*, *12*, 809.

Wang, P., Zhang, L., Liao, Y., Du, J., Xu, M., Zhao, W., et al. (2020). Effect of intratracheal instillation of ZnO nanoparticles on acute lung inflammation induced by lipopolysaccharides in mice. *Toxicological Sciences*, *173*(2), 373–386.

Wilding, L. A., Bassis, C. M., Walacavage, K., Hashway, S., Leroueil, P. R., Morishita, M., et al. (2016). Repeated dose (28-day) administration of silver nanoparticles of varied size and coating does not significantly alter the indigenous murine gut microbiome. *Nanotoxicology*, *10*(5), 513–520.

Wu, J., & Kaufman, R. J. (2006). From acute ER stress to physiological roles of the unfolded protein response. *Cell Death & Differentiation*, *13*(3), 374–384.

Wu, J., Liu, W., Xue, C., Zhou, S., Lan, F., Bi, L., et al. (2009). Toxicity and penetration of TiO_2 nanoparticles in hairless mice and porcine skin after subchronic dermal exposure. *Toxicol Letters*, *191*(1), 1–8.

Wu, H., Yin, J. -J., Wamer, W. G., Zeng, M., & Lo, Y. M. (2014). Reactive oxygen species-related activities of nano-iron metal and nano-iron oxides. *Journal of Food and Drug Analysis*, *22*(1), 86–94.

Yada, R. Y., Buck, N., Canady, R., DeMerlis, C., Duncan, T., Janer, G., et al. (2014). Engineered nanoscale food ingredients: evaluation of current knowledge on material characteristics relevant to uptake from the gastrointestinal tract. *Comprehensive Reviews in Food Science and Food Safety*, *13*(4), 730–744.

Yamashita, K., Yoshioka, Y., Higashisaka, K., Morishita, Y., Yoshida, T., Fujimura, M., et al. (2010). Carbon nanotubes elicit DNA damage and inflammatory response relative to their size and shape. *Inflammation*, *33*(4), 276–280.

Yang, N., Chen, X., Ren, T., Zhang, P., & Yang, D. (2015a). Carbon nanotube based biosensors. *Sensors and Actuators B: Chemical*, *207*, 690–715.

Yang, X., Shao, H., Liu, W., Gu, W., Shu, X., Mo, Y., et al. (2015b). Endoplasmic reticulum stress and oxidative stress are involved in ZnO nanoparticle-induced hepatotoxicity. *Toxicology Letters*, *234*(1), 40–49.

Yin, L., Yuvienco, C., & Montclare, J. K. (2017). Protein based therapeutic delivery agents: Contemporary developments and challenges. *Biomaterials*, *134*, 91–116.

Yu, K. -N., Yoon, T. -J., Minai-Tehrani, A., Kim, J. -E., Park, S. J., Jeong, M. S., et al. (2013). Zinc oxide nanoparticle induced autophagic cell death and mitochondrial damage via reactive oxygen species generation. *Toxicology in Vitro*, *27*(4), 1187–1195.

Yu, Y., Duan, J., Yu, Y., Li, Y., Liu, X., Zhou, X., et al. (2014). Silica nanoparticles induce autophagy and autophagic cell death in HepG2 cells triggered by reactive oxygen species. *Journal of Hazardous Materials*, *270*, 176–186.

Yu, K. -N., Chang, S. -H., Park, S. J., Lim, J., Lee, J., Yoon, T. -J., et al. (2015a). Titanium dioxide nanoparticles induce endoplasmic reticulum stress-mediated autophagic cell death via mitochondria-associated endoplasmic reticulum membrane disruption in normal lung cells. *PloS One*, *10*(6), e0131208.

Yu, K. -N., Sung, J. H., Lee, S., Kim, J. -E., Kim, S., Cho, W. -Y., et al. (2015b). Inhalation of titanium dioxide induces endoplasmic reticulum stress-mediated autophagy and inflammation in mice. *Food and Chemical Toxicology*, *85*, 106–113.

Yu, Y., Zhu, T., Li, Y., Jing, L., Yang, M., Li, Y., et al. (2019). Repeated intravenous administration of silica nanoparticles induces pulmonary inflammation and collagen accumulation via JAK2/STAT3 and TGF-β/Smad3 pathways in vivo. *International Journal of Nanomedicine*, *14*, 7237.

Zeng, Y., Zhu, Z., Du, D., & Lin, Y. (2016). Nanomaterial-based electrochemical biosensors for food safety. *Journal of Electroanalytical Chemistry*, *781*, 147–154.

Zhang, R., Piao, M. J., Kim, K. C., Kim, A. D., Choi, J. -Y., Choi, J., et al. (2012). Endoplasmic reticulum stress signaling is involved in silver nanoparticles-induced apoptosis. *The International Journal of Biochemistry and Cell Biology*, *44*(1), 224–232.

Zhang, H., Zhhai, Y., Wang, J., & Zhai, G. (2016). New progress and prospects: The application of nanogel in drug delivery. *Materials Science and Engineering: C*, *60*, 560–568.

Zhang, R., Wu, W., Zhang, Z., Lv, S., Xing, B., & McClements, D. J. (2019). Impact of food emulsions on the bioaccessibility of hydrophobic pesticide residues in co-ingested natural products: influence of emulsifier and dietary fiber type. *Journal of Agricultural and Food Chemistry*, *67*(21), 6032–6040.

Zhang, X., Guo, X., Kang, X., Yang, H., Guo, W., Guan, L., et al. (2020). Surface functionalization of pegylated gold nanoparticles with antioxidants suppresses nanoparticle-induced oxidative stress and neurotoxicity. *Chemical Research in Toxicology*, *33*(5), 1195–1205.

Zhao, X., & Liu, R. (2012). Recent progress and perspectives on the toxicity of carbon nanotubes at organism, organ, cell, and biomacromolecule levels. *Environment International*, *40*, 244–255.

Zhao, G., Zhang, T., Sun, H., & Liu, J. -X. (2020). Copper nanoparticles induce zebrafish intestinal defects via endoplasmic reticulum and oxidative stress. *Metallomics*, *12*(1), 12–22.

Zhou, C. H., & Keeling, J. (2013). Fundamental and applied research on clay minerals: from climate and environment to nanotechnology. *Applied Clay Science*, *74*, 3–9.

Zhou, F., Liao, F., Chen, L., Liu, Y., Wang, W., & Feng, S. (2019). The size-dependent genotoxicity and oxidative stress of silica nanoparticles on endothelial cells. *Environmental Science and Pollution Research*, *26*(2), 1911–1920.

Zimmermann, M. B., & Hilty, F. M. (2011). Nanocompounds of iron and zinc: their potential in nutrition. *Nanoscale*, *3*(6), 2390–2398.

Zou, L., et al. (2016). Enhancing the bioaccessibility of hydrophobic bioactive agents using mixed colloidal dispersions: Curcumin-loaded zein nanoparticles plus digestible lipid nanoparticles. *Food Research International*, *81*, 74–82.

Zoufan, P., Baroonian, M., & Zargar, B. (2020). ZnO nanoparticles-induced oxidative stress in *Chenopodium murale* L, Zn uptake, and accumulation under hydroponic culture. *Environmental Science and Pollution Research*, 1–13.

CHAPTER 3

In vitro assays for predicting the safety of food-based nanomaterials

Zahra Emam-Djomeh[a], Ali Ebrahimi Pure[b], Monir Ebrahimi Pure[c]
[a]Transfer Phenomena Laboratory (TPL), Functional Food Research Core, Department of Food Science and Technology, College of Agriculture and Natural Resources, University of Tehran, Tehran, Iran
[b]Department of Food Science and Technology, College of Agriculture and Natural Resources, University of Tehran, Tehran, Iran
[c]Department of Chemistry, Alzahra University, Tehran, Iran

3.1 Introduction

The widespread applications of nanotechnology in different aspects of our life had passed the borders of just being useful, involved in our daily life including our foods. The application of nanotechnology in our food system could be indirect and from exposure of raw food material to nano scale substances or much closer like in food packaging or even food ingredients and additives. There are many types of nanoparticles involved in our foods, which most of them have inorganic bases like iron oxide, silver, zinc oxide, etc. (Deshmukh, Patil, Mullani, & Delekar, 2019; Pugazhendhi et al., 2019); these particles have different sizes, shapes, and surface properties which affects their functionality in living organs and cells. For instance, silver nanoparticles had been found to be absorbed by the ingestion tract which might propose safety risks and health issues for consumers (Gaillet & Rouanet, 2015) while still new applications are proposed for silver nanoparticles in food (Gil-Sánchez et al., 2019). There are more than only inorganic nanoparticles, actually organic based and synthetic nano substances play wider applications and roles in our food system. Different lipids, proteins, and polymers are very well known for their application in encapsulation, stabilizing, and delivering food-derived bioactive substances (He, Deng, & Hwang, 2019; Saifullah, Islam Shishir, Ferdowsi, Tanver Rahman, & Van Vuong, 2019; Vaze et al., 2019).

Consuming and digestion could be the easiest and most facile route for a toxic agent to enter our body system; therefore and because of the importance of food safety issues and regarding the properties of nanoscaled

Safety and Regulatory Issues of Nanoencapsulated Food Ingredients
http://dx.doi.org/10.1016/B978-0-12-815725-1.00003-3

materials, safety evaluation of nano-involved foods require much more developed assessments than regular food analytical assays. While these assays had been out of the scopes of expertise for a food technologist before, these days, regarding the prospects of nanotechnology involvement in different industries, it sounds a vital need and even essential for whom studying food safety issues to be aware of them.

This chapter discusses the in vitro assays, which are introduced and developed for investigation and monitoring the major safety concerns related to nano-involved foods. The first step should be the declaration of food safety concerns related to nanotechnology. Principals of food safety introduce different ways to prevent common contaminating agents, causing safety concerns in our food, and these routes are (Santacruz, 2016):

- Through equipment, utensils, and surfaces in contact with foodstuff, which are not adequately cleaned.
- Personal hygiene, especially hand washing.
- Different processing on food, including storing, chilling, heating, etc. which environment, temperature, and equipment might cause contamination in food.
- Effective pest controls.
- Individual reflects particular food substances; allergies, intolerances, etc.

As mentioned earlier, due to the widespread application of nanotechnology in different aspects of our life, nanoscale substances might find a way to our foods in all the earlier-mentioned levels. For example, from the residuals of detergents or antifungal liquids which nanoparticles are being used in their formula and are being used for cleaning surfaces and equipment or even from soaps being used to wash hands, or from nano-antibacterial agents like TiO_2 (Weir, Westerhoff, Fabricius, Hristovski, & von Goetz, 2012; Zhu, Cai, & Sun, 2018) and silver nanoparticles (He & Hwang, 2016) which are already being commonly used in food industries, including poultry industry; or by migrating from the nano-packaging materials used for foods packaging (Garcia, Shin, & Kim, 2018).

There is no consensus about the safety of nanosized materials, but some of them have been found to be toxic for human and animal, mainly acting as oxidant agents (Karakoti, Singh, Dowding, Seal, & Self, 2010; Jiang et al., 2019) and antimicrobial (De Azeredo, 2013; Azeredo et al., 2018). Moreover, there are also mathematical models being developed for *in silico* prediction of the safety of nanotechnology involved products (Richarz et al., 2015; Puzyn et al., 2018). Here, in this chapter, we are introducing *in-vitro* assays for safety evaluation and prediction of nano involved food products.

3.2 Effective characteristics

The significant concerns related to the safety of nano involved food are because of the small size scale of nanoparticles which are easier to migrate from food packaging or pass through the cell wall into living cells of our body. Depending on the composition and processing conditions, these particles size might vary from few nanometers to a few hundred. Smaller nanoparticles are more likely to be dissolved and digested; also, a reduction in particles size increases the surface area. Therefore, interactions between environmental compounds, including digestion enzymes, etc. will be increased, leading to different chemical or biological circumstances (McClements et al., 2016).

Not only the size scale of nanoparticles but also their chemical composition, interfacial properties, aggregation state, and dimensions are of high importance in considering the safety issues. The chemical composition of nanoparticles could help to predict and define how the particles might go under changes inside the body. Enzymatic activity of digestion tract and different chemical conditions throughout this track (mainly pH) might significantly affect the organic and inorganic part of nanoparticles differently. Also delivery and transmission of particles may be different regarding their chemical compositions (Fröhlich & Fröhlich, 2016; De Matteis, 2017). On the other hand, the high antimicrobial activity of some nanoparticles which might be used as food preservatives can potentially influence the microbiome (Jiang et al., 2018). Considering the particles chemical composition, particle size, and environmental conditions, food-grade nanoparticles might exist in the form of individual particles or clustered and aggregated together forming much bigger particles than evaluated based on the individual particle (Wang, Zhang, Wang, & Liang, 2018). Environmental conditions, including pH, ionic strength, and interactions between other ingredients might also influence this aggregation states.

Application of nanoparticles in innovative food packaging with desirable characteristics are emerging, but the control of migration of this particle into the food is a concern. For example, silver nanoparticles are well known for their antimicrobial properties while they have been reported to cause toxicity in mammalian Germline Stem Cells (Braydich-Stolle, Hussain, Schlager, & Hofmann, 2005) or reduce mitochondrial function in the liver (Hussain, Hess, Gearhart, Geiss, & Schlager, 2005; Yue et al., 2019). Therefore, there is safety concern regarding the migration of these nanoparticles from packaging materials into foods (Metak, Nabhani, & Connolly, 2015). There have been efforts to model the kinetic and thermodynamic of silver nanoparticles migration (Kim et al., 2019). It has been demonstrated that this migration is not only related to the particles and packaging material but also correlates with chemical properties of packaged food.

Besides the advantages of nanoparticles, their high capacity, reactivity, and small size, they can transmit between organs in the body and pass the cell wall therefore potentially cause abnormality or death to the cells (Khan, Saeed, & Khan, 2017). Before toxicological evaluation of nanoparticles, potential availability and cell uptake should be demonstrated. Due to the small size and limited amount of these materials, frequently used analytical techniques like spectrophotometer or high-performance liquid chromatography are not very applicable, mainly when migration is investigated; therefore more accurate and advanced techniques are required.

3.3 Investigation methods

3.3.1 Inductively coupled plasma mass spectrometry (ICP-MS)

ICP-MS is a highly accurate mass spectrometry method providing the possibility of detection of metals and several non-metals at concentrations as low as one part per 10^{15} (part per quadrillion, ppq) (Beauchemin, 2017). This method is commonly used to quantify the nanoparticles in food, especially useful for investigations on nanoparticles migration from nano-engineered food packaging to the food, or from nano-involved food to the body organs (Li et al., 2017; Moreno-Olivas, Tako, & Mahler, 2018; Dairi, Ferfera-Harrar, Ramos, & Garrigós, 2019).

Detection of ultra-trace elements using ICP-MS is achieved by ionizing the sample by inductively coupled plasma, which needs a preparation before and the most common methods are using analytical nebulizer to convert the liquid to aerosol (Suddendorf & Boyer, 1980) or using sonication (Candás-Zapico, Kutscher, Montes-Bayón, & Bettmer, 2018; Moreno-Olivas et al., 2018), and then using a mass spectrometer to do separations and quantifications.

3.3.2 Microscopic observation of cell uptake

Nano microscopic methods are the most promising techniques for verification of nanoscaled particles, the size analysis provided by them can be used to predict and standardize nanoparticles migration from food packaging into the food (Huang, Li, & Zhou, 2015; Makaremi et al., 2017). Scanning Electron Microscopy (SEM) and Transmission Electron Microscopy

(TEM) are two of the most commonly used microscopic equipment useful for evaluation of particles size distribution which is a critical point in safety quality assessments of nano involved foods (Bajpai et al., 2019; Fabiano, Reverberi, & Varbanov, 2019) and surface morphology of particles (Shtay, Keppler, Schrader, & Schwarz, 2019). As mentioned before, a concern related to the safety of nanoparticles is the ease of their internalization into human cells which might be positive and desirable occurrence in case of nano drugs but absolutely unwanted in many cases, like nano-preservatives migrated from packaging (Jokar, Correia, & Loeschner, 2018; Bumbudsanpharoke, Choi, Park, & Ko, 2019).

SEM and TEM are also applicable in cell morphology studies, but sample preparations are more complicated comparing to that of Atomic Force Microscopy (AFM), therefore limits the imaging in the natural environment of biological samples. With developments in AFM, it had become possible to observe biological cells in their natural environment (Morris, Gunning, & Kirby, 2009). Beside investigation of nanoparticles in food (Marinello, La Storia, Mauriello, & Passeri, 2018), AFM topography imaging tool provides real-time observation of cell boundary and geometry. If the specific cell's geometry starts to change after addition of nanoparticles to the environment, it can be considered as approval for cell internalization of particles into that specific cell (Fu, Ding, Sun, & Fu, 2018), as shown in Fig. 3.1.

However, there are other microscopy techniques like SEM, TEM (Lopez-Chaves et al., 2018), confocal microscopy (Borker, Patole, Moghe, & Pokharkar, 2016), or optical dark-field (DFM) (Rosman et al., 2012) which can be used to investigate cell uptake mechanism and distribution

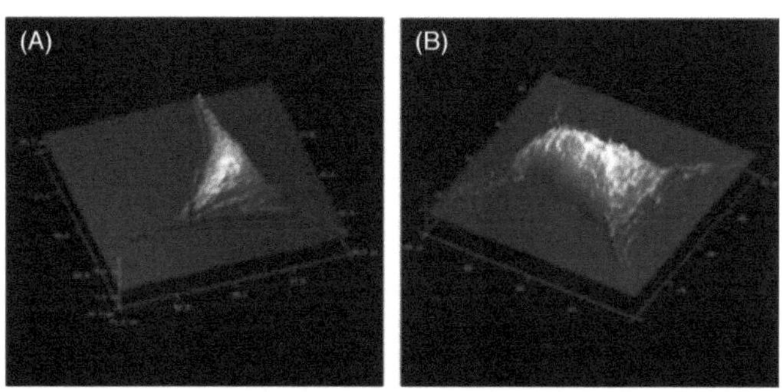

Figure 3.1 ***AFM cell morphology observation.*** (A) AFM images of normal SH-SY5Y cell; (B) AFM images of SH-SY5Y cell with the nanocapsules. *Image from (Fu et al., 2018)*

of nanoparticles size, but it is only AFM which provides the observation in natural environment and condition (Matczuk et al., 2018).

3.3.3 Cell function evaluation

Beside earlier-mentioned methods, the critical stage in safety assessment of nano involved foods is their impact on cell viability and functionality, which there are several methods for evaluating them.

3.3.3.1 xCELLigence system

An xCELLigence system is a technological machine-based analyzing system allowing us to monitor cell phenotypic changes in real-time. There are microelectrodes integrated on the bottom of the cell culture which is called E-Plates that provides a measurement of electrical impedance caused by living cells (Ke, Wang, Xu, & Abassi, 2011; Zhang, Jiang, Hao, & Qu, 2019a). The systems work by applying a weak electrical current to the sensors where the level of AC excitation is in the lower mV range; thus, a current in μA range flows. Cell index (CI), which is a dimensionless parameter that can be calculated from relative changes in impedance (Z) from the following equation:

$$CI = (Zi - Z0)\Omega / 15\ \Omega \tag{3.1}$$

Here Z0 is the base impedance when there are no cells and cell index is 0 at this point, Zi is the impedance of the cells at the time. xCELLigence measures the cell impedance and represent it as cell index to time graph, which is informative for evaluation of changes in cell culture (Doornbos & Heitman, 2018), as illustrated in Fig. 3.2.

Different cells exhibit different impedance profile differing in adhesion, morphology, and growth rates, which can be used as a fingerprint, reflecting the overall behavior of the cell. Changes in cell behaviors like genetic changes or contaminations cause fluctuation in impedance profile that can be used for evaluation of nanomaterials effect on living cells.

Despite the traditional assays xCELLigence as a modern cell viability technique, does not use markers and labels to do the quantifications (Yan et al., 2018). This method has been proven to be useful and reliable for evaluating compounds effects on cellular behavior or changes, for example, DNA damages. Generally, if the compound affects the cytoskeleton of the cells, the cellular impedance would be more than the case in which intracellular pathways are affected by compound (Yan et al., 2018).

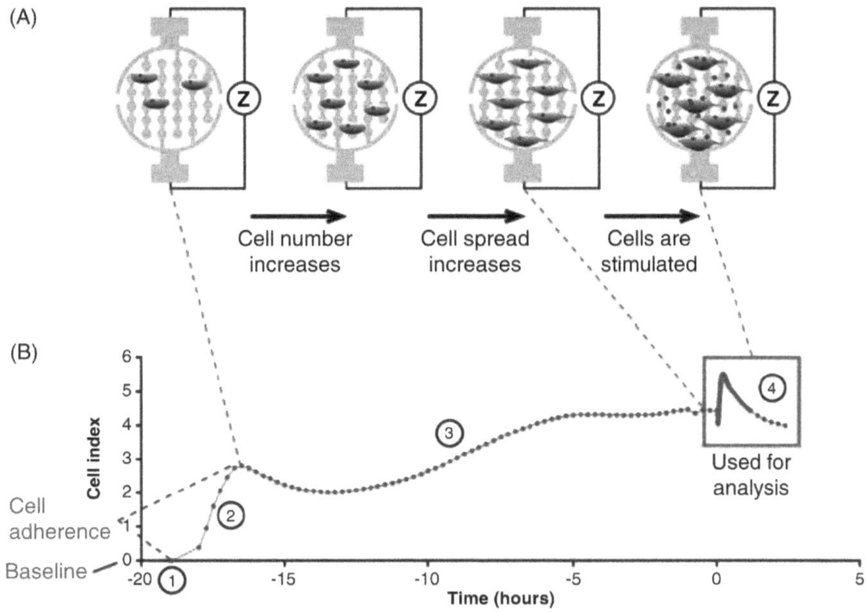

Figure 3.2 ***Cell index to time graph representing cell impedance profile measure by xCELLigence (Doornbos & Heitman, 2018).***

Płaczek et al. (2019) used three independent cytotoxicity assays including xCELLigence, MTT, and flow cytometry analysis to evaluate the safety of liposome, emulsion, and aqueous lecithin on human embryonic kidney 293 (HEK-293) and human promyelocytic leukemia (HL-60) cell lines. They have used the xCELLigence system as a confirmation on MTT assay to make sure about the accuracy of investigations. The difference between results from MTT/cytometry analysis and xCELLigence has been described to be because of xCELLigence higher sensitivity to cell changes than other methods (Erskine, Henle, & Knutson, 2012; Peper et al., 2014).

Park et al. (2017) have used the xCELLigence system combined with other technics to study different pathway of cell death caused by Iron nanoparticles. They could notice that that cytotoxic level of iron nanoparticles may be similar in different cell lines, while different uptake levels had been measured (Park et al., 2015).

3.3.3.2 Flow cytometry analysis

Flow cytometry analysis is an effective system useful for the measurement of physical and chemical properties of cells or particles. The system uses a light source to detect the cells which are individually passing in front of it in a flowing bed (Givan, 2011; Picot, Guerin, Le Van Kim, & Boulanger, 2012). This technique had been applied to investigating a variety of immune system functions, helpful for diagnosing disorders (Delmonte & Fleisher, 2019).

Although the functionality of this instrument provides the possibility to be used for analysis of non-biological structures like liposomes too (Bryła, Juzwa, Weiss, & Lewandowicz, 2017), or detection of food allergens like glycinin (Wang et al., 2015), but the main application of flow cytometry is for biological samples like blood cells, microorganism, and cancer cells. The analysis done by flow cytometry system can give us different results including: cell counting, sorting and characteristics, detection of microorganism and biomarkers (like fluorescents), detection of specific proteins, and several health issues like cancer cells in blood (Goubran, Hart, Othman, & Seghatchian, 2018; Hudson, Burack, & Bennett, 2018; Silva, Magalhães, Freire, & Delerue-Matos, 2018; Delmonte & Fleisher, 2019).

A suspension containing cells pass through a laser beam, where homogeneity of suspension and uniformity of cell flow is a must for accurate measurement (Cossarizza et al., 2017). Therefore, the impact of a secondary factor, like nanoparticle, on cell structure, counting, etc. could be evaluated by comparing a suspension affected by nanoparticles with a control sample. Kadri, E, Devanthi, Overton, and Gkatzionis (2017) have used flow cytometry system to investigate the safety of oil-in-water nanoemulsions on survival and growth of bacteria. As the results obtained from flow cytometry did not show any damage to the bacterial membrane, they have concluded the testing samples being safe on objected microorganisms.

3.3.3.3 Tetrazolium salts assay

Tetrazolium salts and their related formazans, which are chromogenic chemicals are widely being used for accessing cell metabolism viability (Stockert, Horobin, Colombo, & Blázquez-Castro, 2018). Generally, the mechanism of these assays is based on the reduction of tetrazolium salts by coenzymes, enzymes, or other chemicals from the metabolism of biologically active cells which leads to the production of formazan products that are whether colorful or fluorescent where then can be quantified by spectroscopic methods. Therefore, tetrazolium salts assay is introduced as a rapid colorimetric method for cell growth and survival evaluation (Mosmann, 1983).

There are several different tetrazolium salts, but there are two mono-tetrazolium salts more commonly used; 3-(4,5-dimethyl-2-thiazolyl)-2,5-

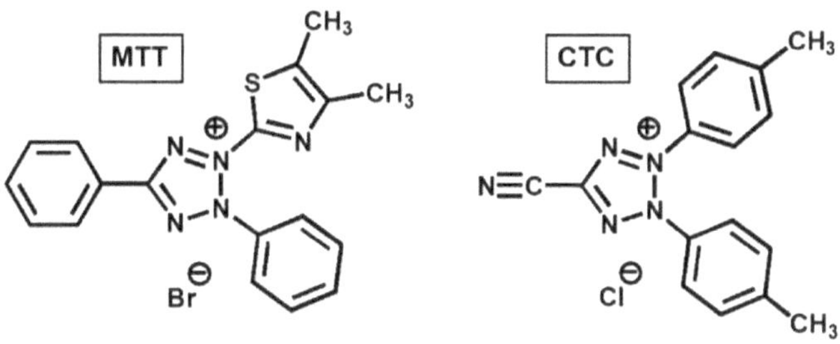

Figure 3.3 ***Chemical structure of MTT and CTC.***

diphenyl-2H-tetrazolium bromide (MTT was also known as thiazolyl blue) is the most widely used tetrazolium salt for viability of eukaryote cells and 5-cyano-2,3-di-(p-tolyl)-tetrazolium chloride (CTC) with fluorescence properties, for bacteria (Stockert & Blazquez-Castro, 2017), as shown in Fig. 3.3.

One of the most widely used in vitro assays for cell viability, proliferation, cytotoxicity (Mosmann, 1983), chemo- and radiosensitivity studies (Carmichael, DeGraff, Gazdar, Minna, & Mitchell, 1987; Merlin et al., 1992) is MTT, which was first introduced by Pearse (PEARSE, 1957) for detection of dehydrogenase activity in living cells. MTT assay is more straightforward and taking shorter time as compared to alternative methods. The chemical pathway is described as the dehydrogenase-catalyzed oxidation that causes the generation of protons which will then be picked up by related coenzyme; the reduced coenzyme causes reduction in some electron carriers like phenazine methosulfate which then transfers the electrons to the tetrazolium salt that causes the production of specific formazan as the final product of the reaction. Formazan compounds from MTT and CTC are water-insoluble and lipophilic, thus can only be dissolved in organic solvents. It has been found that in an MTT viability test, this tetrazolium salt is reduced by NAD(P)H coenzyme and glycolytic enzymes from endoplasmic reticulum, thus MTT reduction in living cells could be considered as a measure of glycolytic NAD(P)H production rate (Berridge, Herst, & Tan, 2005; Lukina et al., 2018; Ma et al., 2018).

This chemical pathway, which occurs in metabolically active cells, causes the yellow color MTT to turn into violet-blue, water-insoluble formazan (λ max = 562 nm). To briefly introduce the workflow, a known concentration of living cells (usually 2×10^3) are seeded into well plates followed by incubation and addition of testing samples at the known concentration(s). The cells will be cultivated after a specific period, and

MTT solution will be added to cell media; if there are any metabolically active cells, MTT reduction will start. After the reduction of MTT, the produced formazan products could be extracted using an appropriate solvent for further colorimetric assessments. DMSO has been suggested as a suitable solvent for this case (Carmichael et al., 1987). Like other spectroscopy methods with a simple proportional equation here, the percentage of cell viability could be evaluated:

$$\text{Cell viability percentage} = \frac{A1 - A2}{A0} \times 100 \tag{3.2}$$

Where A0 is the absorbance of blank, A1 is the absorbance of the treatment group, and A2 is the absorbance of the sample blank.

Although MTT is the most commonly used tetrazolium salt, there are more of this type of chemicals generally known as tetrazolium salts. WST-1 (Tan & Berridge, 2000; Brzicova et al., 2019), XTT (Roehm, Rodgers, Hatfield, & Glasebrook, 1991; Yi, Son, Yoo, Park, & Koo, 2019), and MTS (Dunigan, Waters, & Owen, 1995; Brzicova et al., 2019) are also in this list but less chosen by researchers because their high cell-impermeability makes limitation for their cellular uptake.

A negative aspect of this spectroscopic method is that in some cases, nanoparticles might interfere with dye absorbance, which then may cause overestimation or underestimation on actual results. Not all nanoparticles cause this, but carbon nanoparticles are being reported to cause such interfering (Monteiro-Riviere & Inman, 2006; Casey et al., 2007; Kalman, Merino, Fernández-Cruz, & Navas, 2019). Another critical point which should be considered before performing MTT assay is that vitamin E isomers can cause reduction of MTT salt to purple formazan compounds; therefore it should be noted if there are any vitamin E isomers in the subjecting samples and cell lines (Lim, Loh, Ting, Bradshaw, & Allaudin, 2015).

Besides this, some other assays including neutral red uptake, resazurin reduction, and sulforhodamine B are being reported to be more accurate methods for evaluation of small variations in cell numbers comparing to MTT (van Tonder, Joubert, & Cromarty, 2015). The review article by Stockert et al. mainly focuses on the impact of tetrazolium salts and formazan products in cell biology, giving a good insight into this assay (Stockert et al., 2018).

Considering earlier-mentioned characteristics, tetrazolium salts assay in general and especially MTT assay are potentially useful, fast, inexpensive, and non-radioactive in vitro methods for safety assessments of nano involved food products, being applied by many researchers in different studies (Cheng et al., 2018; Karimi, Sadeghi, & Kokini, 2018; Rout, Liu, & Wu, 2018; Batra, Pawar, & Bahl, 2019).

Hao et al. (2017) had performed MTT assay for evaluating the toxicity of chitosan-coated nano-liposomes of quercetin on HepG2 cells (human liver cancer) and compared it with the toxicity of free quercetin. With this method, they had observed none anti-proliferative activity of samples and were able to measure approximately 5% difference in cell viability between different samples treated by free or encapsulated quercetin (Hao et al., 2017). It was also used by (Hosseini, Nahvi, & Zandi, 2019) to evaluate the safety of nano-bio-packaging they had designed for foods on the L929 mouse fibroblast cell line, where the safety of the nanofibrous mat had been confirmed (Hosseini et al., 2019).

MTT assay was even performed for the evaluation of cytotoxicity of C-dots (Huang, Hung, Weng, Chen, & Lai, 2019; Shi et al., 2019). Zhao et al. had synthesized C-dots from garlic and performed MTT assay to evaluate their cytotoxicity on A549 cells (adenocarcinomic human alveolar basal epithelial cells) where they could observe 100% and 85% at cell viability under 0.5 mg/mL and 1 mg/mL C-dots concentrations (Zhao et al., 2015).

3.3.3.4 Resazurin reduction assay

Resazurin reduction assay is another oxidation-reduction indicator and colorimetric method for the evaluation of cell viability (Bueno et al., 2002; Niu, Liu, Xie, Gao, & Wang, 2019). Resazurin, usually as the sodium salt, has a blue color which will be irreversibly converted to pink fluorescent resorufin after reduction. In the presence of living cells, the blue resazurin is internalized to the cell, and the reduced products are then freely released from cell (O'Brien, Wilson, & Pognan, 2000); this happens both in aerobic and anaerobic respiration (Chen, Ortiz, Xiao, Steele, & Stuckey, 2015).

The fluorescent resorufin could be measured with the wavelengths of 530 nm for excitation and 590 nm for emission, and results would be expressed as a relative fluorescent unit (RFU):

$$\Delta RFU = \frac{RFU\ sample - RFU\ control}{RFU\ control} \times 100\% \tag{3.3}$$

where RFU control indicates the fluorescent signal of the resazurin solution.

Resazurin is also called CellTiter-Blue (CTB) or AlamarBlue, characteristics of which such as stability, being non-toxic, and high color perception due to concentration changes, made it a very preferred staining test for cell viability evaluations (Van den Driessche, Rigole, Brackman, & Coenye, 2014). Besides these, being a fast and inexpensive assay made it suitable for many different investigations. Data obtained from resazurin reduction assay could be used for the correlation between cell death and drug dosage therefore evaluation of minimum inhibitory concentrations of drugs (Kumar, Giribhattanavar, Sagar, & Patil, 2018) or antimicrobials (Sharma et al., 2018) or nanoparticles (Castelo Branco Melo et al., 2018; Ananda, Manukumar, Krishnamurthy, Nagendra, & Savitha, 2019; Larayetan, Ojemaye, Okoh, & Okoh, 2019).

The very early application of this method backs to 1935 which was for evaluation of milk sanitary condition, correlating bacterial growth with milk contamination quality based on colorimetric data (Ramsdell, Johnson, & Evans, 1935). As mentioned earlier, this method could be performed in case of both aerobic and anaerobic respiration systems which makes it a proper assay for the evaluation of probiotics viability estimation (Jung et al., 2017) under different conditions affected by additives, operations, or storage time. Resazurin reduction assay is also applicable in the evaluation of nanoparticles toxicological effect on cells. Kermani et al. (2019) performed this method to evaluate the effect of their food-grade synthesized nanoparticles which were a mixture of oil, water, and polymer (HA/βTCP) on fibroblast cells (NIH/3T3) viability during 7 days to confirm its safety.

In a similar study, Freitas et al. (2019) have evaluated the toxicological effect of encrypted antimicrobial and antitumoral peptides derived from soybean by-products on healthy mouse bone marrow and L929 fibroblast cell lines and human glioblastoma U-87 MG cell line and antimicrobial effect on foodborne pathogens including *S. aureus*, coagulase-negative *S. saprophyticus*, and *A. hydrophila* using resazurin assay.

Cell death caused by sodium alginate-silver nanoparticles which could be used as antibacterial agent in food packaging materials had been evaluated by Shao et al. (2018) who used resazurin assay for the evaluation of minimum inhibitory concentrate of nanoparticles on bacterial pathogens, confirming intense antibacterial activity of their nanoparticles on subjected microorganisms in comparison with standard antibiotics (Fig. 3.4). No bacterial growth makes the color blue while pink and purple mean that there are viable cells. Distilled water was used as a blank control (C); different concentration of nanoparticles are placed in S1, S2, and S3. The positive controls containing 100 μg/mL of antibiotic (tetracycline) is placed in T,

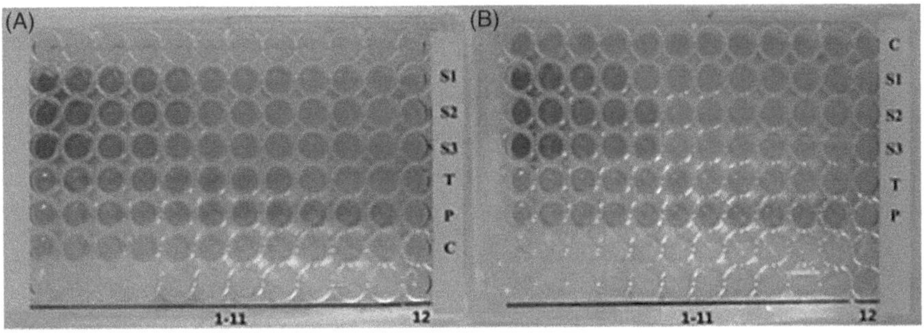

Figure 3.4 Antibacterial activity of Na-Alg-Ag nanoparticles against (A) *Staphylococcus aureus* (KCTC1928); (B) *Escherichia coli* O157: H7 (ATCC25922) using minimal inhibitory concentration (MIC) test.

and Penicillin-Streptomycin in P. column 12 has been left empty (Shao et al., 2018).

3.3.3.5 Neutral red assay

Neutral red (IUPAC name: 3-Amino-7-dimethylamino-2-methylphenazine hydrochloride) assay is a colorimetric method for quantitative estimation of cell viability (Repetto, del Peso, & Zurita, 2008) and a common in vitro toxicity evaluation test (Jain et al., 2018). The method is based on staining cells' lysosomes red (Winckler, 1974).

The principals of neutral red assay for cytotoxicity evaluation was first introduced in 1984 when Dr. Ellen Borenfreund described that living cells uptake the neutral red into their lysosome; therefore there is a direct correlation between neutral red uptake and cell viability in the medium; thus the method had been introduced as a simple quantitative assay for cell viability evaluation (Borenfreund & Puerner, 1985). Lysosome, as a cellular organ, has the critical role of keeping degradative enzymes stored and isolated from other cell organs, therefore any damage or malfunctioning of lysosomes, accumulation of toxic chemicals might cause it, may lead to oxidative damages caused by the release of stored enzymes and chemical that in some cases causes cell death (Moore, Viarengo, Donkin, & Hawkins, 2007). In the case of nanoparticles crossing lysosome membrane, the same situation may lead to cell death. The neutral red assay is a potential method for evaluation of cytotoxicity of nanoparticles or other chemicals migrated from food packaging into the food (Souza & Fernando, 2016).

The spectrophotometric method was introduced in 1990 (Babich & Borenfreund, 1990) and later on developed to microscopic observation level in 2009 for the very first application of neutral red assay for evaluation of the toxicity of nanoparticles by Moore et al. (2009). It was even more developed and miniaturized by Hu et al. (2015) for reading more samples and replications in a shorter time using plate reader. In a recent study, Hu et al. (2019) have combined neutral red with C-dot to develop a colorimetric and fluorescent method for higher sensitivity detection of nitrite in food samples. The schematic overview of the neutral red assay is as shown in Fig. 3.5.

3.3.3.6 LDH cytotoxicity assay

Unlike previously described colorimetric assays which were directly related to living cells, lactate dehydrogenase (LDH) assay quantifies cytotoxicity of chemicals through the spectrophotometrically measurement of LDH, which is released from damaged cells. The stable cytoplasmic enzyme LDH which is present in all cells will be rapidly released due to the damage occurred to the plasma membrane of cells which might be the cause of a process, chemical or nanoparticle (Wang et al., 2012; Chen et al., 2013).

The chemical pathway is through catalytically conversion of lactate to pyruvate by the lactate dehydrogenase along which NAD^+ will be reduced to NADH (Kamalieva, Ishmukhametov, Batasheva, Rozhina, &

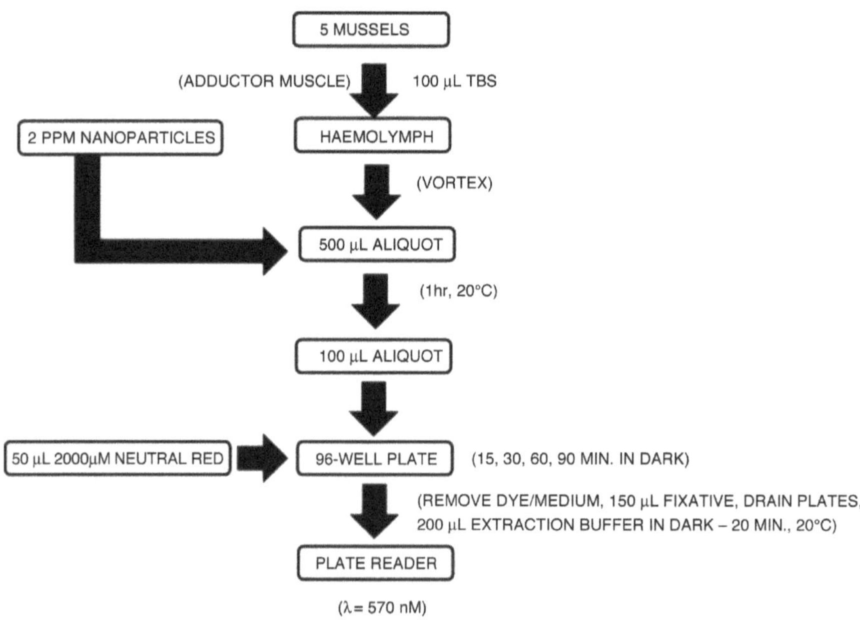

Figure 3.5 ***Schematic overview of NRTT assay (Hu et al., 2015).***

Fakhrullin, 2018). The NADH and H^+ will then catalyze the reduction process of a tetrazolium salt to formazan.

Cell type and age would influence the amount of LDH in the cell, thus the LDH cytotoxicity assay is carried out in two steps, first to standardization by using purified LDH standard and then using a titration curve for cell-derived LDH in which the cell number is known; an LDH units per cell number could be calculated for each cell type which can be used in further toxicity experiments (Decker & Lohmann-Matthes, 1988).

This method had been used in many toxicological studies on a nanomaterial, from simple nanoscale particles to liposomes, emulsions, and mixtures (Nassimi et al., 2010; Płaczek et al., 2019; Zhang et al., 2019a). In a recent study, Zhang et al., 2019b have studied the toxicity effect of nanomaterial ingested through food; according to their findings from LDH assay results, food-grade TiO_2 nanoparticles, which is appreciated for its antimicrobial activities, might be potentially a risk for epithelium. In another study on the toxicity of food-derived chemicals, the cytotoxic effect of rare sugar-glycerol-lipid on IM-9 human myeloma multiple cells had been evaluated using LDH assay, which the toxic dose of the compounds could be estimated based on the obtained results (Hirata, Uchida, Nakajima, Maekawa, & Mizuki, 2018).

3.4 Conclusion

The considerably increasing interests on the application of nanotechnologies in different aspects of our life, including our food and agriculture system, also raise concerns related to the safety of this almost novel and widely applicable products. While safety evaluations and ensuring about long-term side effects of these products on biological individuals and environments need a longer time than fast pace appearance of innovations in nano involved food products; therefore focuses on developments of safety evaluation assays had also become fast-paced to excel the accuracy and reduce in time consumption of evaluation assays. However, there is still no "the best" method for this case, but a combination of approved methods can almost evaluate the safety concerns of nano products in few hours to few days. The extended application and undeniable importance of nanotechnology involvement in our food system, cause a mandatory need for awareness about safety evaluation assays for both food researchers and industries.

References

Ananda, A. P., Manukumar, H. M., Krishnamurthy, N. B., Nagendra, B. S., & Savitha, K. R. (2019). Assessment of antibacterial efficacy of a biocompatible nanoparticle PC@ AgNPs against *Staphylococcus aureus*. *Microbial Pathogenesis, 126*, 27–39. doi: 10.1016/j.micpath.2018.10.029.

Azeredo, H. M. C., de Otoni, C. G., Assis, O. B. G., Corrêa, D. S., de Moura, M. R., & Mattoso, L. H. C. (2018). Nanoparticles and antimicrobial food packaging. In *Reference Module in Food Science*Elsevierdoi: 10.1016/B978-0-08-100596-5.21874-X.

Babich, H., & Borenfreund, E. (1990). Applications of the neutral red cytotoxicity assay to in vitro toxicology. *AtlaAlternatives to Laboratory Animals, 18*, 129–144.

Bajpai, V. K., Park, I., Lee, J., Shukla, S., Nile, S. H., Chun, H. S., et al. (2019). Antioxidant and antimicrobial efficacy of a biflavonoid, amentoflavone from *Nandina domestica* in vitro and in minced chicken meat and apple juice food models. *Food Chemistry, 271*, 239–247. doi: 10.1016/j.foodchem.2018.07.159.

Batra, H., Pawar, S., & Bahl, D. (2019). Curcumin in combination with anti-cancer drugs: A nanomedicine review. *Pharmacological Research, 139*, 91–105. doi: 10.1016/j.phrs.2018.11.005.

Beauchemin, D. (2017). Inductively coupled plasma mass spectrometry methods. In *Encyclopedia of Spectroscopy and Spectrometry* (pp. 236–245.). Elsevierdoi: 10.1016/B978-0-12-409547-2.11222-3.

Berridge, M. V., Herst, P. M., & Tan, A. S. (2005). Tetrazolium dyes as tools in cell biology: New insights into their cellular reduction. *Biotechnology Annual Review, 11*, 127–152. doi: 10.1016/S1387-2656(05)11004-7.

Borenfreund, E., & Puerner, J. A. (1985). A simple quantitative procedure using monolayer cultures for cytotoxicity assays (HTD/NR-90). *Journal of Tissue Culture Methods, 9*(1), 7–9. doi: 10.1007/BF01666038.

Borker, S., Patole, M., Moghe, A., & Pokharkar, V. (2016). Engineering of glycyrrhizin capped gold nanoparticles for liver targeting: in vitro evaluation and in vivo biodistribution study. *RSC Advances, 6*(51), 44944–44954. doi: 10.1039/C6RA05202A.

Braydich-Stolle, L., Hussain, S., Schlager, J. J., & Hofmann, M. -C. (2005). In vitro cytotoxicity of nanoparticles in mammalian germline stem cells. *Toxicological Sciences, 88*(2), 412–419. doi: 10.1093/toxsci/kfi256.

Bryła, A., Juzwa, W., Weiss, M., & Lewandowicz, G. (2017). Lipid nanoparticles assessment by flow cytometry. *International Journal of Pharmaceutics, 520*(1–2), 149–157. doi: 10.1016/j.ijpharm.2017.01.047.

Brzicova, T., Sikorova, J., Milcova, A., Vrbova, K., Klema, J., Pikal, P., et al. (2019). Nano-TiO_2 stability in medium and size as important factors of toxicity in macrophage-like cells. *Toxicology in Vitro, 54*, 178–188. doi: 10.1016/j.tiv.2018.09.019.

Bueno, C., Villegas, M. L., Bertolotti, S. G., Previtali, C. M., Neumann, M. G., & Encinas, M. V. (2002). The excited-state interaction of resazurin and resorufin with aminesin aqueous solutions. photophysics and photochemical reaction¶. *Photochemistry and Photobiology, 76*(4), 385–390. doi: 10.1562/0031-8655(2002)0760385TESIOR2.0.CO2.

Bumbudsanpharoke, N., Choi, J., Park, H. J., & Ko, S. (2019). Zinc migration and its effect on the functionality of a low density polyethylene-ZnO nanocomposite film. *Food Packaging and Shelf Life, 20*, 100301. doi: 10.1016/j.fpsl.2019.100301.

Candás-Zapico, S., Kutscher, D. J., Montes-Bayón, M., & Bettmer, J. (2018). Single particle analysis of TiO_2 in candy products using triple quadrupole ICP-MS. *Talanta, 180*, 309–315. doi: 10.1016/j.talanta.2017.12.041.

Carmichael, J., DeGraff, W. G., Gazdar, A. F., Minna, J. D., & Mitchell, J. B. (1987). Evaluation of a tetrazolium-based semiautomated colorimetric assay: assessment of chemosensitivity testing. *Cancer Research, 47*(4), 936–942. Available at: http://www.ncbi.nlm.nih.gov/pubmed/3802100.

Casey, A., Herzog, E., Davoren, M., Lyng, F. M., Byrne, H. J., & Chambers, G. (2007). Spectroscopic analysis confirms the interactions between single walled carbon nanotubes and various dyes commonly used to assess cytotoxicity. *Carbon*, *45*(7), 1425–1432. doi: 10.1016/j.carbon.2007.03.033.

Castelo Branco Melo, N. F., de MendonçaSoares, B. L., Marques Diniz, K., Ferreira Leal, C., Canto, D., Flores, M. A. P., et al. (2018). Effects of fungal chitosan nanoparticles as eco-friendly edible coatings on the quality of postharvest table grapes. *Postharvest Biology and Technology*, *139*, 56–66. doi: 10.1016/j.postharvbio.2018.01.014.

Chen, X.-X., Cheng, B., Yang, Y.-X., Cao, A., Liu, J.-H., Du, L.-J., et al. (2013). Characterization and preliminary toxicity assay of nano-titanium dioxide additive in sugar-coated chewing gum. *Small*, *9*(9–10), 1765–1774. doi: 10.1002/smll.201201506.

Chen, J. L., Ortiz, R., Xiao, Y., Steele, T. W., & Stuckey, D. C. (2015). Rapid fluorescence-based measurement of toxicity in anaerobic digestion. *Water Research Journal*, *75*, 123–130. doi: 10.1016/j.watres.2015.02.035.

Cheng, Y., Zhao, P., Wu, S., Yang, T., Chen, Y., Zhang, X., et al. (2018). Cisplatin and curcumin co-loaded nano-liposomes for the treatment of hepatocellular carcinoma. *International Journal of Pharmaceutics*, *545*(1–2), 261–273. doi: 10.1016/j.ijpharm.2018.05.007.

Cossarizza, A., Chang, H. -D., Radbruch, A., Akdis, M., Andrä, I., Annunziato, F., et al. (2017). Guidelines for the use of flow cytometry and cell sorting in immunological studies. *European Journal of Immunology*, *47*(10), 1584–1797. doi: 10.1002/eji.201646632.

Dairi, N., Ferfera-Harrar, H., Ramos, M., & Garrigós, M. C. (2019). Cellulose acetate/AgNPs-organoclay and/or thymol nano-biocomposite films with combined antimicrobial/antioxidant properties for active food packaging use. *International Journal of Biological Macromolecules*, *121*, 508–523. doi: 10.1016/j.ijbiomac.2018.10.042.

De Azeredo, H. M. C. (2013). Antimicrobial nanostructures in food packaging. *Trends in Food Science and Technology*, *30*(1), 56–69. doi: 10.1016/j.tifs.2012.11.006.

De Matteis, V. (2017). Exposure to inorganic nanoparticles: routes of entry, immune response, biodistribution and in vitro/in vivo toxicity evaluation. *Toxics*, *5*(4), 29. doi: 10.3390/toxics5040029.

Decker, T., & Lohmann-Matthes, M. -L. (1988). A quick and simple method for the quantitation of lactate dehydrogenase release in measurements of cellular cytotoxicity and tumor necrosis factor (TNF) activity. *Journal of Immunological Methods*, *115*(1), 61–69. doi: 10.1016/0022-1759(88)90310-9.

Delmonte, O. M., & Fleisher, T. A. (2019). Flow cytometry: Surface markers and beyond. *Journal of Allergy and Clinical Immunology*, *143*(2), 528–537. doi: 10.1016/j.jaci.2018.08.011.

Deshmukh, S. P., Patil, S. M., Mullani, S. B., & Delekar, S. D. (2019). Silver nanoparticles as an effective disinfectant: A review. *Materials Science and Engineering: C*, *97*, 954–965. doi: 10.1016/j.msec.2018.12.102.

Doornbos, M. L. J., & Heitman, L. H. (2018). Label-free impedance-based whole cell assay to study GPCR pharmacology. In *Methods in Cell Biology* (2nd ed.). Elsevier Inc.doi: 10.1016/bs.mcb.2018.08.003.

Dunigan, D. D., Waters, S. B., & Owen, T. C. (1995). Aqueous soluble tetrazolium/formazan MTS as an indicator of NADH- and NADPH-dependent dehydrogenase activity. *BioTechniques*, *19*(4), 640–649. Available from: http://www.ncbi.nlm.nih.gov/pubmed/8777059.

Erskine, C. L., Henle, A. M., & Knutson, K. L. (2012). Determining optimal cytotoxic activity of human Her2neu specific CD8 T cells by comparing the Cr51 release assay to the xCELLigence system. *Journal of Visualized Experiments*, *66*, e3683. doi: 10.3791/3683.

Fabiano, B., Reverberi, A. P., & Varbanov, P. S. (2019). Safety opportunities for the synthesis of metal nanoparticles and short-cut approach to workplace risk evaluation. *Journal of Cleaner Production, 209*, 297–308. doi: 10.1016/j.jclepro.2018.10.161.

Fröhlich, E., & Fröhlich, E. (2016). Cytotoxicity of nanoparticles contained in food on intestinal cells and the gut microbiota. *International Journal of Molecular Sciences, 17*(4), 509. doi: 10.3390/ijms17040509.

Freitas, C. S., Vericimo, M. A., da Silva, M. L., da Costa, G. C. V., Pereira, P. R., Paschoalin, V. M. F., et al. (2019). Encrypted antimicrobial and antitumoral peptides recovered from a protein-rich soybean (*Glycine max*) by-product. *Journal of Functional Foods, 54*, 187–198. doi: 10.1016/j.jff.2019.01.024.

Fu, C., Ding, C., Sun, X., & Fu, A. (2018). Curcumin nanocapsules stabilized by bovine serum albumin-capped gold nanoclusters (BSA-AuNCs) for drug delivery and theranosis. *Materials Science and Engineering C., 87*, 149–154. doi: 10.1016/j.msec.2017.12.028.

Gaillet, S., & Rouanet, J. -M. (2015). Silver nanoparticles: Their potential toxic effects after oral exposure and underlying mechanisms – A review. *Food and Chemical Toxicology, 77*, 58–63. doi: 10.1016/j.fct.2014.12.019.

Garcia, C. V., Shin, G. H., & Kim, J. T. (2018). Metal oxide-based nanocomposites in food packaging: Applications, migration, and regulations. *Trends in Food Science and Technology, 82*, 21–31. doi: 10.1016/j.tifs.2018.09.021.

Gil-Sánchez, I., Monge, M., Miralles, B., Armentia, G., Cueva, C., Crespo, J., et al. (2019). Some new findings on the potential use of biocompatible silver nanoparticles in winemaking. *Innovative Food Science and Emerging Technologies, 51*, 64–72. doi: 10.1016/j.ifset.2018.04.017.

Givan, A. L. (2011). Flow cytometry: an introduction. *Methods in Molecular Biology, 699*, 1–29. doi: 10.1007/978-1-61737-950-5_1.

Goubran, H., Hart, C., Othman, I., & Seghatchian, J. (2018). Flow cytometry and immune thrombocytopenic purpura. *Transfusion and Apheresis Science, 57*(6), 800–803. doi: 10.1016/j.transci.2018.10.018.

Hao, J., Guo, B., Yu, S., Zhang, W., Zhang, D., Wang, J., et al. (2017). Encapsulation of the flavonoid quercetin with chitosan-coated nano-liposomes. *LWT - Food Science and Technology, 85*, 37–44. doi: 10.1016/j.lwt.2017.06.048.

He, X., & Hwang, H. M. (2016). Nanotechnology in food science: Functionality, applicability, and safety assessment. *Journal of Food and Drug Analysis. Elsevier Ltd, 24*(4), 671–681. doi: 10.1016/j.jfda.2016.06.001.

He, X., Deng, H., & Hwang, H. (2019). The current application of nanotechnology in food and agriculture. *Journal of Food and Drug Analysis, 27*(1), 1–21. doi: 10.1016/j.jfda.2018.12.002.

Hirata, K., Uchida, T., Nakajima, Y., Maekawa, T., & Mizuki, T. (2018). Chemical synthesis and cytotoxicity of neo-glycolipids; rare sugar-glycerol-lipid compounds. *Heliyon, 4*(10), e00861. doi: 10.1016/j.heliyon.2018.e00861.

Hosseini, S. F., Nahvi, Z., & Zandi, M. (2019). Antioxidant peptide-loaded electrospun chitosan/poly(vinyl alcohol) nanofibrous mat intended for food biopackaging purposes. *Food Hydrocolloids, 89*, 637–648. doi: 10.1016/j.foodhyd.2018.11.033.

Hu, W., Culloty, S., Darmody, G., Lynch, S., Davenport, J., Ramirez-Garcia, S., et al. (2015). Neutral red retention time assay in determination of toxicity of nanoparticles. *Marine Environmental Research, 111*, 158–161. doi: 10.1016/j.marenvres.2015.05.007.

Hu, X., Shi, J., Shi, Y., Zou, X., Tahir, H. E., Holmes, M., et al. (2019). A dual-mode sensor for colorimetric and fluorescent detection of nitrite in hams based on carbon dots-neutral red system. *Meat Science, 147*, 127–134. doi: 10.1016/j.meatsci.2018.09.006.

Huang, J. Y., Li, X., & Zhou, W. (2015). Safety assessment of nanocomposite for food packaging application. *Trends in Food Science and Technology, 45*(2), 187–199. doi: 10.1016/j.tifs.2015.07.002.

Huang, C. -C., Hung, Y. -S., Weng, Y. -M., Chen, W., & Lai, Y. -S. (2019). Sustainable development of carbon nanodots technology: natural products as a carbon source and applications to food safety. *Trends in Food Science and Technology*, *86*, 144–152. doi: 10.1016/j.tifs.2019.02.016.

Hudson, C. A., Burack, W. R., & Bennett, J. M. (2018). Emerging utility of flow cytometry in the diagnosis of chronic myelomonocytic leukemia. *Leukemia Research*, *73*, 12–15. doi: 10.1016/j.leukres.2018.08.015.

Hussain, S. M., Hess, K. L., Gearhart, J. M., Geiss, K. T., & Schlager, J. J. (2005). In vitro toxicity of nanoparticles in BRL 3A rat liver cells. *Toxicology in Vitro*, *19*(7), 975–983. doi: 10.1016/j.tiv.2005.06.034.

Jain, A. K., Singh, D., Dubey, K., Maurya, R., Mittal, S., & Pandey, A. K. (2018). Models and methods for in vitro toxicity. In *In Vitro Toxicology* (pp. 45–65.). Elsevierdoi: 10.1016/B978-0-12-804667-8.00003-1.

Jiang, Z., Jacob, J. A., Li, J., Wu, X., Wei, G., Vimalanathan, A., et al. (2018). Influence of diet and dietary nanoparticles on gut dysbiosis. *Microbial Pathogenesis*, *118*, 61–65. doi: 10.1016/j.micpath.2018.03.017.

Jiang, Z., Shan, K., Song, J., Liu, J., Rajendran, S., Pugazhendhi, A., et al. (2019). Toxic effects of magnetic nanoparticles on normal cells and organs. *Life Sciences*, *220*, 156–161. doi: 10.1016/j.lfs.2019.01.056.

Jokar, M., Correia, M., & Loeschner, K. (2018). Behavior of silver nanoparticles and ions in food simulants and low fat cow milk under migration conditions. *Food Control*, *89*, 77–85. doi: 10.1016/j.foodcont.2018.01.023.

Jung, M. Y., Lee, J., Park, B., Hwang, H., Sohn, S. -O., Lee, S. H., et al. (2017). Applicability of a colorimetric method for evaluation of lactic acid bacteria with probiotic properties. *Food Microbiology*, *64*, 33–38. doi: 10.1016/j.fm.2016.12.008.

Kadri, H., El, Devanthi, P. V. P., Overton, T. W., & Gkatzionis, K. (2017). Do oil-in-water (O/W) nano-emulsions have an effect on survival and growth of bacteria? *Food Research International*, *101*, 114–128. doi: 10.1016/j.foodres.2017.08.064.

Kalman, J., Merino, C., Fernández-Cruz, M. L., & Navas, J. M. (2019). Usefulness of fish cell lines for the initial characterization of toxicity and cellular fate of graphene-related materials (carbon nanofibers and graphene oxide). *Chemosphere*, *218*, 347–358. doi: 10.1016/j.chemosphere.2018.11.130.

Kamalieva, R. F., Ishmukhametov, I. R., Batasheva, S. N., Rozhina, E. V., & Fakhrullin, R. F. (2018). Uptake of halloysite clay nanotubes by human cells: Colourimetric viability tests and microscopy study. *Nano-Structures and Nano-Objects*, *15*, 54–60. doi: 10.1016/j.nanoso.2018.03.009.

Karakoti, A., Singh, S., Dowding, J. M., Seal, S., & Self, W. T. (2010). Redox-active radical scavenging nanomaterials. *Chemical Society Reviews*, *39*(11), 4422. doi: 10.1039/b919677n.

Karimi, M., Sadeghi, R., & Kokini, J. (2018). Human exposure to nanoparticles through trophic transfer and the biosafety concerns that nanoparticle-contaminated foods pose to consumers. *Trends in Food Science and Technology*, *75*, 129–145. doi: 10.1016/j.tifs.2018.03.012.

Ke, N., Wang, X., Xu, X., & Abassi, Y. A. (2011). The xCELLigence system for real-time and label-free monitoring of cell viability. *Methods in Molecular Biology*, 33–43. doi: 10.1007/978-1-61779-108-6_6.

Kermani, F., Kargozar, S., Tayarani-Najaran, Z., Yousefi, A., Beidokhti, S. M., & Moayed, M. H. (2019). Synthesis of nano HA/βTCP mesoporous particles using a simple modification in granulation method. *Materials Science and Engineering: C*, *96*, 859–871. doi: 10.1016/j.msec.2018.11.045.

Khan, I., Saeed, K., & Khan, I. (2017). Nanoparticles: Properties, applications and toxicities. *Arabian Journal of Chemistry*, *12*, 908–931. doi: 10.1016/j.arabjc.2017.05.011.

Kim, M. H., Kim, T. H., Ko, J. A., Ko, S., Oh, J. M., & Park, H. J. (2019). Kinetic and thermodynamic studies of silver migration from nanocomposites. *Journal of Food Engineering, 243*, 1–8. doi: 10.1016/j.jfoodeng.2018.08.028.

Kumar, K., Giribhattanavar, P., Sagar, C., & Patil, S. (2018). A rapid and simple resazurin assay to detect minimum inhibitory concentrations of first-line drugs for *Mycobacterium tuberculosis* isolated from cerebrospinal fluid. *Journal of Global Antimicrobial Resistance, 12*, 157–161. doi: 10.1016/j.jgar.2017.09.012.

Larayetan, R., Ojemaye, M. O., Okoh, O. O., & Okoh, A. I. (2019). Silver nanoparticles mediated by Callistemon citrinus extracts and their antimalaria, antitrypanosoma and antibacterial efficacy. *Journal of Molecular Liquids, 273*, 615–625. doi: 10.1016/j.molliq.2018.10.020.

Li, L., Zhao, C., Zhang, Y., Yao, J., Yang, W., Hu, Q., et al. (2017). Effect of stable antimicrobial nano-silver packaging on inhibiting mildew and in storage of rice. *Food Chemistry, 215*, 477–482. doi: 10.1016/j.foodchem.2016.08.013.

Lim, S. -W., Loh, H. -S., Ting, K. -N., Bradshaw, T. D., & Allaudin, Z. N. (2015). Reduction of MTT to purple formazan by vitamin E isomers in the absence of cells. *Tropical Life Sciences Research, 26*(1), 111–120. Available at: http://www.ncbi.nlm.nih.gov/pubmed/26868595.

Lopez-Chaves, C., Soto-Alvaredo, J., Montes-Bayon, M., Bettmer, J., Llopis, J., & Sanchez-Gonzalez, C. (2018). Gold nanoparticles: Distribution, bioaccumulation and toxicity. In vitro and in vivo studies. *Nanomedicine: Nanotechnology, Biology, and Medicine, 14*(1), 1–12. doi: 10.1016/j.nano.2017.08.011.

Lukina, M. M., Dudenkova, V. V., Ignatova, N. I., Druzhkova, I. N., Shimolina, L. E., Zagaynova, E. V., et al. (2018). Metabolic cofactors NAD(P)H and FAD as potential indicators of cancer cell response to chemotherapy with paclitaxel. *Biochimica et Biophysica Acta (BBA) - General Subjects, 1862*(8), 1693–1700. doi: 10.1016/j.bbagen.2018.04.021.

Ma, Z.-J., Yan, H., Wang, Y.-J., Yang, Y., Li, X.-B., Shi, A.-C., et al. (2018). Proteomics analysis demonstrating rosmarinic acid suppresses cell growth by blocking the glycolytic pathway in human HepG2 cells. *Biomedicine and Pharmacotherapy, 105*, 334–349. doi: 10.1016/j.biopha.2018.05.129.

Makaremi, M., Pasbakhsh, P., Cavallaro, G., Lazzara, G., Aw, Y. K., Lee, S. M., et al. (2017). Effect of morphology and size of halloysite nanotubes on functional pectin bionanocomposites for food packaging applications. *ACS Applied Materials & Interfaces, 9*(20), 17476–17488. doi: 10.1021/acsami.7b04297.

Marinello, F., La Storia, A., Mauriello, G., & Passeri, D. (2018). Atomic Force microscopy techniques to investigate activated food packaging materials. *Trends in Food Science and Technology, 87*, 84–93. doi: 10.1016/j.tifs.2018.05.028.

Matczuk, M., Ruzik, L., Aleksenko, S. S., Keppler, B. K., Jarosz, M., & Timerbaev, A. R. (2018). Analytical methodology for studying cellular uptake, processing and localization of gold nanoparticles. *Analytica Chimica Acta. Elsevier B.V., 1052*, 1–9. doi: 10.1016/j.aca.2018.10.027.

McClements, D. J., DeLoid, G., Pyrgiotakis, G., Shatkin, J. A., Xiao, H., & Demokritou, P. (2016). The role of the food matrix and gastrointestinal tract in the assessment of biological properties of ingested engineered nanomaterials (iENMs): State of the science and knowledge gaps. *NanoImpact, 3-4*, 47–57. doi: 10.1016/j.impact.2016.10.002.

Merlin, J. -L., Azzi, S., Lignon, D., Ramacci, C., Zeghari, N., & Guillemin, F. (1992). MTT assays allow quick and reliable measurement of the response of human tumour cells to photodynamic therapy. *European Journal of Cancer, 28*(8–9), 1452–1458. doi: 10.1016/0959-8049(92)90542-A.

Metak, A. M., Nabhani, F., & Connolly, S. N. (2015). Migration of engineered nanoparticles from packaging into food products. *LWT - Food Science and Technology. Elsevier Ltd, 64*(2), 781–787. doi: 10.1016/j.lwt.2015.06.001.

Monteiro-Riviere, N. A., & Inman, A. O. (2006). Challenges for assessing carbon nanomaterial toxicity to the skin. *Carbon*, *44*(6), 1070–1078. doi: 10.1016/j.carbon.2005.11.004.

Moore, M. N., Viarengo, A., Donkin, P., & Hawkins, A. J. S. (2007). Autophagic and lysosomal reactions to stress in the hepatopancreas of blue mussels. *Aquatic Toxicology*, *84*(1), 80–91. doi: 10.1016/j.aquatox.2007.06.007.

Moore, M. N., Readman, J. A. J., Readman, J. W., Lowe, D. M., Frickers, P. E., & Beesley, A. (2009). Lysosomal cytotoxicity of carbon nanoparticles in cells of the molluscan immune system: An in vitro study. *Nanotoxicology*, *3*(1), 40–45. doi: 10.1080/17435390802593057.

Moreno-Olivas, F., Tako, E., & Mahler, G. J. (2018). ZnO nanoparticles affect intestinal function in an: In vitro model. *Food and Function. Elsevier*, *9*(3), 1475–1491. doi: 10.1039/c7fo02038d.

Morris, V. J., Gunning, A. P., & Kirby, A. R. (2009). In *Atomic Force Microscopy for Biologists* (2nd ed.). Imperial College Press.

Mosmann, T. (1983). Rapid colorimetric assay for cellular growth and survival: Application to proliferation and cytotoxicity assays. *Journal of Immunological Methods*, *65*(1–2), 55–63. doi: 10.1016/0022-1759(83)90303-4.

Nassimi, M., Schleh, C., Lauenstein, H. D., Hussein, R., Hoymann, H. G., Koch, W., et al. (2010). A toxicological evaluation of inhaled solid lipid nanoparticles used as a potential drug delivery system for the lung. *European Journal of Pharmaceutics and Biopharmaceutics*, *75*(2), 107–116. doi: 10.1016/j.ejpb.2010.02.014.

Niu, R., Liu, Y., Xie, X., Gao, X., & Wang, S. (2019). Resazurin reduction assays for screening the synergistic antibacterial effects of flavonoids. *Toxicon*, *158*, S81–S82. doi: 10.1016/j.toxicon.2018.10.278.

O'Brien, J., Wilson, I., & Pognan, F. (2000). Investigation of the Alamar Blue (resazurin) fluorescent dye for the assessment of mammalian cell cytotoxicity. *European Journal of Biochemistry*, *267*(17), 5421–5426. 10.1046/j.1432-1327.2000.01606.x.

Płaczek, M., Wątróbska-Świetlikowska, D., Stefanowicz-Hajduk, J., Drechsler, M., Ochocka, J. R., & Sznitowska, M. (2019). Comparison of the in vitro cytotoxicity among phospholipid-based parenteral drug delivery systems: Emulsions, liposomes and aqueous lecithin dispersions (WLDs). *European Journal of Pharmaceutical Sciences*, *127*, 92–101. doi: 10.1016/j.ejps.2018.10.018.

Park, E. -J., Lee, G. -H., Shim, J., Cho, M. -H., Lee, B. -S., Kim, Y. -B., et al. (2015). Comparison of the toxicity of aluminum oxide nanorods with different aspect ratio. *Archives of Toxicology*, *89*(10), 1771–1782. doi: 10.1007/s00204-014-1332-5.

Park, E. -J., Chae, J. -B., Kang, S., Lyu, J., Jeong, U., Yeom, C., et al. (2017). Nano-sized iron particles may induce multiple pathways of cell death following generation of mistranscripted RNA in human corneal epithelial cells. *Toxicology in Vitro*, *42*, 348–357. doi: 10.1016/j.tiv.2017.04.029.

PEARSE, A. G. E. (1957). Intracellular localisation of dehydrogenase systems using monotetrazolium salts and metal chelation of their formazans. *Journal of Histochemistry and Cytochemistry*, *5*(5), 515–527. doi: 10.1177/5.5.515.

Peper, J. K., Schuster, H., Löffler, M. W., Schmid-Horch, B., Rammensee, H. -G., & Stevanovic´, S. (2014). An impedance-based cytotoxicity assay for real-time and label-free assessment of T-cell-mediated killing of adherent cells. *Journal of Immunological Methods*, *405*, 192–198. doi: 10.1016/j.jim.2014.01.012.

Picot, J., Guerin, C. L., Le Van Kim, C., & Boulanger, C. M. (2012). Flow cytometry: retrospective, fundamentals and recent instrumentation. *Cytotechnology*, *64*(2), 109–130. doi: 10.1007/s10616-011-9415-0.

Pugazhendhi, A., Shobana, S., Nguyen, D. D., Banu, J. R., Sivagurunathan, P., Chang, S. W., et al. (2019). Application of nanotechnology (nanoparticles) in dark fermentative hydrogen production. *International Journal of Hydrogen Energy*, *44*(3), 1431–1440. doi: 10.1016/j.ijhydene.2018.11.114.

Puzyn, T., Jeliazkova, N., Sarimveis, H., Marchese Robinson, R. L., Lobaskin, V., Rallo, R., et al. (2018). Perspectives from the nanosafety modelling cluster on the validation criteria for (Q)SAR models used in nanotechnology. *Food and Chemical Toxicology*, *112*, 478–494. doi: 10.1016/j.fct.2017.09.037.

Ramsdell, G. A., Johnson, W. M. T., & Evans, F. R. (1935). Investigation of resazurin as an indicator of the sanitary condition of milk. *Journal of Dairy Science*, *18*(11), 705–717. doi: 10.3168/jds.S0022-0302(35)93193-9.

Repetto, G., del Peso, A., & Zurita, J. L. (2008). Neutral red uptake assay for the estimation of cell viability/cytotoxicity. *Nature Protocols*, *3*(7), 1125–1131. doi: 10.1038/nprot.2008.75.

Richarz, A. -N., Madden, J. C., Marchese Robinson, R. L., Lubiński, Ł., Mokshina, E., Urbaszek, P., et al. (2015). Development of computational models for the prediction of the toxicity of nanomaterials. *Perspectives in Science*, *3*(1–4), 27–29. doi: 10.1016/j.pisc.2014.11.015.

Roehm, N. W., Rodgers, G. H., Hatfield, S. M., & Glasebrook, A. L. (1991). An improved colorimetric assay for cell proliferation and viability utilizing the tetrazolium salt XTT. *Journal of Immunological Methods*, *142*(2), 257–265. doi: 10.1016/0022-1759(91)90114-U.

Rosman, C., Pierrat, S., Henkel, A., Tarantola, M., Schneider, D., Sunnick, E., et al. (2012). A new approach to assess gold nanoparticle uptake by mammalian cells: combining optical dark-field and transmission electron microscopy. *Small*, *8*(23), 3683–3690. doi: 10.1002/smll.201200853.

Rout, B., Liu, C. -H., & Wu, W. -C. (2018). Increased anti-biofilm efficacy of toluidine blue on *Staphylococcus* species after nano-encapsulation. *Photodiagnosis and Photodynamic Therapy*, *21*, 190–200. doi: 10.1016/j.pdpdt.2017.12.007.

Saifullah, M., Islam Shishir, M. R., Ferdowsi, R., Tanver Rahman, M. R., & Van Vuong, Q. (2019). Micro and nano encapsulation, retention and controlled release of flavor and aroma compounds: A critical review. *Trends in Food Science & Technology*, *86*, 230–251. doi: 10.1016/j.tifs.2019.02.030.

Santacruz, S. (2016). What is food safety? *Australian Institute of Food Safety*. Available from: https://www.foodsafety.com.au/resources/articles/what-is-food-safety (Accessed: 1 December 2018).

Shao, Y., Wu, C., Wu, T., Yuan, C., Chen, S., Ding, T., et al. (2018). Green synthesis of sodium alginate-silver nanoparticles and their antibacterial activity. *International Journal of Biological Macromolecules*, *111*, 1281–1292. doi: 10.1016/j.ijbiomac.2018.01.012.

Sharma, A., Sharma, N. K., Srivastava, A., Kataria, A., Dubey, S., Sharma, S., et al. (2018). Clove and lemongrass oil based non-ionic nanoemulsion for suppressing the growth of plant pathogenic Fusarium oxysporum f.sp. lycopersici. *Industrial Crops and Products*, *123*, 353–362. doi: 10.1016/j.indcrop.2018.06.077.

Shi, X., Wei, W., Fu, Z., Gao, W., Zhang, C., Zhao, Q., et al. (2019). Review on carbon dots in food safety applications. *Talanta*, *194*, 809–821. doi: 10.1016/j.talanta.2018.11.005.

Shtay, R., Keppler, J. K., Schrader, K., & Schwarz, K. (2019). Encapsulation of (-)-epigallocatechin-3-gallate (EGCG) in solid lipid nanoparticles for food applications. *Journal of Food Engineering*, *244*, 91–100. doi: 10.1016/j.jfoodeng.2018.09.008.

Silva, N. F. D., Magalhães, J. M. C. S., Freire, C., & Delerue-Matos, C. (2018). Electrochemical biosensors for Salmonella: State of the art and challenges in food safety assessment. *Biosensors and Bioelectronics*, *99*, 667–682. doi: 10.1016/j.bios.2017.08.019.

Souza, V. G. L., & Fernando, A. L. (2016). Nanoparticles in food packaging: Biodegradability and potential migration to food—A review. *Food Packaging and Shelf Life*, *8*, 63–70. doi: 10.1016/j.fpsl.2016.04.001.

Stockert, J. C., & Blazquez-Castro, A. (2017). *Fluorescence microscopy in life sciences*. Bentham Science Publishersdoi: 10.2174/97816810851801170101.

Stockert, J. C., Horobin, R. W., Colombo, L. L., & Blázquez-Castro, A. (2018). Tetrazolium salts and formazan products in cell biology: Viability assessment, fluorescence imaging, and labeling perspectives. *Acta Histochemica. Elsevier*, *120*(3), 159–167. doi: 10.1016/j.acthis.2018.02.005.

Suddendorf, R.F., & Boyer, K.W. (1980). Mechanical device to produce a finely dispersed aerosol, U.S. Patent 4,206,160.

Tan, A. S., & Berridge, M. V. (2000). Superoxide produced by activated neutrophils efficiently reduces the tetrazolium salt, WST-1 to produce a soluble formazan: a simple colorimetric assay for measuring respiratory burst activation and for screening anti-inflammatory agents. *Journal of Immunological Methods*, *238*(1–2), 59–68. doi: 10.1016/S0022-1759(00)00156-3.

Van den Driessche, F., Rigole, P., Brackman, G., & Coenye, T. (2014). Optimization of resazurin-based viability staining for quantification of microbial biofilms. *Journal of Microbiological Methods*, *98*, 31–34. doi: 10.1016/j.mimet.2013.12.011.

van Tonder, A., Joubert, A. M., & Cromarty, A. (2015). Limitations of the 3-(4,5-dimethylthiazol-2-yl)-2,5-diphenyl-2H-tetrazolium bromide (MTT) assay when compared to three commonly used cell enumeration assays. *BMC Research Notes*, *8*(1), 47. doi: 10.1186/s13104-015-1000-8.

Vaze, N., Pyrgiotakis, G., Mena, L., Baumann, R., Demokritou, A., Ericsson, M., et al. (2019). A nano-carrier platform for the targeted delivery of nature-inspired antimicrobials using Engineered Water Nanostructures for food safety applications. *Food Control*, *96*, 365–374. doi: 10.1016/j.foodcont.2018.09.037.

Wang, G., Zhang, J., Dewilde, A. H., Pal, A. K., Bello, D., Therrien, J. M., et al. (2012). Understanding and correcting for carbon nanotube interferences with a commercial LDH cytotoxicity assay. *Toxicology*, *299*(2–3), 99–111. doi: 10.1016/j.tox.2012.05.012.

Wang, Y., Deng, R., Zhang, G., Li, Q., Yang, J., Sun, Y., et al. (2015). Rapid and sensitive detection of the food allergen glycinin in powdered milk using a lateral flow colloidal gold immunoassay strip test. *Journal of Agricultural and Food Chemistry*, *63*(8), 2172–2178. doi: 10.1021/jf5052128.

Wang, Y., Zhang, Y., Wang, J., & Liang, X. -J. (2018). Aggregation-induced emission (AIE) fluorophores as imaging tools to trace the biological fate of nano-based drug delivery systems. *Advanced Drug Delivery Reviews*, *143*, 161–176. doi: 10.1016/j.addr.2018.12.004.

Weir, A., Westerhoff, P., Fabricius, L., Hristovski, K., & von Goetz, N. (2012). Titanium dioxide nanoparticles in food and personal care products. *Environmental Science and Technology*, *46*(4), 2242–2250. doi: 10.1021/es204168d.

Winckler, J. (1974). Vital staining of lysosomes and other cell organelles of the rat with neutral red (author's transl). *Progress in Histochemistry and Cytochemistry*, *6*(3), 1–91. Available from: http://www.ncbi.nlm.nih.gov/pubmed/4142096.

Yan, G., Du, Q., Wei, X., Miozzi, J., Kang, C., Wang, J., et al. (2018). Application of real-time cell electronic analysis system in modern pharmaceutical evaluation and analysis. *Molecules*, *23*(12), 3280. doi: 10.3390/molecules23123280.

Yi, G., Son, J., Yoo, J., Park, C., & Koo, H. (2019). Emulsan-based nanoparticles for in vivo drug delivery to tumors. *Biochemical and Biophysical Research Communications*, *508*(1), 326–331. doi: 10.1016/j.bbrc.2018.11.106.

Yue, L., Zhao, W., Wang, D., Meng, M., Zheng, Y., Li, Y., et al. (2019). Silver nanoparticles inhibit beige fat function and promote adiposity. *Molecular Metabolism*, *22*, 1–11. doi: 10.1016/j.molmet.2019.01.005.

Zhang, X., Jiang, X., Hao, Z., & Qu, K. (2019a). Advances in online methods for monitoring microbial growth. *Biosensors and Bioelectronics*, *126*, 433–447. doi: 10.1016/j.bios.2018.10.035.

Zhang, Z., Zhang, R., Xiao, H., Bhattacharya, K., Bitounis, D., Demokritou, P., et al. (2019b). Development of a standardized food model for studying the impact of food matrix effects on the gastrointestinal fate and toxicity of ingested nanomaterials. *NanoImpact, 13*, 13–25. doi: 10.1016/j.impact.2018.11.002.

Zhao, S., Lan, M., Zhu, X., Xue, H., Ng, T. -W., Meng, X., et al. (2015). Green synthesis of bifunctional fluorescent carbon dots from garlic for cellular imaging and free radical scavenging. *ACS Applied Materials and Interfaces*, 7(31), 17054–17060. doi: 10.1021/acsami.5b03228.

Zhu, Z., Cai, H., & Sun, D. -W. (2018). Titanium dioxide (TiO_2) photocatalysis technology for nonthermal inactivation of microorganisms in foods. *Trends in Food Science and Technology*, *75*, 23–35. doi: 10.1016/j.tifs.2018.02.018.

CHAPTER 4

In vivo assays for predicting the safety of food-based nanomaterials

Rezvan Shaddel[a], Safoura Akbari-Alavijeh[b], Seid Mahdi Jafari[c]
[a]Department of Food Science and Technology, Faculty of Agriculture and Natural Resources, University of Mohaghegh Ardabili, Ardabil, Iran
[b]Department of Food Science and Technology, College of Agriculture, Isfahan University of Technology, Isfahan, Iran
[c]Department of Food Materials and Process Design Engineering, Gorgan University of Agricultural Sciences and Natural Resources, Gorgan, Iran

4.1 Introduction

Nanotechnology as a modern approach in the food industry has its own benefits along with the safety concerns. The nanoparticles bring unique physical, chemical, and biological properties due to their nanoscale size and large surface-to-volume ratio. However, these properties may generate the toxicity effect as nanoparticles can easily penetrate and disperse through the biological barriers in human body. Nanomaterials can enter the human body and circulatory system via different routes such as skin permeation, respiratory system, and oral ingestion. So, there needs to be comprehensive investigations on use of nanoparticles in foods considering all unpredictable and unexpected bio-kinetic outcomes of their exposure to human health. Furthermore, better understanding of the in-flow and out-flow of nanostructures is essential to manage their toxicity (Dasgupta & Ranjan, 2018). Nevertheless, the harmful biological responses of nanoparticles in food products have not been confirmed yet; therefore, international and governmental organizations are legislating and applying the standards in using nanostructures in foods (Jafari, Katouzian, & Akhavan, 2017c). Several international agencies have elaborated standard protocols or guides for risk evaluation of nanofoods such as European Food Safety Authority that published a guide for risk assessments in food products, which include the characterization and identification of nanomaterials and the related risks, both *in vitro* and *in vivo* (Rubio et al., 2019).

Nanotechnology has provided a wide field for development of nano-delivery systems as an ideal carrier of sensitive bioactive compounds. The

Safety and Regulatory Issues of Nanoencapsulated Food Ingredients
http://dx.doi.org/10.1016/B978-0-12-815725-1.00004-5

nanosized particles are able to cross the intestinal epithelium which is a double-edged sword; it may enhance the absorption and bioavailability of encapsulated bioactives which can trigger the health benefits, on the other hand the unpredictable phenomena inside the living cells brings about health concerns (Jafari, Esfanjani, & Katouzian, 2017b; Jafari et al., 2017c).

Characterization and risk assessment of nanoparticles in a biological system is a serious challenge. Anyway, different assays, both *in vitro* and *in vivo*, have been proposed to anticipate the toxicity of nanomaterials in human body (Rashidi & Khosravi-Darani, 2011; Jafari et al., 2017a). The main factors related to toxicity assessment are toxic dosage, biodistribution, biodegradability, entrance route, tissue accumulation, and the health impacts related to the certain nanoparticles after exposure or ingestion which all depends on the properties of nanoparticle such as the shape, size, charge, and composition (Rubio et al., 2019).

This chapter focuses on different *in vivo* assays applied for investigation of the safety of food-based nanomaterials.

4.2 Biological fate of food-based nanomaterials

There are raising concerns about the toxicity effect of nanoparticles on human body; so, it would be very important to know the biological fate of nanoparticles following the exposure to consumers. Previously, the absorption process of micro/nano materials through the gastrointestinal tract (GIT) was announced for the first time (Florence, 1997). So, the investigation of the toxicokinetics of nanostructures is not a new research field. Nevertheless, there is still a serious need for evaluating the absorption, metabolism, distribution, and excretion of orally consumed nanomaterials (Card et al., 2011).

After oral consumption, nanostructures have to go through the enzymatic environment and neutral pH of mouth, which contains enzymes (e.g., amylase), saliva, and electrolytes. Subsequently, the nanostructures pass the esophagus into the stomach with acidic condition (pH = 2–3), digestive enzymes (e.g., lipase and pepsin), and electrolytes. Going through the pylorus sphincter, nanostructures arrive the small bowel (pH = 5–7) with digestive enzymes such as amylase, bile salts, phospholipids, electrolytes, proteases, and pancreatic lipase. The nanostructures with repulsive surface charge or anionic are able to diffuse through the negatively charged mucus layer and can reach to the enterocytes. All non-absorbed nanostructures may reach to the colon (pH = 6–7) and be exposed to the undigested food and colon microbiota (Card et al., 2011; McClements & Xiao, 2017; Vita, Royse, & Pullen, 2019).

As previously mentioned, the potential risks of nanoparticles are mainly due to the high surface-to-volume ratio which makes them highly reactive; as well as, their nanoscale size which may facilitate their transfer through the cell membranes and may accumulate in different biological systems (Jafari et al., 2017c; Dasgupta & Ranjan, 2018). Furthermore, the nanomaterials possibly cause cell disruption, which may lead to the cell death. Alteration of redox balance and tissue inflammation, are two examples of the toxic effects (Arora, Rajwade, & Paknikar, 2012).

Three main biological fates have been proposed for nanoparticles in food matrix after oral ingestion (Jafari et al., 2017c). First, the nanoparticles are completely digested and absorbed though the gastrointestinal tract. In this case, all chemicals used in the nanoencapsulation process such as surfactants should be evaluated via safety assays. Second, the nanoparticles are partially degraded and the encapsulated bioactive compound is slightly released. Hence, different types of conjugates will be developed between core-shell materials and diverse biokinetic actions may be observed. These conjugates can penetrate to different organs and tissues, which may lead to immunological responses by immune system. Third, the nanoparticles are totally stable through the gastrointestinal tract and the encapsulated bioactive compound is not released which makes two options possible; excretion of the intact nanoparticles from the digestive system or passing of the nanoparticles through the epithelial cells and entering the bloodstream (Katouzian & Jafari, 2016; Jafari et al., 2017c).

Regarding the special features of nanostructures, some physiological and immunological responses are expected including inflammation, mitochondrial disorders, protein denaturation, corona generation, phagocytosis, and ultimately DNA alteraion and cell death (Simón-Vázquez et al., 2014). These interactions can be beneficial for the consumer in some cases. For instance, using nanostructure as an antigen carrier for vaccines to trigger the immunological reactions. Nevertheless, other interactions can be hazardous (Jafari et al., 2017c).

Nanoparticles may alter the physiological mechanisms via conformational modifications of proteins. One of the suggested routes of the alteration is the synthesis of protein corona with different pharmacokinetic and biodistribution profile. Protein corona generation may occur via the binding of cellular proteins to nanomaterials due to the alteration of their action and stability. These proteins can modify the physiological responses such as endocytosis and exocytosis, and phagocytosis process (Peng et al., 2014; Wolfram et al., 2014). Pros and cons of protein corona are not well understood. Although the protein can be utilized as a carrier to deliver the nanoparticles to the specific sites of body, but they can also be rapidly absorbed to spleen or liver and subsequently be eliminated by reticuloendothelial system (Jafari et al., 2017c). These characteristics might be hazardous when the long life of the nanoparticles and their retention are required.

Oxidative stress is another proposed route for toxic effect of nanoparticles in cells via generation of reactive oxygen species (ROS). Direct exposure of nanoparticles with cellular ingredients is required for the generation of ROS. Nanoparticles with extremely active bonds can take the electrons from the electron donors and further oxidation process may damage the cellular organelles like mitochondria. High doses of ROS in cells may trigger inflammation, lipid peroxidation, and modifications in gene sequences (Arora et al., 2012; Manke, Wang, & Rojanasakul, 2013; Jafari et al., 2017c).

Stimulation of innate and adaptive immune systems via activating different cell species may develop unwanted immune responses such as cytotoxicity and inflammation. The physicochemical properties of the nanoparticles define the type of the responses. Innate immune system involves monocytes, neutrophils, macrophages, and dendritic cells while adaptive immune system involves Type B and T lymphocytes which may trigger definite responses to the nanoparticles' exposure. Furthermore, both innate and adaptive immune systems are able to secrete active agents such as cytokines (interleukin (IL)-2, IL-6) and antibodies which are responsible for inflammation (Giese & Marx, 2014; Jafari et al., 2017c).

4.3 Possible health risks of food-based nanomaterials

4.3.1 GIT disruption and malfunction

GIT as the main route of entry for orally administrated nanostructures has three possible means including normal enterocytes, M-cells, and paracellular or transcellular transport. Absorption may happen through the M-cells involving endocytosis, phagocytosis, and clathrin-coated vesicles and pits. Endocytosis may also occur via uptake of nanostructures through the apical side of enterocytes into the lymph. The other possible way for uptake of the nanomaterials is the paracellular or transcellular pathway by passage between the intestinal cells. This pathway depends on the impaired tight junctions between the intestinal cells (Card et al., 2011; Vita et al., 2019).

Nanostructures predominantly trigger oxidative damage; impair the cellular redox homeostasis; cause lysosomal disrupt, which may all result in an inflammatory response and induce an impaired molecular pattern in gut cells. Moreover, nanoparticles can induce an immune-stimulating molecular pattern (Khanna et al., 2015; Nakayama, 2018). This may cause adverse health effects specifically for the cases with inflammatory bowel diseases (IBDs). Herein, Lomer, Thompson, and Powell (2002) announced that the TiO_2 accumulated epithelial cells were responsible for the inflammation in patients with Crohn's disease. Crohn's disease and colon cancer are two main GIT disorders with approved relationship with the nanoparticles entry into the GIT (Maksimović, Omanović-Mikličanin, & Badnjević, 2019).

GIT function may be disturbed by small-sized nanomaterials because of their large surface area for adsorption of GIT components such as digestive or metabolic enzymes. Denaturation of enzyme after adsorption onto the surfaces of nanoparticles may lead to a lower catalytic activity and further reduction of the extent or rate of digestion process in GIT. This would be more probable about the indigestible organic nanoparticles or inorganic ones (McClements & Xiao, 2017). It has been also reported that the mixture of TiO_2 nanoparticles with lipid droplets may lead to a slight reduction of lipid digestion.

Another adverse health effect of nanomaterials is disruption of GIT structures such as microvilli of enterocytes and intercellular tight junction, which may promote the alteration of their abnormal nutrient absorption and protective function (Fröhlich & Fröhlich, 2016). For example, Chalew and Schwab (2013) and Martirosyan, Bazes, and Schneider (2014) reported that TiO_2 and Ag nanoparticles ruptured the tight junctions. Koeneman et al. (2010) also announced that the microvilli of Caco-2 cells collapsed by exposure to 10 μg/mL of TiO_2 nanoparticles and an increment in intracellular-free calcium was triggered. Furthermore, Faust et al. (2014) revealed that presence of TiO_2 nanoparticles in the candy coating of chewing gum resulted in disruption of the brush border of enterocytes at the concentration of 350 ng/mL. Other studies have also confirmed that TiO_2, Ag, ZnO, SiO_2, and Fe_3O_4 may accumulate at the surface of cells in different cell lines *in vitro*, as well as in cytoplasm, M-cells, subcellular surfaces of enterocytes, macrophages, and Peyer's patches within the GIT (Vita et al., 2019). However, not all of nanoparticles have the adverse effect on the epithelial barrier function; thus *in vitro* administration of Ag nanoparticles did not induce loss of barrier integrity on Caco-2 cells (Vila et al., 2018).

Nanostructures in GIT may also stimulate immune responses which is connected with gut-associated lymphoid tissue (GALT) (Őrfi & Szebeni, 2016; McClements & Xiao, 2017). Nanosilica particles used as preservative in the food can stimulate lymphocytes including Th1, Th2, and Th17 which is considered as an probable reason for food allergy (Toda and Yoshino 2016). Shumakova et al. (2015) also confirmed that oral consumption of silicon dioxide nanoparticles in rats may lead to a reduction in Th cells and leukocytes and further increase in cytotoxic lymphocytes and plasma TNF-alpha.

4.3.2 Tissue accumulation

Ingested nanomaterials might be adsorbed and accumulated in different organs and tissues of human body. Nanoparticles can transfer through the mucosae and be absorbed by different active or passive transport mechanisms. The process directly depends on the physicochemical characteristics of nanomaterials such as composition, dimension, morphology, aggregation, and interfacial characteristics (McClements & Xiao, 2017). Recently, studies have reported the transfer of TiO_2, Ag, and carbon nanoparticles from GIT into the blood circulation, and further to other organs like liver, spleen, brain, and reproductive system (Maksimović et al., 2019). Long-term toxicity of nanomaterials can be due to their accumulation in specific tissues which is more probable for the non-digested nanoparticles through the GIT (McClements & Xiao, 2017).

Another exposure rout, which may cause toxicity by accumulation, is inhalation. Dose and size of nanomaterials and the physique of the lungs, the exposure time as well as the breathing pattern may affect the deposition and retention of nanomaterials in different parts of respiratory systems. Nanoparticles smaller than 100 nm elevate the risk of pulmonary diseases compared to the larger ones. Furthermore, the inhaled nanomaterials can be absorbed through the lymph and blood stream and finally reach to the other organs or tissues such as spleen, bone marrow, heart, lymph nodes, etc. and may trigger serious diseases (Maksimović et al., 2019). There is a direct relationship between the inhalation of nanomaterials and pulmonary diseases such as emphysema, asthma, bronchitis, lung cancer, and some neurodegenerative diseases like Alzheimer's and Parkinson's diseases (Chau, Wu, & Yen, 2007). Heart diseases, arrhythmia, autoimmune diseases, arteriosclerosis, and a broad range of malfunctioning in brain, liver, heart, kidney, lung, etc. are possible by the entrance of the nanomaterials into the blood stream. In this regard, titanium oxide, silver, and silicon dioxide nanoparticles have been proved to be capable of passing the physiological barriers posing the toxicological effects. Additionally, zinc oxide nanoparticles may damage DNA and are remarkably cytotoxic to hepatic and renal cells (Maksimović et al., 2019).

4.3.3 Cellular malfunction

Nanoparticles may induce cytotoxicity, cellular malfunction, and ultimately cell death through different mechanisms such as mitochondrial damage, apoptosis, oxidative stress, and cell wall disruption, etc. (Dasgupta & Ranjan, 2018). Inorganic nanoparticles are more susceptible to promote these mechanisms due to their absorption in intestinal cells, since most of the organic nanoparticles can be easily digested before absorption (McClements & Xiao, 2017).

The development of reactive oxygen species (ROS) such as hydroxyl radicals, hydrogen peroxide, singlet oxygen, and superoxide can trigger the oxidative stress in cells. ROS can potentially interact with protein, nucleic acid, and lipid components of cellular structure and subsequently damage to nucleus, cell membrane, and organelles. Further, some biochemical pathways like gene expression, DNA replication, and ATP production may be adversely affected. ZnO, Ag, and SiO_2 nanoparticles have been detected as the potent inorganic nanoparticles for interaction with the cellular components resulting in the cell malfunction (Choi et al., 2015; Gaillet & Rouanet, 2015; Jafari et al., 2017a; Mc Clements, Li, & Xiao, 2015; Wu et al., 2014). Mitochondria as the main organelle in regulation of programmed cell death (apoptosis) as well as normal cellular function are the intracellular center of ROS production within cells. In apoptosis process, caspases as a group of proteases in cytoplasm induce the production of some proteins leading to the increased production of ROS in cells (Ricci, Gottlieb, & Green, 2003). Wang et al. (2014) reported that the combination of zinc oxide nanoparticles plus ascorbic acid induced a higher activity of caspase-3 in comparison to the control group which revealed the synergistic cytotoxic effect of this combination and the related mitochondrial signal pathway (Wang et al., 2014).

Innate and adaptive immune systems are responsible for the immunological responses as the other type of cell reaction to the absorbed nanomaterial. Since the nanoparticles arrive the blood stream, the complement system (innate humoral immunity) will be activated and then the nanomaterials surface will be marked by C3b and iC3b opsonic complement fragments which could be subsequently removed by phagocytic cells.

Neutrophils, dendritic cells, macrophages, and monocytes are responsible for innate immune responses while lymphocytes (Type T and B) are activated in adaptive immune response (Jafari et al., 2017c). Furthermore, the secreted antibodies and cytokines (interleukin (IL)-1β, IL-6) may be responsible for inflammation responses. Bettini et al. (2017) announced that titanium oxide nanoparticles was identified in immune cells of Peyer's patches in orally exposed rats for 1 week. In all TiO_2-treated rats a reducing trend in secretion of Thelper (Th)-1 IFN-γ was shown. Whereas, secretion of splenic Th1/Th17 raised considerably which led to inflammatory responses (Bettini et al., 2017).

4.4 Detection, characterization, and safety assessment of food-based nanomaterials

Detection and characterization of food-based nanoparticles in different cell cultures or tissue samples require separation or pre-treatment steps to isolate the target nanoparticle from the interfering components. Of note, the results obtained from sampling can only partially reflect the natural and original condition. So, scientists try to keep the sample preparation steps as minimal (Bouwmeester et al., 2014).

Nanotoxicology aims to find the relationships between the physicochemical properties of nanoparticles and their potential toxic effects. However, there is still challenge to determine the physicochemical parameters with the most relevance to the adverse health effects. In this regard, systematic investigations have reported the following ranking for the relevancy (Orts-Gil, Natte, & Österle, 2013): 100% surface area, 96% elemental composition, 89% surface chemistry, 86% particle size, 86% particle size distribution, 86% surface charge, 71% agglomeration state, and 61% crystalline structure. The chemical composition of nanoparticles is essential for understanding the adverse health effects of nanoparticles. Nevertheless, the surface properties of the nanostructures show a different behavior compared to the non-nanoforms of the material with similar chemical composition (Ribeiro et al., 2017; Gupta, Kumar, & Kumar, 2019).

Fig. 4.1 represents the physicochemical properties of nanoparticles relevant for toxicological assays.

Regarding chemistry and coatings, nanoparticles with different characteristics and surface alteration use different routes of absorption and trigger different cellular responses. Photo reactivity of nanoparticles is another important factor in forming their potential to elicit toxicity by oxidative

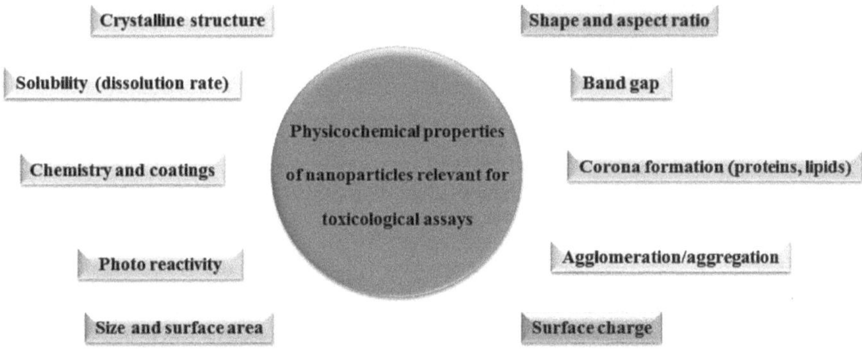

Figure 4.1 ***Physicochemical properties of nanoparticles relevant for toxicological assays (Ribeiro et al., 2017).***

stress pathways. The level of conduction band energy can also be used in order to estimate the toxicological potentials at cellular levels (Oberdörster et al., 2005; Zhang et al., 2012; Ribeiro et al., 2017). Size and surface area is another effective property of nanoparticles. Smaller nanoparticles of the same material are potentially more toxic than the larger ones specifically in pulmonary diseases. Furthermore, surface charge of nanoparticles may impact on their interactions with cells and organs as well as corona formation. Corona formation (proteins, lipids) can determine the nanoparticles uptake and their distribution in cells and tissues. Moreover, nanoparticles with a high aspect ratio (tubular shape) tend to be more toxic. On the other hand, agglomeration or aggregation of nanoparticles influence on their size distribution leading to a modified translocation across the cellular barriers. Crystalline structure may also affect the other properties of the nanomaterials such as reactivity and zeta potential with adverse health effects. Solubility and dissolution rate of ions/molecules releasing from the structure of nanomaterials over time show their potential toxic effects in biological systems (Oberdörster et al., 2005; Dekkers et al., 2016; Ribeiro et al., 2017).

There are several analytical methods available for characterization of the properties of nanoparticles in cells and tissues such as (Gupta et al., 2019): High-Resolution Transmission Electron Microscopy (HRTEM); *In situ* Transmission Electron Microscopy (TEM); Inductively-Coupled Mass Spectroscopy (ICP-MS); Energy-Dispersive Dispersive X- Ray Spectroscopy in an Electron Microscope (EDXS);Atomic Force Microscopy (AFM); Confocal Raman Spectroscopy (CRS); X-ray Fluorescence Microscopy (XRF); Environmental Scanning Electron Microscopy (ESEM); Magnetic

Resonance Imaging (MRI); Fluorescence Microscopy (FLM); and Confocal Light Microscopy (CLM).

TEM is the best standard technique for assessment of shape, size distribution, and morphology; meanwhile, this method is time consuming, expensive, and an unsuitable technique for systematic monitoring (Ribeiro et al., 2017). ICP-MS-based techniques and also flow cytometry can estimate and quantify the nanoparticles internalization into the cells (Oberdörster et al., 2005). All techniques based on microscopic observations including EDXS, AFM, XRF, ESEM, FLM, and CLM beside the imaging methods such as MRI facilitate the study of the alterations in physicochemical properties, biodistribution and bioaccumulation of nanoparticles in tissues, cells, and organs.

Chusuei et al. (2013) studied the cytotoxicity of Cr_2O_3, TiO_2, Mn_2O_3, ZnO, CuO, and Fe_2O_3. They introduced four determinants of cytotoxicity to design a risk assessment procedure including:

1. Quantified relative number of available binding sites on the surface of nanoparticles by X-ray photoelectron spectroscopy was applied to estimate the biomolecular interactions on the nanoparticle surface.
2. Point-of-zero charge (PZC) determined the surface charge of nanoparticles in lysosomal and cytosolic compartments.
3. Measurement of band-gap energy to estimate the electron abstraction from nanoparticles leading to oxidative stress and cell death.
4. Metal ion dissolution from the nanoparticles via ultracentrifugation or ultrafiltration coupled with plasma mass spectrometry.

Overall, toxicity assessment in a complex biological system is a big challenge. Meanwhile, various model systems, both *in vitro* and *in vivo* have been introduced to anticipate the possible health risks of the food-based nanomaterials in human body (Jafari et al., 2017b). Further sections discuss about the available techniques for both *in vitro* and *in vivo* assays of nanomaterials in food.

4.4.1 *In vitro* toxicity assays

In vitro assays are commonly in priority due to their simplicity, controllability, and low cost to evaluate the potential toxicity of the nanomaterials. Fig. 4.2 represents the advantages and disadvantages of *in vitro* and *in vivo* techniques for nanomaterials toxicity assays. *In vitro* assays can give us useful data for designing appropriate *in vivo* assays. Additionally, sufficient information can be obtained by the *in vitro* studies for bulk nanomaterials while the *in vivo* assays are complicated due to the different interactions in the

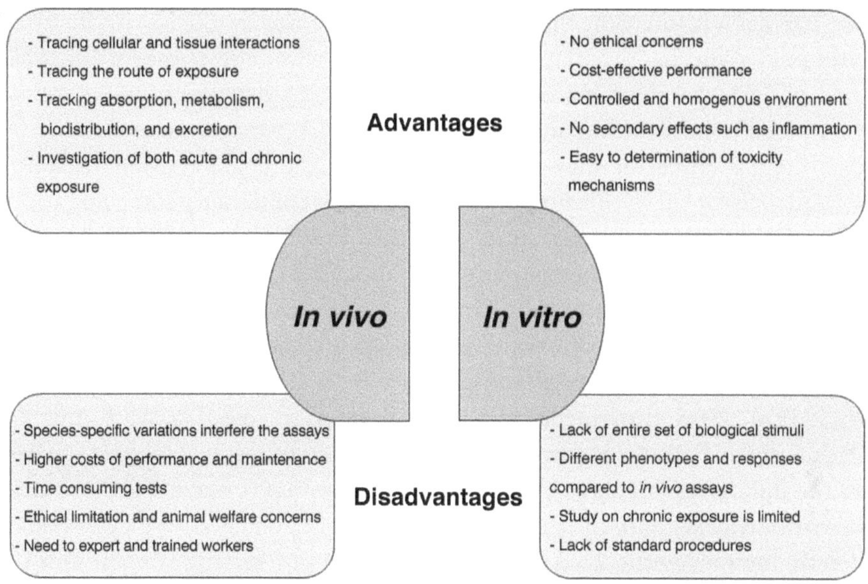

Figure 4.2 ***Advantages and disadvantages of using in vitro and in vivo toxicity assays to evaluate possible health risks of food-based nanoparticles (Greish et al., 2012).***

biological systems; so, *in vitro* studies are economically ideal for large-scale production (Jafari et al., 2017c). For this purpose, it is needed to follow the standard protocols introduced by regulatory organizations and institutes.

In vitro assays can be done by both chemicals and cell cultures. Among chemicals, the GIT fluid components have been mostly applied to simulate the GIT condition in terms of enzymatic degradation and pH fluctuation (Jafari et al., 2017a). On the other hand, using cell culture provides substantial advantages due to utilizing the human origin cells and the obtained results can reveal the biologic and genetic happenings inside the human body (Calvo et al., 2016).

Basically, *in vitro* tests concentrate on the uptake process and the rate of penetration of nanoparticles, but ignore the possible effects of digestion process on their potent adverse health effects. For instance, Malaekeh-Nikouei et al. (2013) investigated the cytotoxicity of the nanoliposomes loaded by safranal as an aromatic substance with antitumor effect. HeLa, MCF7 plus L929 cell lines were applied in order to evaluate apoptosis by staining with propidium iodide; Furthermore, the MTT (3-(4,5-dimethylthiazol-2-yl)-2-yl)-2,5-diphenyl tetrazlium) assay was done to detect the cytotoxicity characteristics. Their results showed that entrapped safranal had higher cytotoxic effect on MCF7 and HeLa cell lines in comparison to free safranal. On the other hand, Hamzeh and Sunahara (2013) studied the cytotoxicity and genotoxicity of TiO_2 nanoparticles. They observed that all types of TiO_2 samples reduced cell viability in a size-dependent manner and mass-based concentration and suggested that TiO_2 nanoparticles differed in toxicological characteristics as well as uncoated and coated nanoparticles.

Today, the need for finding the rapid and efficient toxicity screening tools with mechanistic profiling features feels than ever before, which is due to a large tendency for commercialization of fabricated nanomaterials in laboratory scales. Oxidative stress and genotoxic potential of nanoparticles have been identified as mechanisms of nanoparticle-induced toxicity in which the oxidative stress is the predominant one (Attia, Nounou, & Shalaby, 2018; Sruthi et al., 2018; Brown et al., 2019). Changes in the redox balance, inflammation, and cell death as a result of enhanced levels of reactive oxygen species are the indicators which could be obtained by oxidative stress. Furthermore, the nanomaterials and/or the produced oxidants upon oxidative stress could directly interact with DNA and subsequent, demonstrating the amount of nanoparticle-induced toxicity (Kermanizadeh et al., 2015b). Comet assay (Møller et al., 2015) and Toxtracker assay (Hendriks et al., 2016) are two main *in vitro* techniques for assessing genotoxicity of nanoparticles. The comet assay as a sensitive method uses single cell gel electrophoresis (SCGE) for detecting strand damages in the DNA of a cell while the ToxTracker assay uses green fluorescent protein (GFP)-tagged biomarkers for identification of DNA damage, oxidative stress, protein damage, and general cellular stress upon exposure (Møller et al., 2015; Hendriks et al., 2016).

Intestinal and liver sourced cells are vital target organs for toxicity assessment by ToxTracker assay after oral exposure (Brown et al., 2019). More recently, multiple criteria decision analysis method has been developed as an emerging route for *in vitro* toxicity assessment, working of which is based on physicochemical features, bioavailability, bioaccumulation, and toxic potential of nanoparticles (Bouwmeester et al., 2014; Hristozov et al., 2014b; Hristozov et al., 2014a). NanoRiskCat tool is an example of multiple criteria decision analysis method which uses a systematic approach for finding both exposure and hazard potentials of products with nanoparticles using five color-coded ranks (Hansen, Baun, & Alstrup-Jensen, 2011; Bouwmeester et al., 2014).

Overall, *in vitro* assays by chemicals or cell cultures are preliminary studies to assay possible toxic effects of food-based nanomaterials (Chapter 3). Further, *in vivo* studies are necessary to confirm the obtained results by the *in vitro* study before the subsequent clinical studies.

4.4.2 *In vivo* toxicity assays

Despite the tremendous researches which have done on risk assessment of nanoparticle usage, the need for further studies can still be seen. Lack of enough knowledge of probable detrimental effects of orally exposure nanomaterials, the pathways of their exposure, the mechanism of their uptake into the body, toxicokinetics that are the parameters affecting the nanoparticles distribution and toxicodynamics which are the ways of the nanoparticles interactions with the body's biological systems is still challengeable (Brown et al., 2019). Upon digestion, the translocation of particles is a possible scenario for the interactions with tissues in the GIT (Panyala et al., 2017; Richard, 2017) in which the particles have been observed in the blood capillaries and the lymphatics as sub-mucosal tissues, translocating to, for example, the liver as one of the other secondary organs (Kermanizadeh et al., 2015a). Nanoparticles upon arrival to the liver are taken up via phagocytosis of Kupffer cells, which is likely to cause several responses such as activation of reactive oxygen species formation and oxidative stress reactions in the hepatocytes, or inflammatory tissue feedbacks (Fernandezurrusuno et al., 1995). As a consequence, it is essential to assess both *in vitro* and *in vivo* toxicity of the nanoparticles.

A wide range of live animal models such as *Drosophila melanogaster*, *Caenorhabditis elegans*, Zebrafish, mice, rat, rabbit, and dog can be applied for *in vivo* studies (Rubió, Motilva, & Romero, 2013; Calvo et al., 2016). The live model systems provide physiological information about the biological fate of nanoparticles as well as the possible health risks (Jafari et al., 2017a). *In vivo* assays involve the quantification of the absorbed nanomaterials through the GIT. Furthermore, the plasma concentration methods plus atomic absorption assessment can be used to trace the nanoparticles (Navarro & Wood, 2003; Jafari et al., 2017a). Application of imaging and radioisotopes studies are necessary to evaluate the metabolization, absorption, and excretion rate of the nanomaterials. Further, tissue sampling after the knocking down and surgery of the model organism can be utilized following by staining, imaging, or other techniques to get more information about the absorption and accumulation of nanoparticles in specific organs and tissues (Jafari et al., 2017a).

So, Jang, and Han (2008) studied the potential toxic effects of SiO_2 micro/nanoparticles on dieted mice with the dose of 140 g SiO_2/kg for 10 weeks. They indicated that feeding with SiO_2 microparticles (30 μm) resulted in fatty liver symptoms while the content of both micro and nanoparticles in the liver were almost equal, indicating their difficulty for getting through the GIT barriers.

Yang et al. (2013) assessed the long-term toxicity of graphene oxide (GO) nanocomponents and polyethylene glycol (PEG) functionalized GO by intraperitoneal and oral injection in mouse models. By intraperitoneal injection, high accumulation of PEGylated GO, but not GO was observed in the reticuloendothelial system, specifically in the liver and spleen via a size and coating surface dependent manner. However, oral administrated PEGylated GO derivatives could not be adsorbed via organs and were excreted rapidly. Conclusively, they confirmed that delivery routes, surface coatings, and size distribution of nanomaterials directly influence on their toxicological effects.

4.5 Model organisms for *in vivo* studies of food-based nanomaterials

Live model systems which have been utilized for *in vivo* study of the nanomaterials toxicity mostly belong to murine group including mice or rat. Common *in vivo* models in this field are zebrafish, *Drosophila melanogaster*, and *C. elegans* (Ting et al., 2014; Calvo et al., 2016). Fig. 4.3 present different model organisms applied for *in vitro* and *in vivo* toxic characterization of nanoparticles and their functions (Yang et al., 2017).

Murine models are mammals with the most similarity to humans, which provide more confidence in extrapolation of the observations. On the other hand, the high taxonomical similarity brings ethical issues. Moreover, they are large and long-lived that makes it hard and costly to handle for most

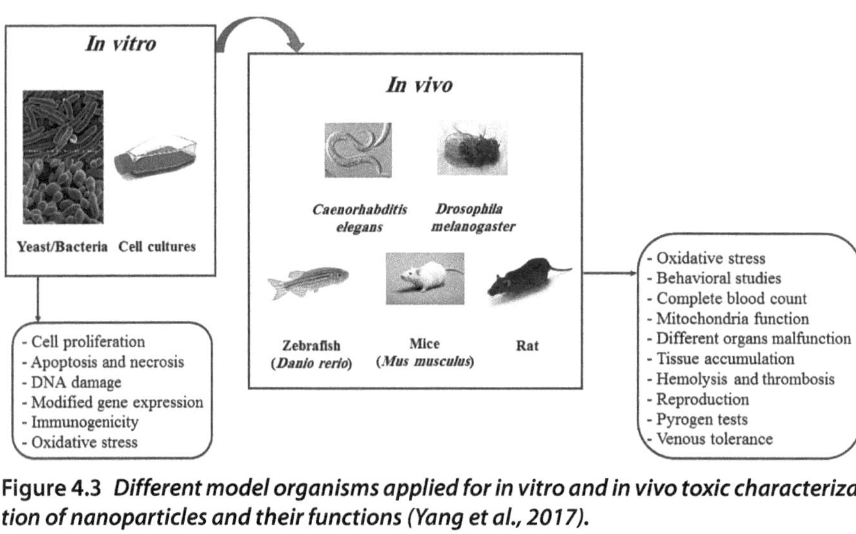

Figure 4.3 ***Different model organisms applied for in vitro and in vivo toxic characterization of nanoparticles and their functions (Yang et al., 2017).***

of researchers and food industries. This is why most of researchers prefer to use more affordable and smaller model systems such as zebrafish, *Drosophila,* and *C. elegans* (Calvo et al., 2016). Zebrafish was another attractive model organism during the last decade in food and pharmaceutical industries. This model has a complete circulatory apparatus, high fecundity, low cost, fast and characterized development, embryo transparency, gene manipulation accessibility, and also a greatly conserved immune system. Furthermore, it is easy-to-handle and small even in adult form. However, the long lifespan (3–4 years) limits working with this model. Nevertheless, zebrafish has widely been applied to assess the safety of nanomaterials for human body (Lin et al., 2014).

Drosophila is another favorable model system due to its small-sized body and remarkable genetic and biologic resemblance, which makes it a popular model in food and pharma field of study. The biggest challenge in working with *Drosophila* is its flying ability. Moreover, its active breeding may lead to hygiene issues in laboratories (Calvo et al., 2016). Alaraby et al. (2015) studied the biological effects of ZnO nanoparticles exposure via *Drosophila melanogaster*. They observed the internalization process of ZnO nanoparticles through the intestinal barriers; however, they revealed that ZnO nanoparticles did not trigger toxicity or genotoxicity in *Drosophila*, although the alteration of gene expression was happened.

*C. elegans*is a small-sized (about 1 mm) free-living nematode with 1000 somatic cells feeding bacteria originated from soil environments. It was the first multicellular model organism with a completely sequenced genome (Brenner, 1974). For the first time, RNAi (RNA interference) was described by this model system, which make it a desirable genetic model. *C. elegans* has also a complete nervous system and digestive tract with a known neural map. This model can easily grow on an agar plate at ambient temperature feeding on a non-pathogenic *E. coli* culture with a short life cycle of 3 weeks. A broad range of studies such as aging, neural disorders, cell apoptosis, and development, molecular biology, and metabolic syndrome can be done by this model. Its transparent body facilitates using green fluorescent proteins as marker to track all types of nanomaterials through the digestion and absorption process inside the body. The generation of new mutants of *C. elegans* is very simple by RNAi and too many mutant strains are available. The earlier-mentioned properties of the model organism facilitate dietary interventions as well as the mechanistic studies. The only requirement to assess a food-based nanomaterial is that it should be mixed well with the growth medium and let the *C. elegans* consume it. Afterwards, definite phenotypic and genotypic factors should be followed up to check all expected toxic effect of the nanomaterials. Gonzalez-moragas et al. (2017) assessed the biodistribution of citrate-capped Au nanoparticles with the size range of 11–150 nm by *C. elegans* as a model organism. Their results showed that Au nanoparticles did not effectively cross intestinal and dermal barriers and did not significantly influence the intestinal barrier's' integrity. However, they claimed that exposure to 11 nm Au nanoparticles resulted in a higher toxicity compared to the larger ones.

4.6 Different approaches for risk assessment by *in vivo* assays

Compared to *in vitro* researches, which are essential for implementing any *in vivo* assays, the *in vivo* studies are complicated because of the reactions built within the biological systems (Jafari et al., 2017c). It is important to obtain comprehensive information on the purity and physicochemical characteristics of studied nanoparticles before any toxicity assays (Oberdörster et al., 2005; Sayes & Warheit, 2009) in order to obtain a strong correlation between the findings and toxicity of the fabricated nanomaterials. The rat, zebrafish, *C. elegans*, and mice models are the common model systems which have been applied to assess the toxicity of nanoparticles (Table 4.1). The interactions between cells or tissues and nanoparticles is a critical issue that give possibility to find information for the identification of the accumulation site and toxicological profiles within a specific organ in order to determine any future biological application of such structures. *In vivo* risk assessment could be identified by several parameters as follows:

1. Histopathological analysis for assessment of nanoparticles distribution at different organ sites (liver, kidney, spleen, lungs, heart, aorta, and brain) by various instruments such as optical and electron microscopy, confocal microscopy, and micro X-ray fluorescence imaging of stained organs: In this method, the organs (liver, kidney, …) from treated and control model are fixed in formalin solution and then in paraffin blocks. The glass slides including sliced tissues are examined by light microscopy or other appropriate instruments for the histopathological alterations after hematoxylin and eosin (HE) staining (Kim et al., 2009; Chupani et al., 2018).
2. Acute toxicity and sublethal effects of nanoparticles under a dissecting microscope for zebrafish embryos model studies: Zebrafish embryos are fed by different concentrations of nanoparticles in an incubator at

Table 4.1 Common *in vivo* model systems for toxicity and safety assessment of food-based nanomaterials.

Nanoencapsulation method	Core/shell	Type of nano/micro structure	Model system	Techniques for toxicity and safety assessment	Study outcomes	References
Cation-exchange method	Poly(lactic) acid/modified clay	Nanocomposite	Wistar rat	• Histopathological analysis by optical and electron microscopy of stained organs • Clinical biochemistry parameters of blood samples	No apparent evidence of damage which shows that this nanocomposite is a good choice in the food packaging industry	Maisanaba et al. (2014)
Dialysis method	Curcumin/poly(dimethylaminoethyl methacrylate) - poly(e-caprolactone) -poly(dimethylaminoethyl methacrylate)	Nanoparticle	Wistar rat	• Clinical hematology analysis and biochemical assay of blood samples • Histopathology of livers segments by a light microscope connected to a digital camera	No significant toxicity reactions or pathological liver changes which indicate the lack of toxicity of cationic micelles and their biocompatibility as potential drug delivery systems	Tzankova et al. (2016)

(*Continued*)

Table 4.1 Common *in vivo* model systems for toxicity and safety assessment of food-based nanomaterials. (*Cont.*)

Nanoencapsulation method	Core/shell	Type of nano/micro structure	Model system	Techniques for toxicity and safety assessment	Study outcomes	References
Solution mixing and freeze drying method	Fluorescein isothiocyanate (FITC)/Halloysite nanotubes (HNTs)	Nanotube	Zebrafish embryos	• Acute toxicity and sublethal effects of HNTs to zebrafish embryos and larva under a dissecting microscope • Assessment of HNTs distribution in the zebrafish under a fluorescence microscope	• No significant acute toxicity and sublethal effects, including the percent survival, morphological malformation and cardiac toxicity during the development period of zebrafish by HNTs $\leq$ 25 mg/mL. • Enrichment the surface of chorion by HNTs, ingestion and then excretion by zebrafish larvae	Long et al. (2018)

Co precipitation route	Curcumin/Chitosan and Titanium dioxide	Nanoparticle	Sprague dawley rat	• Biodistribution studies of the nanomaterial at different organ sites (liver, kidney, spleen, lungs, heart, aorta, and brain) • Toxicity and hematological parameters • Genotoxicity study by comet assay on blood cells	Efficient drug delivery system with improved half-life and nontoxicity	Sherin et al. (2017)
Solution mixing and freeze drying method	*Raw nanoparticles: -/* Halloysite nanotubes (HNTs) and chitin nanocrystals (ChNCs) *Rhodamine-labeled nanoparticles:* Rhodamine B (Aladdin)/ Halloysite nanotubes (HNTs) and chitin nanocrystals (ChNCs)	Nanotube Nanocrystal	*Caenorhabditis elegans*	• Lifespan assay • Movement assays • Fertility assay • Development analysis • ROS production • Oil Red O staining (ORO) and quantification • Quantitative RT-PCR assay • Intestinal barrier function assay	No obvious toxicity for both ChNCs and HNTs below the concentration of 50 mg/mL which shows promising application in biomedical areas	Zhao et al. (2019)

(*Continued*)

Table 4.1 Common *in vivo* model systems for toxicity and safety assessment of food-based nanomaterials. (*Cont.*)

Nanoencapsulation method	Core/shell	Type of nano/micro structure	Model system	Techniques for toxicity and safety assessment	Study outcomes	References
N/A	-/Zinc oxide nanoparticles (ZnO-NPs)	Nanoparticle	Mice	• Biochemical assays of serum and blood • Zinc content analysis in Tissues • Gene expression of the liver and kidneys post exposure of ZnO NPs, vitamin C (Vc) and ZnO NPs plus Vc	The significant increase in toxicity in the *in vivo* experiments over repeated oral exposure to ZnO NPs plus Vc, indicating the necessity of synergistic toxicological evaluation and safety assessment of nanofood in a complex system	Wang et al. (2014)
N/A	Silver nanoparticles (Ag)/PVP-coated	Nanoparticle	Sprague-Dawley rats	• Determination of total silver content in tissues, organs and gastrointestinal content • Determination of silver nanoparticles (AgNPs) in tissues and gastrointestinal content. • Blood biochemical analysis • Immunotoxicity analysis	• Persistent retention of silver in brain and testis after 8 weeks postdosing by dose range of 90 mg/kg • No significant evidence of hepatotoxicity or immunotoxicity of the silver exposure	van der Zande et al. (2012)

N/A	-/Zinc oxide nanoparticles (ZnO-NPs)	Nanoparticle	Swiss albino mice	• Zn content analysis in tissues • Biochemical assays in serum • Lipid peroxidation (LPO) assay • Histopathological analysis • Cell viability assay using trypan blue • Genotoxicity assessment: Single cell gel electrophoresis (Comet) & Fpg-modified Comet assays • Terminal deoxynucleotide-transferase (TdT)-mediated dUTP nick end labeling (TUNEL) assay using "*In situ* Cell Death Detection Kit, Fluorescein"	• Pathological lesions in the liver by dose range of 300 mg/kg for 14 d due to accumulation of nanoparticles • DNA damage and subsequent apoptosis in liver cells by increasing oxidative stress	Sharma et al. (2012)
N/A	-/Zinc oxide nanoparticles (ZnO-NPs)	Nanoparticle	Sprague-Dawley rats	• Assessment of blood samples and ZnO concentration by measuring zinc level with inductively coupled plasma-atomic emission spectroscopy (ICP-AES)	• Persistent increased levels in systemic circulation • No observed adverse effect level values may be less than 125 mg/kg via oral ingestion	Chung et al. (2013)

(*Continued*)

Table 4.1 Common *in vivo* model systems for toxicity and safety assessment of food-based nanomaterials. (*Cont.*)

Nanoencapsulation method	Core/shell	Type of nano/micro structure	Model system	Techniques for toxicity and safety assessment	Study outcomes	References
N/A	-/Titanium dioxide nanoparticles (TiO_2-NPs)	Nanoparticle	Sprague-Dawley rats	• Organ coefficients • Hematological parameters • Blood biochemical assay • Histological examination	• Lung, kidney, and heart injuries • White blood cells and red blood cells changes • TiO_2 NPs combined with glucose could cause more obvious toxic effects which indicate the necessity of control the dietary glucose and TiO_2 NPs intake among children	Chen et al. (2015)
N/A	-/Titanium dioxide nanoparticles (TiO_2(E171)- NPs)	Nanoparticle	Wistar rats	• Confocal microscopy and micro X-ray fluorescence imaging of tissue samples • Intestinal permeability measurement	Inflammatory development with preneoplastic lesions in the colon	Bettini et al. (2017)

Solution mixing and evaporation by rotary evaporator	-/Titanium dioxide nanoparticles (TiO_2-NPs)	Nanopar-ticle	ICR mice	• Stomach collection and coefficients of stomach • Titanium content analysis of gastric tissues • Determination of gastric secretion of blood samples • Histopathological evaluation of stomachs • Evaluation of mucosal smooth muscle cell ultrastructure • Western blot assay	Dysfunction of gastric secretion, inflammation, and atrophy upon long-term exposure to TiO_2- NPs	Hong et al. (2017)
N/A	-/ZnO nanoparticles (ZnO-NPs)	Nanopar-ticle	ICR mice	• Symptoms and morality were observed by LD_{50}estimination • Kinetic study of blood and tissues (lung, liver, kidney, and spleen) based on LD_{50} • Tissue accumulation study • Pathological examination • 8-OHdG analysis • Zn assay of blood on tissue samples	• Rapidly removing of ZnO-NPs from the blood and distributing to various organs, following injection • The spleen, liver, lungs are target organs for ZnO-NPs as observed by tissue distribution and pathology studies	Fujihara et al. (2015)

N/A: Not available.

28.51°C. The subsequent percent survival and percent of malformed zebrafish assessments are done with a dissecting microscope every 24 h from 24 to 120 hpf (Long et al., 2018).

3. Clinical hematology analysis and biochemical assay of blood and serum samples: In hematology tests, the blood samples collected from the animals via tail vein puncture are assessed for some parameters such as the hemoglobin (HGB), red blood count (RBC), hematocrit (HCT), mean corpuscular hemoglobin (MCH), mean corpuscular volume (MCV), white blood cell count (WBC), platelet count (PLT), platelet, mean platelet volume using a hematology analyzer (Sharma et al., 2012). In biochemical assay, the serum achieved by centrifugation of the whole blood are measured for its biochemical levels such as alkaline phosphatase (ALP), alanine aminotransferase (ALT), creatinine, bilirubin, blood urea nitrogen (BUN) using an automated hematology analyzer (Sharma et al., 2012) as compared to standard controls.
4. Lifespan assay, movement assays, fertility assay, development analysis, ROS production, Oil Red O staining (ORO) and quantification, quantitative RT-PCR assay, and intestinal barrier function assay for studies using *C. elegans* model (Zhao et al., 2019): In lifespan assay, a number of larvae or young adults are transferred to fresh plates containing the proposed nanoparticles and floxuridine, and scored every day. In movement assay, a previously described larvae or young adults in the lifespan assay were evaluated by tapping plates and monitoring the worms moving whether it is in a continuous, coordinated sinusoidal way (fast movement), or not (non-fast movement). In fertility assay, the number of eggs and hatched larvae are determined by each worm on each day. The hatching rate is calculated as the number of larvae/number of egg. In development analysis, the effects of various concentrations of nanoparticles on development of worms are determined by calculating the percentage of L4 worms in the worm population. In ROS production, previously labeled worms using MitoSOX are picked on the agar pads, then imaged using a fluorescence microscope coupled with Image J. In Oil Red O staining (ORO) and quantification, ORO-stained worms are picked onto agar pads and imaged using a fluorescence microscope coupled with Image-Pro-Plus processing software. In quantitative RT-PCR assay, total RNA of worms are isolated and cDNA are synthesized using total RNA via specific kits based on the protocols. Expression levels of examined genes are analyzed on a real-time system. In intestinal barrier function assay, the

blue food dye-stained worms are picked onto agar pads and imaged using an imager with microscope.

5. Gene expression of the different organ sites post-exposure of nanoparticles: The various gene expressions of the organ sites such as liver fatty acid binding protein (L-FABP), neutrophil gelatinase associated lipocalin (NGAL), and apolipoprotein (APO) post exposure of nanoparticles for a period of time are analyzed by the mRNA reverse transcription-polymerase chain reaction (RT-PCR) method (Wang et al., 2014).
6. Immunotoxicity analysis: Immunotoxicity analysis is done by evaluating the proliferation of T- and B-cells isolated from spleen and mesenteric lymph nodes (MLN) in response tolipopolysaccharides (LPS) or Concanavalin A (ConA) using a Coulter counter (van der Zande et al., 2012).
7. Cell viability assay using trypan blue: The Comet assay are usually used for cell viability assessment. In addition, the trypan blue dye exclusion technique could be used for this purpose in some cases by diluting cells obtained from organs with PBS and mixing with trypan blue dye. The viable cells with intact cell membrane exclude the dye and are counted via a hemocytometer (Sharma et al., 2012).
8. Organ coefficients: After weighing the body and organs, the ratio of the weight of the liver, kidney, spleen, testicle or ovary, and heart (wet weight, mg) is divided by body weight (g) to calculate the organ coefficients (Chen et al., 2015).
9. Intestinal permeability measurement: Intestinal permeability is assessed using a marker of paracellular permeation after administration of Cr-EDTA diluted in saline to model (e.g., rats) by using gavage and subsequent analysis of their urine samples upon a period of time (e.g., 24 h) (Bettini et al., 2017).
10. Symptoms and morality by LD_{50} estimation: Symptoms and morality is observed based on the lethal dose 50 (LD_{50}) estimation by a group of model (e.g., 65 mice) after injection of different concentrations of nanoparticles via the tail vein (Fujihara et al., 2015).
11. Kinetic study of blood and tissues (lung, liver, kidney, spleen, and …) based on LD_{50}: Kinetic study is performed by a group of model (e.g., 6 × 24 mice for each time point) after injection of a single intravenous dose (based on LD_{50}) of nanoparticles via the tail vein and subsequent scarifying the mice (or applied model) at the due time point to collect blood and tissues for nanoparticles assay (Fujihara et al., 2015).

12. Tissue accumulation study: Tissue accumulation study is performed similar to the mentioned kinetic study with some differences in the number of mice (or applied model), the due time point for collection of blood and tissues as well as a single intravenous dose of nanoparticles (Fujihara et al., 2015).
13. 8-OHdG analysis: Both 8-OHdG and creatinine concentrations in urine samples is measured using a competitive immunochromatography analyzer and is expressed as ng 8-OHdG/mg creatinine (Fujihara et al., 2015).

More details of case-by-case studies and techniques for toxicity and safety assessment for each of them have been presented in Table 4.1. For example, Tzankova et al. (2016) used Wistar rat model to evaluate the toxicity of Curcumin/poly(dimethylaminoethyl methacrylate)-poly(e-caprolactone)-poly(dimethylaminoethyl methacrylate) nanoparticles obtained by dialysis method. They utilized clinical hematology analysis and biochemical assay of blood samples as well as histopathology of livers segments by a light microscope connected to a digital camera. According their results, no significant toxicity reactions or pathological liver changes were observed upon single (7.5 mg/kg, i.p.), or repeated (3.5 mg/kg, i.p., 14 days) exposure to empty or curcumin loaded polymeric micelles which indicate the lack of toxicity of cationic micelles and their biocompatibility as potential drug delivery systems (Tzankova et al., 2016).

Many studies about acute, subacute, and subchronic toxicity, after oral exposure in rodents exhibited that acute toxicity may occur at high doses, depending on chemical composition, size, and coating of the nanoparticles (Bouwmeester et al., 2009). For instance, Sharma et al. (2012) evaluated zinc oxide nanoparticles (ZnO nanoparticles) toxicity in Swiss albino mice at dose range of 300 mg/kg for 14 d. They used various *in vitro* and *in vivo* studies including Zn content analysis in tissues, biochemical assays in serum, lipid peroxidation (LPO) assay, histopathological analysis, cell viability assay using trypan blue, genotoxicity assessment (Single cell gel electrophoresis (Comet) & Fpg-modified Comet assays) as well as Terminal deoxynucleotide-transferase (TdT)-mediated dUTP nick end labeling (TUNEL) assay using "In situ Cell Death Detection Kit, Fluorescein." According to their results, pathological lesions in the liver occurred by dose range of 300 mg/kg for 14 d due to accumulation of nanoparticles. Applying this dosage showed DNA damage and subsequent apoptosis in liver cells by increasing oxidative stress (Sharma et al., 2012).

The lack of sufficient understanding about the toxicity of nanoparticles and the inconsistencies between studies are often related to the variability in protocols applied in cell culture and animal researches and the difficulties for analytical detection of nanoparticles in complex biological systems like food matrices and mammalian tissues (McClements & Xiao, 2017). To some extent, it could be solved by the application of alternative models for *in vivo* studies, including terrestrial invertebrates such as insects and nematodes to marine life and freshwater including planarians, crustaceans, molluscs, as well as transparent nematode *C. elegans*, the fruit fly *Drosophila melanogaster*, the moth *Galleria mellonella*, and the water flea *Daphnia* spp. larvae (Wilson-Sanders, 2011; Brandelli, 2019).

4.7 Key issues in development and standardization of *in vivo* assays

Characterization and measurement of nanomaterials are still associated with some shortcomings which may complicate the procedures used to identify nanomaterials in biological matrices. The other challenge is the physicochemical change of nanomaterials during processing, producing, packaging, ingestion, and absorption. These changes include aggregation or agglomeration, transformation, chemical form and surface, dispersibility/solubility, chemical reactivity, and porosity (Jain et al., 2018).

The difficulty and complexity of tracing nanoparticles inside the live organism cause some issues in working with the live organisms through *in vivo* assessments (Gopi et al., 2016). On the other hand, uncorrelated toxicological results by *in vivo* assays represent a serious problem and it would be hard to compare them. In the United States and in Europe, new approaches have been developed to reduce the use of animal models for toxicity tests (Hofmann-Amtenbrink, Grainger, & Hofmann, 2015). So, the computational system biology like fuzzy logic or quantitative structure activity relationships approaches have been proposed to analyze the complex biological system of live organism (Bhattacharya et al., 2011). Following recommendations should be remarked regarding the methodology of *in vivo* studies to reduce the possible limitations (Greish, Thiagarajan, & Ghandehari, 2012; Hofmann-Amtenbrink et al., 2015; Yu et al., 2017):

- Characterization of the administered nanomaterials is crucial for *in vivo* toxicity assessment due to the important role of their physicochemical attributes such as charge, purity, charge density, shape, size, surface area, and density in causing toxicity effects. The results should be presented appropriately and allow the comparison with the other studies.
- Administrated dosage can be designed in the range of the expected exposure or the potential application of the nanomaterial. Single or

multiple escalation doses can be used in definition of the dose range. Additionally, the tissues or cells should not be overloaded by nanomaterials which may cause false results. On the other hand, the time frame of toxicity assessments should be carefully determined based on the possible interactions between the nanomaterials and the biological system. The acute toxicity studies usually span 2 weeks while the chronic studies can take up to 2 years. Overall, dose and time variations are main issues in making comparison between *in vitro* and *in vivo* studies.
- The designed route of the administration should simulate the real exposure of the nanomaterial including ingestion, nasal inhalation, dermal exposure, as well as intravenous administration.
- Expected mechanism of toxicity causing by nanomaterials should be considered which can lead to a thorough investigation of targeted organs, tissues, or cells affected by the nanomaterial. Due to the higher uptake of nanoparticles by macrophages, it would be appropriate to consider the macrophage population in liver and spleen through the safety assessment.
- Among the rodent model organisms, rats are commonly preferred to mice because of their higher similarity in hemodynamics to human. Moreover, in choosing the model organisms, two species are required (e.g., one rodent and one non-rodent) to confirm the pattern of toxicity and to exclude the species-specific toxic effects.
- The establishment of best protocols and standards for *in vivo* assays by regulators and industries has to be initiated to minimize all issues in this regard.

All in all, the challenges in designing an *in vivo* study to evaluate the toxic effects of the nanomaterials should be taken into consideration to maximize the reliability of the tests.

4.8 Conclusion

There are raising concerns about the toxicity effect of nanoparticles on human body; so, it would be very important to know the biological fate of nanoparticles following the exposure to consumers. Most of the studies have been based on the *in vitro* models using chemicals or cell lines. Despite the fact that there is low correlation between the results obtained by *in vitro* and *in vivo* studies, *in vivo* assays are usually preferred due to providing a real physiological environment. Moreover, the biotransformation and excretion of nanoparticles in human body is important to take a closer look, which can definitely affect their toxicity effect. Taken together, this chapter indicated that working with live model organism to evaluate the controlled release and bioavailability of nanoencapsulated bioactive compounds is still in first steps and more works are required in this area before general consumption of the novel functional foods.

References

Alaraby, M., et al. (2015). A comprehensive study of the harmful effects of ZnO nanoparticles using *Drosophila melanogaster* as an in vivo model. *Journal of Hazardous Materials, 296*, 166–174. doi: 10.1016/j.jhazmat.2015.04.053. Elsevier B.V..

Arora, S., Rajwade, J. M., & Paknikar, K. M. (2012). Nanotoxicology and in vitro studies: the need of the hour. *Toxicology and Applied Pharmacology, 258*(2), 151–165. Elsevier.

Attia, H., Nounou, H., & Shalaby, M. (2018). Zinc oxide nanoparticles induced oxidative DNA damage, inflammation and apoptosis in rat's brain after oral exposure. *Toxics, 6*(2), 29. Multidisciplinary Digital Publishing Institute.

Bettini, S., et al. (2017). Food-grade TiO_2 impairs intestinal and systemic immune homeostasis, initiates preneoplastic lesions and promotes aberrant crypt development in the rat colon. *Scientific Reports*, 7, 40373. Nature Publishing Group.

Bhattacharya, S., et al. (2011). Toxicity testing in the 21st century: defining new risk assessment approaches based on perturbation of intracellular toxicity pathways. *Plos One, 6*(6), e20887. doi: 10.1371/journal.pone.0020887.

Bouwmeester, H., et al. (2009). Review of health safety aspects of nanotechnologies in food production. *Regulatory Toxicology and Pharmacology, 53*(1), 52–62. Elsevier.

Bouwmeester, H., et al. (2014). State of the safety assessment and current use of nanomaterials in food and food production. *Trends in Food Science & Technology, 40*(2), 200–210. Elsevier.

Brandelli, A. (2019). The interaction of nanostructured antimicrobials with biological systems: cellular uptake, trafficking and potential toxicity. *Food Science and Human Wellness, 9*, 8–20. Elsevier.

Brenner, S. (1974). The genetics of *Caenorhabditis elegans*. *Genetics*, 77(1), 71–94. doi: 10.1002/cbic.200300625.

Brown, D. M., et al. (2019). The mechanism-based toxicity screening of particles with use in the food and nutrition sector via the ToxTracker reporter system. *Toxicology in Vitro, 61*, 104594. Elsevier.

Calvo, D. R., et al. (2016). Development of novel functional ingredients: Need for testing systems and solutions with *Caenorhabditis elegans*. *Trends in Food Science and Technology, 54*, 197–203. doi: 10.1016/j.tifs.2016.05.006. Elsevier Ltd.

Card, J. W., et al. (2011). An appraisal of the published literature on the safety and toxicity of food-related nanomaterials. *Critical Reviews in Toxicology, 41*(1), 20–49. doi: 10.3109/10408444.2010.524636.

Chalew, T. E. A., & Schwab, K. J. (2013). Toxicity of commercially available engineered nanoparticles to Caco-2 and SW480 human intestinal epithelial cells. *Cell Biology and Toxicology, 29*(2), 101–116. Springer.

Chau, C. F., Wu, S. H., & Yen, G. C. (2007). The development of regulations for food nanotechnology. *Trends in Food Science and Technology, 18*(5), 269–280. doi: 10.1016/j.tifs.2007.01.007.

Chen, Z., et al. (2015). Interaction of titanium dioxide nanoparticles with glucose on young rats after oral administration. *Nanomedicine: Nanotechnology, Biology and Medicine, 11*(7), 1633–1642. Elsevier.

Choi, J., et al. (2015). Toxicity of zinc oxide nanoparticles in rats treated by two different routes: single intravenous injection and single oral administration. *Journal of Toxicology and Environmental Health, Part A*, *78*(4), 226–243.

Chung, H. E., et al. (2013). Toxicokinetics of zinc oxide nanoparticles in rats. In *Journal of Physics: Conference Series* (pp. 12037). IOP Publishing.

Chupani, L., et al. (2018). Chronic dietary toxicity of zinc oxide nanoparticles in common carp (*Cyprinus carpio* L.): tissue accumulation and physiological responses. *Ecotoxicology and Environmental Safety*, *147*, 110–116. Elsevier.

Chusuei, C. C., et al. (2013). Cytotoxicity in the age of nano: the role of fourth period transition metal oxide nanoparticle physicochemical properties. *Chemico-Biological Interactions*, *206*(2), 319–326. Elsevier.

Dasgupta, N., & Ranjan, S. (2018). Nano-food toxicity and regulations. In *An introduction to food grade nanoemulsions* (pp. 151–179). Springer.

Dekkers, S., et al. (2016). Towards a nanospecific approach for risk assessment. *Regulatory Toxicology and Pharmacology*, *80*, 46–59. Elsevier.

Faust, J. J., et al. (2014). Food grade titanium dioxide disrupts intestinal brush border microvilli in vitro independent of sedimentation. *Cell Biology and Toxicology*, *30*(3), 169–188. Springer.

Fernandezurrusuno, R., et al. (1995). Evaluation of liver toxicological effects induced by polyalkylcyanoacrylate nanoparticles. *Toxicology and Applied Pharmacology*, *130*(2), 272–279. Elsevier.

Florence, A. T. (1997). The oral absorption of micro-and nanoparticulates: neither exceptional nor unusual. *Pharmaceutical Research*, *14*(3), 259–266. Springer.

Fröhlich, E. E., & Fröhlich, E. (2016). Cytotoxicity of nanoparticles contained in food on intestinal cells and the gut microbiota. *International Journal of Molecular Sciences*, *17*, 509. doi: 10.3390/ijms17040509.

Fujihara, J., et al. (2015). Distribution and toxicity evaluation of ZnO dispersion nanoparticles in single intravenously exposed mice. *The Journal of Medical Investigation*, *62*(1.2), 45–50. The University of Tokushima Faculty of Medicine.

Gaillet, S., & Rouanet, J. M. (2015). Silver nanoparticles: their potential toxic effects after oral exposure and underlying mechanisms–a review. *Food and Chemical Toxicology*, 77, 58–63. https://doi.org/10.1016/j.fct.2014.12.019.

Giese, C., & Marx, U. (2014). Human immunity in vitro—Solving immunogenicity and more. *Advanced Drug Delivery Reviews*, *69*, 103–122. Elsevier.

Gonzalez-moragas, L., et al. (2017). Acta Biomaterialia in vivo testing of gold nanoparticles using the *Caenorhabditis elegans* model organism. *Acta Biomaterialia*, *53*, 598–609. doi: 10.1016/j.actbio.2017.01.080. Acta Materialia Inc..

Gopi, S., et al. (2016). Biotherapeutic discovery introduction of nanotechnology in herbal drugs and nutraceutical: a review. *Journal of Nanomedicine and Biotherapeutic Discovery*, *6*(2), 1–8. doi: 10.4172/2155-983X.1000143.

Greish, K., Thiagarajan, G., & Ghandehari, H. (2012). In vivo methods of nanotoxicology. In *Nanotoxicity* (pp. 235–253). Springer.

Gupta, A., Kumar, S., & Kumar, V. (2019). Challenges for assessing toxicity of nanomaterials. In *Biochemical toxicology-heavy metals and nanomaterials*IntechOpen.

Hamzeh, M., & Sunahara, G. I. (2013). In vitro cytotoxicity and genotoxicity studies of titanium dioxide (TiO_2) nanoparticles in Chinese hamster lung fibroblast cells. *Toxicology in Vitro*, *27*(2), 864–873. Elsevier.

Hansen, S. F., Baun, A., & Alstrup-Jensen, K. (2011). *NanoRiskCat–a conceptual decision support tool for nanomaterials*. Danish Ministry of the Environment.

Hendriks, G., et al. (2016). The extended ToxTracker assay discriminates between induction of DNA damage, oxidative stress, and protein misfolding. *Toxicological Sciences*, *150*(1), 190–203. Oxford University Press.

Hofmann-Amtenbrink, M., Grainger, D.W., & Hofmann, H. (2015). Nanoparticles in medicine: Current challenges facing inorganic nanoparticle toxicity assessments and standardizations. *Nanomedicine: Nanotechnology, Biology, and Medicine, 11*, 1689–1694. doi: 10.1016/j.nano.2015.05.005. Elsevier B.V..

Hong, F., et al. (2017). Gastric toxicity involving alterations of gastritis-related protein expression in mice following long-term exposure to nano TiO_2. *Food Research International, 95*, 38–45. doi: 10.1016/j.foodres.2017.02.013.

Hristozov, D. R., Zabeo, A., et al. (2014a). A weight of evidence approach for hazard screening of engineered nanomaterials. *Nanotoxicology, 8*(1), 72–87. Taylor & Francis.

Hristozov, D. R., Gottardo, S., et al. (2014b). Application of a quantitative weight of evidence approach for ranking and prioritising occupational exposure scenarios for titanium dioxide and carbon nanomaterials. *Nanotoxicology, 8*(2), 117–131. Taylor & Francis.

Jafari, S. M., et al. (2017a). Bioavailability and release of bioactive components from nanocapsules. In *Nanoencapsulation technologies for the food and nutraceutical industries* (pp. 494–523.). Elsevier Incdoi: 10.1016/B978-0-12-809436-5/00013-6.

Jafari, S. M., Esfanjani, A. F., & Katouzian, I. (2017b). Release, characterization, and safety of nanoencapsulated food ingredients. In *Nanoencapsulation of food bioactive ingredients* (pp. 401–453.). Elsevier Incdoi: 10.1016/B978-0-12-809740-3.00010-6.

Jafari, S. M., Katouzian, I., & Akhavan, S. (2017c). Safety and regulatory issues of nanocapsules. In *Nanoencapsulation technologies for the food and nutraceutical industries* (pp. 545–590). Elsevier.

Jain, A., et al. (2018). Nanomaterials in food and agriculture: an overview on their safety concerns and regulatory issues. *Critical Reviews in Food Science and Nutrition, 58*(2), 297–317. Taylor & Francis.

Katouzian, I., & Jafari, S. M. (2016). Nano-encapsulation as a promising approach for targeted delivery and controlled release of vitamins. *Trends in Food Science and Technology, 53*, 34–48. doi: 10.1016/j.tifs.2016.05.002.

Kermanizadeh, A., Balharry, D., et al. (2015a). Nanomaterial translocation–the biokinetics, tissue accumulation, toxicity and fate of materials in secondary organs–a review. *Critical Reviews in Toxicology, 45*(10), 837–872. Taylor & Francis.

Kermanizadeh, A., Chauché, C., et al. (2015b). The role of intracellular redox imbalance in nanomaterial induced cellular damage and genotoxicity: a review. *Environmental and Molecular Mutagenesis, 56*(2), 111–124. Wiley Online Library.

Khanna, P., et al. (2015). Nanotoxicity: an interplay of oxidative stress, inflammation and cell death. *Nanomaterials, 5*(3), 1163–1180. Multidisciplinary Digital Publishing Institute.

Kim, W. -Y., et al. (2009). Histological study of gender differences in accumulation of silver nanoparticles in kidneys of Fischer 344 rats. *Journal of Toxicology and Environmental Health, Part A, 72*(21–22), 1279–1284. Taylor & Francis.

Koeneman, B. A., et al. (2010). Toxicity and cellular responses of intestinal cells exposed to titanium dioxide. *Cell Biology and Toxicology, 26*(3), 225–238. Springer.

Lin, S., et al. (2014). Zebrafish: an in vivo model for nano EHS studies. *Small, 9*(0), 1608–1618. doi: 10.1002/smll.201202115.Zebrafish.

Lomer, M. C. E., Thompson, R. P. H., & Powell, J. J. (2002). Fine and ultrafine particles of the diet: influence on the mucosal immune response and association with Crohn's disease. *Proceedings of the Nutrition Society, 61*(1), 123–130. Cambridge University Press.

Long, Z., et al. (2018). In vitro and in vivo toxicity evaluation of halloysite nanotubes. *Journal of Materials Chemistry B, 6*(44), 7204–7216. Royal Society of Chemistry.

Maisanaba, S., et al. (2014). In vivo toxicity evaluation of the migration extract of an organo-modified clay-poly(lactic) acid nanocomposite. *Journal of Toxicology and Environmental Health - Part A: Current Issues*, 77(13), 731–746. doi: 10.1080/15287394.2014.890987.

Maksimović, M., Omanović-Mikličanin, E., & Badnjević, A. (2019). Is nanofood safe? In *Nanofood and Internet of nano things* (pp. 87–137). Springer.

Malaekeh-Nikouei, B., et al. (2013). Assessment of cytotoxic properties of safranal and nano-liposomal safranal in various cancer cell lines. *Phytotherapy Research, 27*(12), 1868–1873. Wiley Online Library.

Manke, A., Wang, L., & Rojanasakul, Y. (2013). Mechanisms of nanoparticle-induced oxidative stress and toxicity. In *BioMed research international* (pp. 2013). Hindawi.

Martirosyan, A., Bazes, A., & Schneider, Y. -J. (2014). In vitro toxicity assessment of silver nanoparticles in the presence of phenolic compounds–preventive agents against the harmful effect? *Nanotoxicology, 8*(5), 573–582. Taylor & Francis.

Mc Clements, D. J., Li, F., & Xiao, H. (2015). The nutraceutical bioavailability classification scheme: classifying nutraceuticals according to factors limiting their oral bioavailability. *Annual Review of Food Science and Technology, 6*, 299–327. https://doi.org/10.1146/annurev-food-032814-014043.

McClements, D. J., & Xiao, H. (2017). Is nano safe in foods? Establishing the factors impacting the gastrointestinal fate and toxicity of organic and inorganic food-grade nanoparticles. *NPJ Science of Food, 1*(1), 1–13. doi: 10.1038/s41538-017-0005-1. Springer US.

Møller, P., et al. (2015). Applications of the comet assay in particle toxicology: air pollution and engineered nanomaterials exposure. *Mutagenesis, 30*(1), 67–83. Oxford University Press UK.

Nakayama, M. (2018). Macrophage recognition of crystals and nanoparticles. *Frontiers in Immunology, 9*, 103. Frontiers.

Navarro, M., & Wood, R. J. (2003). Plasma changes in micronutrients following a multivitamin and mineral supplement in healthy adults. *Journal of the American College of Nutrition, 22*(2), 124–132.

Oberdörster, G., et al. (2005). Principles for characterizing the potential human health effects from exposure to nanomaterials: elements of a screening strategy. *Particle and Fibre Toxicology, 2*(1), 8. BioMed Central.

Őrfi, E., & Szebeni, J. (2016). The immune systemof the gut and potential adverse effects of oral nanocarriers on its function. *Advanced Drug Delivery Reviews journal, 15*(106), 402–409. doi: 10.1016/j.addr.2016.09.009.

Orts-Gil, G., Natte, K., & Österle, W. (2013). Multi-parametric reference nanomaterials for toxicology: state of the art, future challenges and potential candidates. *RSC Advances, 3*(40), 18202–18215. Royal Society of Chemistry.

Panyala, A., et al. (2017). Assessment of genotoxicity and biodistribution of nano-and micron-sized yttrium oxide in rats after acute oral treatment. *Journal of Applied Toxicology, 37*(12), 1379–1395. Wiley Online Library.

Peng, Q., et al. (2014). Nanocomplex based on biocompatible phospholipids and albumin for long-circulation applications. *ACS Applied Materials and Interfaces, 6*(16), 13730–13737. ACS Publications.

Rashidi, L., & Khosravi-Darani, K. (2011). The applications of nanotechnology in food industry. *Critical Reviews in Food Science and Nutrition, 51*(8), 723–730. Taylor & Francis.

Ribeiro, A. R., et al. (2017). Challenges on the toxicological predictions of engineered nanoparticles. *NanoImpact, 8*, 59–72. Elsevier.

Ricci, J. -E., Gottlieb, R. A., & Green, D. R. (2003). Caspase-mediated loss of mitochondrial function and generation of reactive oxygen species during apoptosis. *The Journal of Cell Biology, 160*(1), 65–75. Rockefeller University Press.

Richard, J. (2017). Challenges in oral peptide delivery: lessons learnt from the clinic and future prospects. *Therapeutic Delivery, 8*(8), 663–684. Future Science.

Rubio, A. L., et al. (2019). Nanomaterials for food applications: general introduction and overview of the book. In *Nanomaterials for food applications* (pp. 1–9). Elsevier.

Rubió, L., Motilva, M. -J., & Romero, M. -P. (2013). Recent advances in biologically active compounds in herbs and spices: a review of the most effective antioxidant and anti-inflammatory active principles. *Critical Reviews in Food Science and Nutrition, 53*(9), 943–953.

Sayes, C. M., & Warheit, D. B. (2009). Characterization of nanomaterials for toxicity assessment. *Wiley Interdisciplinary Reviews: Nanomedicine and Nanobiotechnology, 1*(6), 660–670. Wiley Online Library.

Sharma, V., et al. (2012). Induction of oxidative stress, DNA damage and apoptosis in mouse liver after sub-acute oral exposure to zinc oxide nanoparticles. *Mutation Research/Genetic Toxicology and Environmental Mutagenesis, 745*(1–2), 84–91. Elsevier.

Sherin, S., et al. (2017). In vitro and in vivo pharmacokinetics and toxicity evaluation of curcumin incorporated titanium dioxide nanoparticles for biomedical applications. *Chemico-Biological Interactions, 275*, 35–46. Elsevier.

Shumakova, A. A., et al. (2015). Toxicological assessment of nanostructured silica. III. Microecological, hematological indices, state of cellular immunity. *Voprosy Pitaniia, 84*(4), 55–65.

Simón-Vázquez, R., et al. (2014). Conformational changes in human plasma proteins induced by metal oxide nanoparticles. *Colloids and Surfaces B: Biointerfaces, 113*, 198–206. Elsevier.

So, S. J., Jang, I. S., & Han, C. S. (2008). Effect of micro/nano silica particle feeding for mice. *Journal of Nanoscience and Nanotechnology, 8*(10), 5367–5371. American Scientific Publishers.

Sruthi, S., et al. (2018). Cellular interactions of functionalized superparamagnetic iron oxide nanoparticles on oligodendrocytes without detrimental side effects: Cell death induction, oxidative stress and inflammation. *Colloids and Surfaces B: Biointerfaces, 170*, 454–462. Elsevier.

Ting, Y., et al. (2014). Common delivery systems for enhancing in vivo bioavailability and biological efficacy of nutraceuticals. *Journal of Functional Foods*, 7, 112–128. doi: 10.1016/j.jff.2013.12.010. Elsevier Ltd.

Toda, T., & Yoshino, S. (2016). Amorphous nanosilica particles block induction of oral tolerance in mice. *Journal of Immunotoxicology, 13*(5), 723–728. Taylor & Francis.

Tzankova, V., et al. (2016). In vitro and in vivo toxicity evaluation of cationic PDMAEMA-PCL-PDMAEMA micelles as a carrier of curcumin. *Food and Chemical Toxicology, 97*, 1–10. Elsevier.

Vila, L., et al. (2018). Assessing the effects of silver nanoparticles on monolayers of differentiated Caco-2 cells, as a model of intestinal barrier. *Food and Chemical Toxicology, 116*, 1–10. Elsevier.

Vita, A. A., Royse, E. A., & Pullen, N. A. (2019). Nanoparticles and danger signals: Oral delivery vehicles as potential disruptors of intestinal barrier homeostasis. *Journal of Leukocyte Biology, 106*(1), 95–103. doi: 10.1002/JLB.3MIR1118-414RR.

Wang, Y., et al. (2014). A combined toxicity study of zinc oxide nanoparticles and vitamin C in food additives. *Nanoscale, 6*(24), 15333–15342. Royal Society of Chemistry.

Wilson-Sanders, S. E. (2011). Invertebrate models for biomedical research, testing, and education. *ILAR Journal, 52*(2), 126–152. Institute for Laboratory Animal Research.

Wolfram, J., et al. (2014). The nano-plasma interface: Implications of the protein corona. *Colloids and Surfaces B: Biointerfaces, 124*, 17–24. Elsevier.

Wu, H., et al. (2014). Reactive oxygen related activities of nano-iron metal and nano-iron oxides. *Journal of Food and Drug Analysis, 22*(1), 86–94. https://doi.org/10.1016/j.jfda.2014.01.007.

Yang, K., et al. (2013). In vivo biodistribution and toxicology of functionalized nano-graphene oxide in mice after oral and intraperitoneal administration. *Biomaterials, 34*(11), 2787–2795. Elsevier.

Yang, Y., et al. (2017). Toxicity assessment of nanoparticles in various systems and organs. *Nanotechnology Reviews, 6*(3), 279–289. De Gruyter.

Yu, X. -H., et al. (2017). Isolation, purification, characterization and immunostimulatory activity of polysaccharides derived from American ginseng. *Carbohydrate Polymers, 156*, 9–18. doi: 10.1016/j.carbpol.2016.08.092. Elsevier Ltd..

van der Zande, M., et al. (2012). Distribution, elimination, and toxicity of silver nanoparticles and silver ions in rats after 28-day oral exposure. *ACS Nano*, *6*(8), 7427–7442. ACS Publications.

Zhang, H., et al. (2012). Use of metal oxide nanoparticle band gap to develop a predictive paradigm for oxidative stress and acute pulmonary inflammation. *ACS Nano*, *6*(5), 4349–4368. ACS Publications.

Zhao, X., et al. (2019). Toxicity evaluation of one-dimensional nanoparticles using caenorhabditis elegans: a comparative study of halloysite nanotubes and chitin nanocrystals. *ACS Sustainable Chemistry & Engineering*, 7(23), 18965–18975. ACS Publications.

CHAPTER 5

Regulatory principles on food nano-particles legislated by European countries

Marjan Daeihamed[a], Masoud Faghih Akhlaghi[b], Seid Mahdi Jafari[c]
[a]Department of Pharmaceutics, School of Pharmacy, Guilan University of Medical Sciences, Rasht, Iran
[b]Department of Medicinal Chemistry, School of Pharmacy, Guilan University of Medical Sciences, Rasht, Iran
[c]Department of Food Materials and Process Design Engineering, Gorgan University of Agricultural Science and Natural Resources, Gorgan, Iran

5.1 Introduction

Nanotechnology has opened new horizons in different scientific and industrial fields and lots of research and investments have been focused in this area. The small particle size and significantly high surface area in the nanometer scale provides new characteristics and novel implications for compounds. Nanomaterials have also entered different fields of food industry to provide rapid and effective manufacturing and transport of food products. Nanotechnology has been used in numerous sectors of food chain from crop growth and pesticides to food production and packaging. However, going through the nanometer scale may amplify some other properties related to their toxicity and exposure, raising serious concerns regarding their safety for human and environmental health. The widespread use of nanotechnology necessitates standardization and regulation to ensure their safety. This issue is even more critical regarding food safety because it deals with everyday life and studies show a tremendous growth in their application.

Regulating any new technology requires exact and enough information about different aspects of the field. Our current understanding clearly reveals that the applications of nanotechnology in food may alter the toxicity profile and needs new safety considerations. However, highly diverse structure of nanomaterials with their unique characteristics makes it difficult to provide a harmonized and standard protocol to follow. Different countries and regulatory agencies have been alert about this global issue and the presence of nanomaterials in food products have been taken into consideration through various guidelines, legislation, and standard protocols but the

Safety and Regulatory Issues of Nanoencapsulated Food Ingredients
http://dx.doi.org/10.1016/B978-0-12-815725-1.00005-7

regulatory frameworks are not enough and needs to be updated according to the received information.

In spite of different limitations, the European Union (EU) was among the first regions, which provided rules, guidelines, and legislation regarding the applications of nanotechnology in food chain. The European Food Safety Authority (EFSA) is responsible for food safety in EU and provides updated information and guidelines in this regard. Different regulatory aspects for applications of nanotechnology in food sector through EU are described in this chapter.

5.2 Potential applications of nanotechnology in European food industries

Nanotechnology has revolutionized food industries from preparation and processing of food to packaging and consumption of products. It provides innovative and novel approaches toward high-speed manufacturing, increased production efficiency, cost-effectiveness, and better consumer satisfaction. Nanotechnology has a tremendous potential to be used in different steps of food production chain including agriculture, veterinary and animal feed, food processing and packaging (He, Deng, & Hwang, 2019, He & Hwang, 2016; Peters et al., 2016).

Several nanotechnology-based products are used in food science and industry worldwide. They are either available in the market or under development to be a commercial product, while there is a substantial increase in the number of research and publications in this field. This rapidly progressing technology has made a remarkable impact on the multi trillion-dollar food-related industries. Therefore, different countries invest considerable funds on nanotechnology with the European Union being among the countries on top of the list (Bugusu, Lay Ma, & Floros, 2011; Joseph & Morrison, 2006).

Food-related nanomaterials involve the application of nanotechnology in agriculture, feed and/or food sector. It was simply defined as "foods that are cultivated, produced, processed or packaged using nanotechnology techniques or tools, or to which manufactured nanomaterials have been added" in a nanoforum report from European Nanotechnology Gateway in 2006 (Joseph & Morrison, 2006) (A European nanotechnology network funded by the European Union to provide information on European nanotechnology).

The potential applications of nanotechnology in food chain production including agriculture, feed and food, has been descried thoroughly in the literature (He et al., 2019; He & Hwang, 2016; Peters et al., 2016). In agriculture, nanotechnology is employed for large scale production and farming of food materials with better quality and safety. Various organic and inorganic nanomaterials are used for increasing the efficiency of pesticides, herbicides, and fertilizers, controlled delivery of nutrients, purification of water and soil by nanofilters, and controlling the environmental conditions for crop growth using nano sensors (Parisi, Vigani, & Rodríguez-Cerezo, 2015; Usman et al., 2020).

Nanomaterials have shown promising and various potentials in the food processing and packaging procedure, as well. They are used as nano-additives to preserve the product from microbial or chemical degradation, or enhance food characteristics such as color, flavor, and texture. Nanoencapsulation can also improve the absorption and bioavailability of nutrients. Some other applications, categorized as food contact materials (FCM), which is defined as materials that come into contact with food during food production, processing, storage, and customer supply, involve novel packaging and edible coating methods resulting in better shelf life and the development of nanosensors to monitor packaged foods during transport and storage (Bajpai et al., 2018; Nile et al., 2020; Ravichandran, 2010).

Feasible applications of nanotechnology in European countries are the same as mentioned earlier (Sodano, Gorgitano, Quaglietta, & Verneau, 2016). EFSA published an inventory of current and future applications of nanotechnology in 2014. This inventory was prepared through an extensive literature search, questioners, and company websites and the list covers a total of 633 nanotechnology applications consisting of 55 different nanomaterials and 12 different applications. About 40% of this list consists of the products available on the market, 20% were under development, and the remaining 40% consists of potential applications published in research fields (Peters et al., 2014). A list of some nanotechnology-based food products is also available on "Nanotechnology in Food" database, created by Center for Food Safety (CFS) (Csf, 2018). However, having precise information regarding available products in the market is not completely practical.

Some of the important applications of nanotechnology in agriculture section among European countries include nanocapsule and nanoemulsion-based delivery of pesticides like Apron MAXX and Primo MAXX and nanomaterial-based delivery of fertilizers and nanoclays and sustained delivery of nutrients for better production of crops. Nanotechnology is also applied for water and soil purification and cleaning via cotton-based nanofibers or metal nanoparticles but they are not anticipated to enter the food chain. Another important section in food chain production is animal feeding. Nanoparticles are employed for dietary supplementation (including copper oxide and zinc oxide) or as bacteria and toxin absorbents in animal feeds. The examples include addition of nanosilver as antibacterial agent in drinking water of chickens and nanoclays to absorbent toxins in animal feed, that is, montmorillonite-based nanoclay for mycotoxin (Peters et al., 2014; Sodano et al., 2016; Csf, 2018; King, Osmond-Mcleod, & Duffy, 2018).

5.3 The development of regulations for food nanotechnology in European countries

Considering the raised safety issues and widespread implications of nanotechnology in food science and industries, proper regulation and legislations is demanded to ensure consumer and environmental safety as well as standard marketing of products. The approved regulations would also serve as official and trusted source for public information and awareness (Siegrist, Cousin, Kastenholz, & Wiek, 2007; Dingman, 2008). The development of legislations usually starts from one or more hints from research and investigations, which is usually suggested by academic parties to a government agency. The proposed suggestions will be known as legislations if passed by the legislative arm of a government and becomes a regulation when it is enforced by regulators (He et al., 2019; Corley, Scheufele, & Hu, 2009).

The evolvement of regulations for food nanomaterials was not as rapid as its corresponding technological developments and commercialization rate. This process faced challenges due to some important scientific uncertainty and data limitations (Corley et al., 2009; Cushen, Kerry, Morris, Cruz-Romero, & Cummins, 2012). The versatile and complex structure of food nanomaterials, lack of proper data about their potential toxicities and biodistribution, problems related to standard characterization of nanomaterials and the appropriate differentiation between natural and engineered nanoparticles, as well as the absence of unified and standard extraction and analytical methodologies for risk assessment were the main obstacles in this regard (Dingman, 2008; Tiede et al., 2008; Cushen et al., 2012; Magnuson, Jonaitis, & Card, 2011).

The European approach was toward regulation and control of nanotechnology-based products within the existing regulatory frameworks and they were considered comprehensive enough to cover nanotechnology-related products as well as conventional foods (Sodano et al., 2016; Amenta et al., 2015). In the first year, just a few manufacturers clearly made nanotechnology claims and the consumers in Europe were not completely aware of presence of nanomaterials in food products (Chau, Wu, & Yen, 2007). However, nanomaterials were shown to have different toxicological properties from their conventional counterparts that were significantly related to their physicochemical properties and nanotechnology-related considerations were necessary within the existing regulations (Coles & Frewer, 2013; Park & Yeo, 2016).

In 2008, a legislative review was adopted by European Commission (EC) through a "Communication on the Regulatory Aspects of Nanomaterials" to check the appropriateness of existing regulations for nanotechnology (European Commission, 2008a). Completed by a report published 1 year later as a scientific opinion on "The Potential Risks Arising from Nanoscience and Nanotechnologies on Food and Feed Safety" from the Scientific Committee of EFSA (EFSA, 2009), it was concluded that legislations required modifications for nanotechnology in food science. In the same year, the European Parliament wanted the Commission to "review all relevant legislation within two years to ensure that legislative provisions and instruments of implementation reflect the particular features of nanomaterials to which workers, consumers and/or the environment may be exposed" (European Parliament, 2009).

After 2 years in 2011, EFSA published a "Guidance on the risk assessment of the application of nanoscience and nanotechnologies in the food and feed chain" (EFSA, 2011), compromised of practical approaches to assess potential risks of food-related nanomaterials. It provides some information for applicants regarding physicochemical characterization and testing approaches to submit dossiers related to application of nanotechnology in food industries. EU and Switzerland were among the first countries that considered nanotechnology-related approaches in food legislation, including information regarding risk assessment and/or legal definitions of the term "nanomaterial" (Amenta et al., 2015).

EFSA is positioned to precisely follow any progress in applications of nanotechnology in food and feed sector and will provide specific guidance when possible through proper co-operation and networking with Member States and the updates are reported annually as "Annual report of the EFSA Scientific Network of Risk Assessment of Nanotechnologies in Food and Feed." The latest guidance was published in 2018 as "Guidance on risk assessment of the application of nanoscience and nanotechnologies in the food and feed chain: Part 1, human and animal health" (EFSA, 2018b), considering new developments since publication of the previous Guidance in 2011. It provides updated perception about physicochemical properties, exposure, and hazard assessment of food nanomaterials.

5.3.1 Different points to be considered for the development of regulations

As described previously, the applications of nanomaterials in different sectors of food production chain results in existence of highly diverse and complex nanomaterials in food industry. The rate and extent of possible toxicities and/or efficacy of nanomaterials depend on different factors like physicochemical characteristics, exposure routes, and rate of exposure. The composition and physicochemical properties of nanoparticles including particle size, shape, crystal structure, and surface properties can highly affect their toxicity outcomes. On the other hand, different step of manufacturing process and complex media of food products can result in various physicochemical characteristics for particles of the same composition. It is therefore necessary to exactly consider and measure the affecting parameters and assess possible toxicities via validated methods in regulatory frameworks (Bouwmeester et al., 2009; Tiede et al., 2008). Regulations are supposed to be transparent and clear to ensure food safety, hence different aspects of diversity and complexity should be considered to provide precise information. Valid and consistent manufacturing of products and environmental safety are other important factors, which can be ensured within the regulatory frameworks. Some important points that are necessary in development of regulations are described in this section.

Particle size is the initial and most important determinant to elucidate whether a product is considered nanomaterial or not. Therefore, a size limit should be clearly defined in food regulations. It is also necessary to consider other contributing factors like measurement methods and apparatus, sampling and sample preparation techniques, as well as type of data interpretation which is often reported as number distribution or volume/mass distribution. It is usually recommended to use more than one technique, one of them being an imaging technique like electron microscopy (Bleeker et al., 2013; Calzolai, Gilliland, & Rossi, 2012; Sharma, Jaiswal, Duffy, & Jaiswal, 2019). Particle shape and porosity, surface area, surface charge, preparation method (top-down or bottom-up approaches) and general structure of particles (like thickness and surface composition) are other important factors to be considered as important physicochemical characteristics of nanomaterials in development of legislations (Jeevanandam, Barhoum, Chan, Dufresne, & Danquah, 2018; Pettitt & Lead, 2013).

As the particle size and other physicochemical characteristics of nanomaterial are likely to change during food processing and storage (i.e., formation of agglomerates), it would be better to consider in which step of production including raw materials, intermediates, or final products the measurement of characteristics should be applied. The possible production of residual harmful components from nanomaterials during processing and manufacture should be considered as well (Chau et al., 2007; Contado, 2015).

In addition to physicochemical characteristics of nanomaterials, the regulatory framework should provide exact information regarding toxicity assessments (i.e., hazard identification and risk assessments), depending on exposure routes and the proposed application of nanomaterial. The food-related nanomaterial can exhibit direct or indirect exposure. Direct exposure is usually related to feed or food additives while indirect exposure occurs via FCMs having migration potential from the packaging materials into food materials. All the applied methods and instruments should meet the requirements of standardization and validation according to guidelines (Warheit, 2018; Bouwmeester et al., 2009; EFSA, 2018b).

Oral route is the main exposure route in nanofoods while dermal and inhalation exposures are also possible in some cases including the application of pesticides and feed additives. The quantity of exposure and possible digestion of nanomaterials in gastrointestinal tract are some determining factors estimating the toxicity. Generally, the potential risk of the toxicity, biodistribution and accumulation of nanomaterials in the body should be evaluated via proper *in vitro* and *in vivo* studies; as described in Chapters 3 and 4, respectively. The results of hazard characterization and exposure assessment are combined to provide quantitative, and in some cases qualitative information named as risk characterization and gives an overall assessment of product safety (Wang, Du, Song, & Chen, 2013; Martirosyan & Schneider, 2014).

5.4 Available legislations and guidance related to nanomaterials in European countries

Although initially food-relevant regulations were considered comprehensive enough for nanotechnology-based products, but some safety concerns necessitated modification of current legislations. The EU was among the leading and first global regions that paid attention to nanotechnology-related issues in food regulations (Amenta et al., 2015). EU regulations that cover applications of nanomaterials in food industry exhibit in both horizontal and vertical food-related regulations. The horizontal regulations are usually applicable to all products while vertical regulations are related to specific food sector (Sodano et al., 2016). These legislations are available online at https://eur-lex.europa.eu.

Under EU regulations, all food products have to meet the general safety requirement under the General Food Law legislation (EC/178/2002). However, there is no reference or definitions directly related to nanomaterials in this document (European Commission, 2002b). Food-related nanomaterials may be related to REACH (Registration, Evaluation, Authorization, and restriction of Chemical substances) regulations, which is concerned with safe use of chemicals manufactured or imported in amounts more than 1 ton per year in EU (European Commission, 2006a). However, REACH generally deals with FCMs and not nanomaterials directly used in food products (Jafari, Katouzian, & Akhavan, 2017). REACH has recently adopted revisions by EU commission regarding registration of dossiers for nanomaterials.

There are also some specific nanotechnology considerations in legislations related to food additives, food enzymes, flavorings, food supplements, novel foods, and FCMs. Regulation No 1333/2008 on food additives (European Commission, 2008c), mentions re-evaluation and pre-market authorization of food additives that are already authorized but change through nanotechnology before commercialization of product. Although a definition of nanomaterial is not specifically given in this document.

According to Article 12 of this regulation:

"When a food additive is already included in a Community list and there is a significant change in its production methods or in the starting materials used, or there is a change in **particle size, for example through nanotechnology**, the food additive prepared by those new methods or materials shall be considered as a different additive and a new entry in the Community lists or a change in the specifications shall be required before it can be placed on the market."

Regulation No 1332/2008 on food enzymes (European Commission, 2008b), states that "a food enzyme already included in the Community list but prepared by different methods or using starting materials significantly different from those included in the risk assessment of the Authority, should be submitted for evaluation by Authority." As changes in the particle size has been clearly indicated as "significantly different" in this documents, it will cover the application of nanomaterials in the food enzymes as well, although it is not directly referred to nanomaterials or nanotechnology.

Regulation No 1334/2008 on flavorings and certain food ingredients with flavorings properties (Article 19) (European Commission, 2008d), has mentioned submission of the necessary data to the Commission if an existing flavor is prepared by production methods or starting materials significantly different from those included in the risk assessment of the Authority.

The most recent and relevant regulation directly considering nanomaterials in food is the Regulation No 2015/2283 on novel foods (European Parliament, 2015). Novel foods are defined as "any food that was not used for human consumption to a significant degree within the Union before 15 May 1997." According to this definition foods containing engineered nanomaterials are considered to be novel food and are exactly defined in Article 3. Moreover, vitamins, minerals, or other substances used in accordance with Directive 2002/46/EC (European Commission, 2002a) and Regulation No 1925/2006 (Regulation on addition of vitamins and minerals and of certain other substances to foods) (European Commission, 2006b), or food for specific groups like infants and young children under Regulation No 609/2013 (European Parliament, 2013) which contain or consist of engineered nanomaterials, are also considered novel foods and should be assessed first in accordance with this Regulation and subsequently in accordance with the relevant specific legislation. According to this regulation, any product containing engineered nanomaterials, which is considered novel food under these definitions require a premarket authorization and proper labeling. The necessary information regarding methods and requirements for risk assessment are available in the "Guidance on risk assessment of the application of nanoscience and nanotechnologies in the food and feed chain" published by EFSA in 2018 (EFSA, 2018b).

Regulation No 609/2013 on food intended for infants and young children, food for special medical purposes, and total diet replacement for weight control (European Parliament, 2013) which is now integrated in the Novel Foods Regulation, has mentioned engineered nanomaterials in Article 2 and according to these regulations substances with a change in particle size, like nanomaterials, are considered to be novel foods and needs premarket authorization and re-evaluation.

Concerning the FCMs there are two regulations namely Regulation No 450/2009 (European Commission, 2009) and Regulation No 10/2011 (European Parliament, 2011a). Migration of these materials is the most important point to take into consideration. The migration should occur below the safety limits and should not bring unfavorable effects on human health or quality of the food. In Article 5 of Regulation No 450/2009 on active and intelligent materials and articles intended to come into contact with food (European Commission, 2009), it is referred to "substances deliberately engineered to particle size which exhibit functional physical and chemical properties that significantly differ from those at a larger scale," and generally a case-by-case risk assessment has been recommended.

Regulation No 10/2011 on plastic materials and articles intended to come into contact with food (European Parliament, 2011a), refers to "nanoparticles" and "substances in nanoform" in Articles 9, 12, and 13. According to this regulation only the substances included in the authorized list of substances in Annex I can be used in the manufacture of plastic materials and "authorizations based on the risk assessment of the conventional particle size of a substance do not cover engineered nanoparticles." It has been stated that "substances in nanoform shall only be used if explicitly authorized and mentioned in the specifications in Annex I," which requires safety assessment by the relevant EFSA Panel.

In addition to the mentioned legislations, there are also soft law frameworks including guidelines, communications, commentaries, recommendations, opinions and reports with a weak legal force which are updated thoroughly and frequently and usually published by EFSA (Sodano et al., 2016). EFSA basically deals with scientific information on any risks regarding food chain and helps regulatory-related decision makers in EU to ensure safety in food chain and proper health of consumers all around Europe. It provided the first guidance regarding nanotechnologies in the food and feed chain in 2011 (EFSA, 2011), which was updated in 2018 (EFSA, 2018b). The guidance contains important information about engineered nanomaterials in food chain and gives exact information for analysis and risk assessment of nanomaterials in food and feed. The results of activities and investigations by involvement of member states are reported as "Annual report of the EFSA Scientific Network of Risk Assessment of Nanotechnologies in Food and Feed" each year (EFSA, 2016; EFSA, 2018a). All the updated information and the mentioned guidelines and reports are available at http://www.efsa.europa.eu/

5.4.1 Definition of nanomaterials

In order to develop transparent and harmonized regulatory approaches for food-related nanotechnology, it is necessary to determine a definition of the terms used in legislations including nanomaterial, engineered nanomaterial, nanoform, and nanoparticle. It will help to understand whether a material is regarded as a nanomaterial in the context of food regulations.

In 2011, a definition of the term "nanomaterial" was defined in a Recommendation adopted by European Commission (European Commission, 2011):

"A natural, incidental or manufactured material containing particles, in an unbound state or as an aggregate or as an agglomerate and where, for 50% or more of the particles in the number size distribution, one or more external dimensions is in the size range 1-100 nm."

Although this definition provided a harmonized approach but "size" is the only characteristic that has been considered for all nanomaterials. However, food-related regulations are generally related to the assessments of hazards and toxicities, and biological toxicity is not strictly related to a specific size limit. Therefore, it demands considering sector specific characteristics and requirements in food legislations (Martirosyan & Schneider, 2014).

In the same year, nanotechnology-related definitions were addressed in some vertical food legislations as well. Regulation No 1169/2011 on Food Information to Consumers (European Parliament, 2011b), provided a definition of "engineered nanomaterials" for labeling purposes. Regulation on Plastic Materials in Contact with Food (Regulation No 10/2011) (European Parliament, 2011a) referred to "nanoparticles" and "nanoform" but there was not any definition of the given terms.

The Novel Foods Regulation (European Parliament, 2015), used the definition proposed by Regulation on Food Information for Consumers for "engineered nanomaterial" to provide consistency in the area of food law:

"Any intentionally produced material that has one or more dimensions of the order of 100 nm or less or that is composed of discrete functional parts, either internally or at the surface, many of which have one or more dimensions of the order of 100 nm or less, including structures, agglomerates, or aggregates, which may have a size above the order of 100 nm but retain properties that are characteristic of the nanoscale. Properties that are characteristic of the nanoscale include: (i) those related to the large specific surface area of the materials considered; and/or (ii) specific physicochemical properties that are different from those of the non nanoform of the same material."

In this definition the size range is not restricted to a rigid value and allows consideration of other additional physicochemical properties to identify nanomaterials. The term "engineered" is also defined by "intentionally produced" which helps to distinguish between natural and engineered nanomaterials.

According the latest EFSA guidance (EFSA, 2018b), food-related "engineered nanomaterials" that are considered for risk assessment within the scope of guidance consists of following materials:

1. Engineered nanomaterials that have particle sizes in the defined range (1–100 nm).
2. Particles having a size above 100 nm which exhibit the characteristic properties of the nanoscale like large specific surface area leading to a significant change in toxicokinetics.
3. Materials that are not engineered as nanomaterial but contain a fraction of particles less than 50% in the number–size distribution with one or more external dimensions in the size range of 1–100 nm. This condition usually occurs during manufacturing of powders and although the median size of the particles is significantly greater than 100 nm, but contains a small fraction is present with at least one dimension below 100 nm.
4. Nanomaterials having the same elemental composition but that occur in different morphological shapes, sizes, crystalline forms and/or surface properties.
5. Nanoscale entities made of natural materials that have been deliberately produced to have nano-enabled properties like encapsulating (bioactive) compounds.

Other pieces of EU legislation relevant to food have introduced terms such as "nanotechnology," "nanomaterial," "nanoparticle," or general references to engineered materials with different chemical or physical properties but do not define these terms.

5.4.2 Exposure and risk assessments

Similar to conventional food, the marketing and authorization of nanofoods require risk assessment to ensure the safety of products for consumers. The process of risk assessment involves identification of potential hazards in the product and analysis of the risks associated with that hazard. The probability of this risk depends on the amount that consumers are exposed to that hazard, which is generally related to the exposure route and level (presented as dose and duration of exposure). Generally, there are three main steps for risk assessment of food products: hazard identification and characterization, exposure assessment, and risk characterization (Warheit, 2018; Magnuson et al., 2011).

The latest EFSA guidance (EFSA, 2018b) provides a practical approach and guide for manufacturers and regulators regarding the risk assessment of engineered nanomaterials in food products. The risk assessment process involves physicochemical characterization of nanomaterial, its interactions with tissues, and potential exposure levels. Physicochemical characterization helps to decide whether it is considered nanofood or not. This information is useful for hazard characterization as well and combined with the exposure assessment, will form the basis for the risk characterization. If the product is not considered "nano" according to physicochemical characterizations or there are sufficient evidence showing that there is no migration from FCMs to the product or the nanomaterial is degraded completely in the body, it is either considered safe or can be treated as a conventional material.

5.4.2.1 Physicochemical characterization of engineered nanomaterials

There are some characteristics of nanomaterials which may influence their toxicity and toxicokinetic behavior. These effects may significantly alter the potential risk of these particles compared to their conventional counterparts and consequently affect the results of risk assessment (Martirosyan & Schneider, 2014; Pettitt & Lead, 2013).

EFSA guidance has mentioned some important characteristics to be considered for engineered nanomaterials. The physicochemical parameters should be assessed by validated methods and analytical techniques and are an essential requirement for all dossiers related to food nanomaterials (EFSA, 2018b).

The physicochemical characterization of nanomaterials will consider following issues:

1. Whether the material is subjected for nanospecific risk assessment,
2. Determination of the physical and chemical properties of nanomaterial,
3. Determination of the characteristics of nanomaterial in the media that is used for *in vitro* and *in vivo* toxicological studies, and
4. Physicochemical characterization of the material in complex matrices for implications in exposure assessment.

Different physical and chemical properties are required for engineered nanomaterials including the composition and purity, particle size, and other physical characteristics like particle shape and porosity, surface area, surface charge, etc. Table 5.1 lists some of the typical and important characteristics which are necessary to be analyzed and reported.

Particle size is the main determining parameter in the assessment of nanomaterials. The EFSA guidance recommends reporting the size of particles, agglomerates, and aggregates based on number and mass distribution. It has been shown that the measuring technique significantly influences the results of size determination because techniques address various size-related

Table 5.1 Required parameters for the physicochemical characterization of food-related nanomaterials. *(Adopted from (EFSA, 2018b) with some modifications).*

Parameter	Description
Composition	Relative amounts of components, purity, and impurities should be reported in mass%, as well as its elemental composition
Particle size	The average particle size of nanomaterials and related agglomerates and aggregates should be reported as number and mass-based size distribution, measured by more than one independent technique, one of them being electron microscopy
Particle shape	Describing the particle shape (supplemented by microscopic images), aspect ratio
Structure	Spatial description of nanomaterial (i.e., core-shell, homogeneous, thickness of elements) which can be presented by a graphical sketch
Surface composition	Explanations of functional groups or coating on particle surface
Surface area	Mass and volume specific surface area and measurement method (i.e., BET)
Surface charge	The zeta potential and measurement conditions like pH and ionic strength should be reported
Porosity	Fraction of volume of voids in total volume (gas adsorption measurements)
Production process	The preparation method of nanomaterials
Physical characteristics	Physical appearance, melting and boiling point, dustiness
Other properties	pH, stability, solubility, degradation rate, dispersibility, chemical reactivity

parameters including hydrodynamic or geometric diameter. Moreover, the experimental conditions like efficient dispersion and the refractive index of media in Dynamic Light Scattering method may affect the results. Consequently, the measurement should be conducted at least by two techniques, one of them being electron microscopy or imaging techniques. In addition to the mentioned physical and chemical parameters, some other characteristics like stability, solubility and degradation rate, dispensability should also be reported (EFSA, 2018b). The physicochemical characteristics must be evaluated in the following conditions:

1. The original form
2. In matrix of food/feed product
3. In the relevant biological media used for *in vitro* and *in vivo* tests
4. In FCM and after migration

The applicants must also provide information regarding validation and quality assurance of methods and the performance parameters (sensitivity and precision) must meet the requirements. A list of standard methods is available in EFSA guidance. Searching through the ISO and CEN databases also provides up-to-date and appropriate methods. In case the standard method was not available, household methods can be devolved but documents regarding the validation process should be submitted.

The "Guidance on risk assessment of the application of nanoscience and nanotechnologies in the food and feed chain" provides comprehensive information regarding analysis of physicochemical parameters, proper methods and how to report the results (EFSA, 2018b).

5.4.2.2 Exposure assessment

In addition to the amount of nanomaterials in food products, the level of exposure is dependent on the type of nanomaterial and its corresponding application (Kuhlbusch, Wijnhoven, & Haase, 2018).

The EFSA guidance have considered direct and indirect exposure of human or animals to food-related nanomaterials depending on their indicated application. Novel food and food additives usually result in the direct exposure while indirect exposure include the use of FCM or transfer by from feed, or pesticide or even as a contaminant (Peters et al., 2016; EFSA, 2018b).

In either direct or indirect exposure, the assessment process should be designed in a way that helps to estimate the exposure level. The EFSA guidance provides a flowchart for the assessment of oral exposure. In applications leading to direct exposure of nanomaterial two questions are necessary to be answered:

1. Does the nanomaterial remain stable in food/feed product or it is degraded?
2. Is the material degraded under the *in vitro* simulated conditions of the gastrointestinal (GI) tract?

In the case of indirect exposure, the following questions should be answered:

1. Does the nanomaterial migrate to food/feed product?
2. Is the material degraded under the *in vitro* simulated conditions of the gastrointestinal (GI) tract?

If the evaluations show degradation of nanomaterial in food or GI tract in a way that no particle in nanoscale exists, or the migration is just related to solutes and not particles, the risk assessment should be followed as EFSA guidance for conventional materials.

A case-by-case analysis is recommended if migration of a nanomaterial occur. In cases where determination of nanomaterials is not possible in complex matrices, the worst-case should be considered as all nanomaterial is present and is ingested and absorbed. The assessment should consider special population that is expected to have high exposures (i.e., infant formulas).

5.4.2.3 Hazard identification and characterization

A stepwise approach has been proposed for hazard identification and characterization in EFSA guidance to avoid unnecessary tests for nanomaterials. First, *in vitro* GI digestion should be evaluated (Step 0). If the nanomaterial does not degrade quickly, a review of the existing data must be conducted (Step 1a) and some *in vitro* tests including stability in lysosomal conditions, *in vitro* genotoxicity and cell toxicity should be done (Step 1b). If the results indicate no potential toxicity, waiving of *in vivo* studies can be reviewed by EFSA based on a case-by-case approach. However, if the results of Step 1 show the presence of toxicity, *in vivo* studies including genotoxicity, 90-day oral toxicity, and biodistribution studies are suggested to further evaluate the toxicity (Step 2). It should be noted that in most cases it is usually needed to go through Step 2, because the results of *in vitro* tests don't lead to conclusive evidence for toxicity. According to the results of Step 2, further toxicological evaluations may be required (Step 3). The additional tests include toxicokinetic studies, reproductive and developmental, immuno and neurotoxicity, carcinogenicity or mutagenicity, endocrine effects, and gut microbiome (EFSA, 2018b). A general description of proposed tests in each step are outlined in Table 5.2.

5.4.2.4 Risk characterization and uncertainty analysis

In this step, the results of hazard characterization are combined with exposure assessment to give a quantitative data. If quantitative assessments are not possible the qualitative information is reported. The output will give us an overall assessment of the safety of the nanomaterial in its intended application considering the validation parameters and the uncertainties related to the risk assessment protocol (Warheit, 2018).

The uncertainty factors for a nanomaterial can be the same as conventional material if validated and nanospecific methods were used. However, in the cases where data is inadequate, like limited toxicological information or lack of validated methods for nanomaterials, additional uncertainty factors may be considered and the limitations should be highlighted (EFSA, 2018b).

Table 5.2 A general description of the tests needed for different steps of hazard identification and characterization of nanomaterials.

Hazard identification step	Proposed tests	Experimental conditions
Step 0	*In vitro* GI digestion	• The experimental conditions should mimic physiologic state of exposure including temperature, pH, mixing rate, presence of enzymes and bile salts. • Reliability and reproducibility should be assessed. • The study is performed at three different concentrations with the middle concentration being representative of human exposure according to daily intake and volume of secretions in GI tract. • The particle number–size distribution and concentration should be analytically determined. • If 12% or less of the material is present as particles after 30 min, the nanomaterial is considered quickly dissolving.
Step 1	Stability in lysosomal fluid	• The *in vitro* dissolution rate profile is evaluated in simulated lysosomal fluid (buffered at pH 4.5–5.0) at least in four different time points up to 72–96 h and at three different concentrations tested.
	In vitro genotoxicity	• Three critical genotoxicity endpoints including gene mutation, structural, and numerical chromosome aberrations should be considered. • Bacterial reverse mutation (Ames) assay and *in vitro* mammalian micronucleus test is recommended for *in vitro* detection of gene mutations and structural and numerical chromosome damage, respectively.
	In vitro cell toxicity	• Specific endpoints to be considered include cytotoxicity/cell viability, induction of oxidative stress, (pro-) inflammation, and gastrointestinal barrier integrity impairment.
Step 2	*In vivo* genotoxicity	• *In vivo* micronucleus test, mammalian alkaline comet assay, and transgenic rodent somatic and germ cell gene mutation assay are recommended for *in vivo* tests. • The bacterial reverse mutation (Ames) assay is not considered suitable for nanomaterials due to limitations in the penetration of particles through the bacterial cell wall and the lack of internalization in bacteria.

(*Continued*)

Table 5.2 A general description of the tests needed for different steps of hazard identification and characterization of nanomaterials. (*Cont.*)

Hazard identification step	Proposed tests	Experimental conditions
	Toxicokinetics (ADME)	• Performing a pilot study in Step 2a is recommended for dose ranging. • An appropriate measurement system should be used. • Tissue distribution should be obtained before dosing, after a short period (14 days) of dosing, and at the end of 90-day toxicity study. Any significant increase in tissue concentration between days 14 and 90 may require further assessment in Step 3.
	In vivo repeated-dose 90-day oral toxicity study	• Cardiovascular and inflammatory parameters and organs involved in MPS are important in repeated-dose studies.
Step 3	Chronic toxicity and carcinogenicity	• Chronic toxicity and carcinogenicity studies are usually performed in rat and testing in a second species would be recommended when it is required to observe the mechanism of toxicity.
	Reproductive and developmental toxicity	• Developmental neurotoxicity, immunotoxicity, and endocrine disruption are specific endpoints to be considered. • Tests should include a prenatal developmental and prenatal and postnatal reproductive developmental toxicity study.
	Immunotoxicity and allergenicity	• Assessing the potential of nanoparticles to carry immunogens/antigens and conducted when the nanomaterial is a potential allergen or contains allergenic entities.
	Neurotoxicity	• If experiments in Step 2 show some potential neurotoxicity, further neurotoxicity testing should be considered on a case-by-case approach.
	Endocrine activity	• The starting point is modified 90-day toxicity test. The additional parameters to emphasis on endocrine endpoints include determination of thyroid hormones, gross necropsy, and histopathology of tissues and assessment of oestrous cycles.
	Gut microbiome	• Evaluating the interactions of nanomaterials with the gut microbiome especially for those that have antimicrobial effects.

5.5 Labeling and reporting

The widespread potentials and proper efficacy of nanotechnology in food industry resulted in rapid entrance of nanotechnology-based products in the market. This rapid entrance and limited data in the first year resulted in consumption of these products by consumers while they were not informed about the addition of nanoparticles. The consumers should be clearly aware of what they consume from both ethical and legislative point of view and proper and transparent labeling should be mandatory.

During the first year, there were no labeling requirements for nanomaterials in food products through EU. The Regulation 1169/2011 on Food Information to Consumers (European Parliament, 2011b) which deals with "the general principles, requirements and responsibilities governing food information, and in particular food labelling" has mentioned nanomaterials according to Novel Food Regulation and stated that "All ingredients present in the form of engineered nanomaterials shall be clearly indicated in the list of ingredients. The names of such ingredients shall be followed by the word 'nano' in brackets." These requirements were restricted to foods sold within the EU from 2014.

The labeling clearly informs the consumers about the contents and is intended to help them have informed choices and there is no need to have any information on risks. However, successful, clear, and mandatory labeling faced some limitations due to the gaps in current technical and regulatory issues which needs to be elucidated.

Exact labeling requires clarification and harmonization of the regulatory definitions of the term "engineered nanomaterials," providing convenient commercialization of products in different countries in the world. Another requirement is implications of reliable techniques to detect and characterize different nanomaterials in food products. Providing a proper measuring method and reference standards is also challenging because nanomaterials have diverse structures and their characterization in complex food matrices may face some limitations. Moreover, the physicochemical properties of nanomaterial may change during food processing like aggregations or aging. It is also necessary to exactly define which nanomaterials can be exempt from labeling requirements, like some food additives that has been used for decades (Grieger, Hansen, Mortensen, Cates, & Kowalcyk, 2016). However, like any other regulatory item, the labeling requirements of nanotechnology-related food products will be updated according to upcoming information in the coming years.

5.6 Challenges for risk assessment and risk management

The applications of nanotechnology in food products is relatively a new field and like any novel technology, development and statements of regulations and legislations require data entry from scientific evaluations, experiments from market and regulatory organizations. Although some standards, guidelines, and legislations were provided in EU, but there are still some gaps and limitations which necessitate further evaluations.

It has been globally accepted that nanomaterials exhibit potential toxicities and their toxicity profile is significantly different from conventional substances. Besides, the toxicological properties of nanomaterials are highly dependent on their physicochemical properties including particle size and shape, solubility, etc. (Martirosyan & Schneider, 2014). It is very important to take these differences into consideration and provide a clear definition of "engineered nanomaterials" in food products to ensure that all nanomaterials are assessed for possible risks. There are also some problems related to distinctions between natural and engineered nanomaterials in food products (Bleeker et al., 2013; Tiede et al., 2008). Different nanomaterials are used in food products with very diverse physicochemical properties, structures, and compositions. This makes it difficult to provide general methods for their characterization or risk assessments and usually a case-by-case evaluation is recommended. Although some analytical methods were suggested for characterization of nanomaterials but further research and work is needed to incorporate standard and validated methods in legislative framework. There are difficulties in measuring highly diverse characteristics, extraction, and analysis of nanomaterials in complex food matrices, and lack of reference standards. Moreover, the properties of nanoparticles may change in food matrices during time, for example, they can react with food components, form aggregations or change in size (Wang et al., 2013; Contado, 2015; Pettitt & Lead, 2013).

There is also a lack of proper data regarding toxicological assessment of nanomaterials. It is not clear whether the current *in vitro* and *in vivo* methods are suitable for nanomaterials. As these tiny particles exhibit variable toxicokinetic properties according to their physicochemical properties, there is a debate on the *in vivo* accumulation and toxicity profile of different nanomaterials (Warheit, 2018).

In EU, most of regulatory aspects of nanomaterials are considered and discussed according to currently available scientific data. The guidelines and legislations are published and updated thoroughly, but there is still a far way to go.

5.7 Conclusions and future remarks

Nanotechnology is increasingly introduced into food industries and had influenced this huge industry from production and processing to packaging and transportation having significant effects on quality of food, changing the nutritional values and shelf life of products. It is estimated that more commercial nanotechnology-related food products will enter the market with an increased rate (Nile et al., 2020; Bajpai et al., 2018; He et al., 2019). Consequently, the human and ecosystem exposure to nanomaterials will also increase dramatically and a standardized regulatory system for applications of nanotechnology in food is demanded. Although the development of legislations was not as fast as it was expected in the initial years, but different countries and world regions have been active in this regard with the EU being among the leading organizations. The European commission and EFSA are responsible for evaluations of food-related safety issues and some guidelines, definitions, and regulatory considerations were provided regarding food-related nanomaterials.

The evolvement of transparent regulations will help the industries and manufacturers in development of safe and standard products. Additionally, the public acceptance of nanotechnology-applied food products will depend on their safety.

Current regulation of nanotechnology has some limitations and faced some challenges including analytical and risk assessment issues. However, updating the existing information will support the development of harmonized regulations and may facilitate global and safe consumption of these products.

References

Amenta, V., Aschberger, K., Arena, M., Bouwmeester, H., Botelho Moniz, F., Brandhoff, P., et al. (2015). Regulatory aspects of nanotechnology in the agri/feed/food sector in EU and non-EU countries. *Regulatory Toxicology and Pharmacology*, *73*, 463–476.

Bajpai, V. K., Kamle, M., Shukla, S., Mahato, D. K., Chandra, P., Hwang, S. K., et al. (2018). Prospects of using nanotechnology for food preservation, safety, and security. *Journal of Food and Drug Analysis*, *26*, 1201–1214.

Bleeker, E. A., De Jong, W. H., Geertsma, R. E., Groenewold, M., Heugens, E. H., Koers-Jacquemijns, M., et al. (2013). Considerations on the EU definition of a nanomaterial: science to support policy making. *Regulatory Toxicology and Pharmacology*, *65*, 119–125.

Bouwmeester, H., Dekkers, S., Noordam, M.Y., Hagens, W. I., Bulder, A. S., De Heer, C., et al. (2009). Review of health safety aspects of nanotechnologies in food production. *Regulatory Toxicology and Pharmacology*, *53*, 52–62.

Bugusu, B., Lay Ma, U.V., & Floros, J. D. (2011). Products and their commercialization. In L. J. Frewer, W. Norde, A. Fischer, & F. Kampers (Eds.), *Nanotechnology in the Agri-Food Sector*.

Calzolai, L., Gilliland, D., & Rossi, F. (2012). Measuring nanoparticles size distribution in food and consumer products: a review. *Food Additives and Contaminants. Part A, Chemistry, Analysis, Control, Exposure and Risk Assessment, 29*, 1183–1193.
Chau, C. -F., Wu, S. -H., & Yen, G. -C. (2007). The development of regulations for food nanotechnology. *Trends in Food Science and Technology, 18*, 269–280.
Coles, D., & Frewer, L. J. (2013). Nanotechnology applied to European food production—A review of ethical and regulatory issues. *Trends in Food Science and Technology, 34*, 32–43.
Contado, C. (2015). Nanomaterials in consumer products: a challenging analytical problem. *Frontiers in Chemistry, 3*, 48.
Corley, E. A., Scheufele, D. A., & Hu, Q. (2009). Of risks and regulations: how leading US nanoscientists form policy stances about nanotechnology. *Journal of Nanoparticle Research, 11*, 1573–1585.
Csf, C. F. F. S. (2018). *Nanotechnology in our food: Interactive database of consumer food products containing nanomaterials. Available from: https://www.centerforfoodsafety.org/nanotechnology-in-food.*
Cushen, M., Kerry, J., Morris, M., Cruz-Romero, M., & Cummins, E. (2012). Nanotechnologies in the food industry—Recent developments, risks and regulation. *Trends in Food Science and Technology, 24*, 30–46.
Dingman, J. (2008). Nanotechnology: its impact on food safety. *Journal of Environmental Health, 70*, 47–50.
Efsa. (2009). The potential risks arising from nanoscience and nanotechnologies on food and feed safety. *EFSA Journal*, 7, 958.
Efsa. (2011). Guidance on the risk assessment of the application of nanoscience and nanotechnologies in the food and feed chain. *EFSA Journal, 9*, 2140.
Efsa. (2016). Annual report of the EFSA Scientific Network of Risk Assessment of Nanotechnologies in Food and Feed for 2016. *EFSA Supporting Publications, 13*, 1145E.
Efsa. (2018a). Annual report of the EFSA scientific network of risk assessment of nanotechnologies in food and feed for 2017. *EFSA Supporting Publications, 15*, 1393E.
Efsa. (2018b). Guidance on risk assessment of the application of nanoscience and nanotechnologies in the food and feed chain: Part 1, human and animal health. *EFSA Journal, 16*, e05327.
European Commission. (2002a). Directive 2002/46/EC of the European Parliament and of the Council of 10 June 2002 on the Approximation of the Laws of the Member States Relating to Food Supplements. *Official Journal of the European Union L, 183*, 51–57.
European Commission (2002b). Regulation (EC) No 178/2002 of the European Parliament and of the Council of 28 January 2002 Laying Down the General Principles and Requirements of Food Law, Establishing the European Food Safety Authority and Laying Down Procedures in Matters of Food Safety. *Official Journal L, 031*, 1–24.
European Commission. (2006a). Regulation (EC) No 1907/2006 of the European Parliament and of the Council of 18 December 2006 concerning the Registration, Evaluation, Authorisation and restriction of Chemicals (REACH), establishing a European Chemicals Agency, amending Directive 1999/45/EC and repealing Council Regulation (EEC) No 793/93 and Commission Regulation (EC) No 1488/94 as well as Council Directive 76/769/EEC and Commission Directives 91/155/EEC, 93/67/EEC, 93/105/EC and 2000/21/EC. *Official Journal L, 396*(1), 1–849.
European Commission. (2006b). Regulation (EC) No 1925/2006 of the European Parliament and of the Council of 20 December on the Addition of Vitamins and Minerals and of Certain Other Substances to Foods. *Official Journal of the European Union L, 404*, 26–38.
European Commission, 2008a. Communication from the commission to the European parliament, the council and the european economic and social committee regulatory aspects of nanomaterials. Available from: https://eur-lex.europa.eu/LexUriServ/LexUriServ.do?uri=COM: 2008:0366:FIN:EN:PDF.

European Commission. (2008b). Regulation (EC) No 1332/2008 of the European Parliament and of the Council of 16 December 2008 on Food Enzymes. *Official Journal of the European Union L, 354*, 7–15.

European Commission. (2008c). Regulation (EC) No 1333/2008 of the European Parliament and of the Council of 16 December 2008 on Food Additives. *Official Journal of the European Union L, 354*, 16–33.

European Commission. (2008d). Regulation (EC) No 1334/2008 of the European Parliament and of the Council of 16 December 2008 on flavourings and certain food ingredients with flavouring properties for use in and on foods. *Official Journal of the European Union L, 354*, 34–50.

European Commission. (2009). Commission Regulation (EC) No 450/2009 of 29 May 2009 on active and intelligent materials and articles intended to come into contact with food. *Official Journal of the European Union L, 135*, 3–11.

European Commission. (2011). Commission Recommendation of 18 October 2011 on the definition of nanomaterial Text with EEA relevance. *Official Journal of the European Union L, 275*, 38–40.

European Parliament, 2009. European Parliament resolution on regulatory aspects of nanomaterials (2008/2208(INI)). Available from: https://www.europarl.europa.eu/sides/getDoc.do?pubRef=-//EP//TEXT+TA+P6-TA-2009-0328+0+DOC+XML+V0//EN.

European Parliament. (2011a). Commission Regulation (EU) No 10/2011 of 14 January 2011 on plastic materials and articles intended to come into contact with food. *Official Journal of the European Union L, 328*, 20–29.

European Parliament. (2011b). Regulation (EU) No 1169/2011 of the European Parliament and of the Council of 25 October 2011 on the provision of food information to consumers. *Official Journal of the European Union L, 304*, 18–63.

European Parliament. (2013). Regulation (EU) No 609/2013 of the European Parliament and of the Council of 12 June 2013 on Food Intended for Infants and Young Children, Food for Special Medical Purposes, and Total Diet Replacement for Weight Control and Repealing Council Directive 92/52/EEC, Commission Directives 96/8/EC, 1999/21/EC, 2006/125/EC and 2006/141/EC, Directive 2009/39/EC of the European Parliament and of the Council and Commission Regulations (EC) No 41/2009 and (EC) No 953/2009. *Official Journal of the European Union L, 181*, 35–56.

European Parliament. (2015). Regulation (EU) No 2015/2283 of the European Parliament and of the Council on novel foods, amending Regulation (EU) No 1169/2011 of the European Parliament and of the Council and repealing Regulation (EC) No 258/97 of the European Parliament and of the Council and Commission Regulation (EC) No 1852/2001. *Official Journal of the European Union L, 327*, 1–22.

Grieger, K. D., Hansen, S. F., Mortensen, N. P., Cates, S., & Kowalcyk, B. (2016). International implications of labeling foods containing engineered nanomaterials. *Journal of Food Protection, 79*, 830–842.

He, X., Deng, H., & Hwang, H. M. (2019). The current application of nanotechnology in food and agriculture. *Journal of Food and Drug Analysis, 27*, 1–21.

He, X., & Hwang, H. -M. (2016). Nanotechnology in food science: Functionality, applicability, and safety assessment. *Journal of Food and Drug Analysis, 24*, 671–681.

Jafari, S. M., Katouzian, I., & Akhavan, S. (2017). 15 - Safety and regulatory issues of nanocapsules. In S. M. Jafari (Ed.), *Nanoencapsulation technologies for the food and nutraceutical industries*Academic Press.

Jeevanandam, J., Barhoum, A., Chan, Y. S., Dufresne, A., & Danquah, M. K. (2018). Review on nanoparticles and nanostructured materials: history, sources, toxicity and regulations. *Beilstein Journal of Nanotechnology, 9*, 1050–1074.

Joseph, T., Morrison, M., 2006. Nanotechnology in agriculture and food. A Nanoforum Report. Available from: www.nanoforum.org.

King, T., Osmond-Mcleod, M. J., & Duffy, L. L. (2018). Nanotechnology in the food sector and potential applications for the poultry industry. *Trends in Food Science and Technology, 72*, 62–73.

Kuhlbusch, T. a. J., Wijnhoven, S. W. P., & Haase, A. (2018). Nanomaterial exposures for worker, consumer and the general public. *NanoImpact, 10*, 11–25.

Magnuson, B. A., Jonaitis, T. S., & Card, J. W. (2011). A brief review of the occurrence, use, and safety of food-related nanomaterials. *Journal of Food Science, 76*, R126–R133.

Martirosyan, A., & Schneider, Y. J. (2014). Engineered nanomaterials in food: implications for food safety and consumer health. *International Journal of Environmental Research and Public Health, 11*, 5720–5750.

Nile, S. H., Baskar, V., Selvaraj, D., Nile, A., Xiao, J., & Kai, G. (2020). Nanotechnologies in food science: applications, recent trends, and future perspectives. *Nano-Micro Letters, 12*, 45.

Parisi, C., Vigani, M., & Rodríguez-Cerezo, E. (2015). Nano Today. *Agricultural nanotechnologies: What are the current possibilities?, 10*, 124–127.

Park, H. -G., & Yeo, M. -K. (2016). Nanomaterial regulatory policy for human health and environment. *Molecular and Cellular Toxicology, 12*, 223–236.

Peters, R., Brandhoff, P., Weigel, S., Marvin, H., Bouwmeester, H., Aschberger, K., et al. (2014). Inventory of Nanotechnology applications in the agricultural, feed and food sector. *EFSA Supporting Publications, 11*, 621E.

Peters, R. J. B., Bouwmeester, H., Gottardo, S., Amenta, V., Arena, M., Brandhoff, P., et al. (2016). Nanomaterials for products and application in agriculture, feed and food. *Trends in Food Science and Technology, 54*, 155–164.

Pettitt, M. E., & Lead, J. R. (2013). Minimum physicochemical characterisation requirements for nanomaterial regulation. *Environment International, 52*, 41–50.

Ravichandran, R. (2010). Nanotechnology applications in food and food processing: innovative green approaches opportunities and uncertainties for global market. *International Journal of Green Nanotechnology: Physics and Chemistry, 1*, P72–P96.

Sharma, S., Jaiswal, S., Duffy, B., & Jaiswal, A. K. (2019). Nanostructured materials for food applications: spectroscopy, microscopy and physical properties. *Bioengineering (Basel), 6*, 26.

Siegrist, M., Cousin, M. E., Kastenholz, H., & Wiek, A. (2007). Public acceptance of nanotechnology foods and food packaging: the influence of affect and trust. *Appetite, 49*, 459–466.

Sodano, V., Gorgitano, M. T., Quaglietta, M., & Verneau, F. (2016). Regulating food nanotechnologies in the European Union: Open issues and political challenges. *Trends in Food Science and Technology, 54*, 216–226.

Tiede, K., Boxall, A. B., Tear, S. P., Lewis, J., David, H., & Hassellov, M. (2008). Detection and characterization of engineered nanoparticles in food and the environment. *Food Additives and Contaminants. Part A, Chemistry, Analysis, Control, Exposure and Risk Assessment, 25*, 795–821.

Usman, M., Farooq, M., Wakeel, A., Nawaz, A., Cheema, S. A., Rehman, H. U., et al. (2020). Nanotechnology in agriculture: Current status, challenges and future opportunities. *Science of the Total Environment, 721*, 137778.

Wang, H., Du, L. J., Song, Z. M., & Chen, X. X. (2013). Progress in the characterization and safety evaluation of engineered inorganic nanomaterials in food. *Nanomedicine (Lond), 8*, 2007–2025.

Warheit, D. B. (2018). Hazard and risk assessment strategies for nanoparticle exposures: how far have we come in the past 10 years? *F1000Res, 7*, 376.

CHAPTER 6

Regulatory principles on food nanoparticles legislated by Asian and Oceanian countries

Zahra Mohammadi, Seid Mahdi Jafari
Department of Food Materials and Process Design Engineering, Gorgan University of Agricultural Sciences and Natural Resources, Gorgan, Iran

6.1 Introduction

Nanotechnology is a growing and developing field whose role in improving human life quality in today's world cannot be ignored. This rapid expansion has paved the way for innovation in many industrial sectors, such as food processing, food contact materials, and agricultural production. Numerous renowned companies are looking for more beneficial ways to enhance their products' quality and optimize their production manners. By 2015, more than 300 companies were producing nano-based food and beverage products (Momin & Joshi, 2015). According to the common terminology, nanotechnology is the science of cognition and control of matter in dimensions of approximately 1–100 nm. This unique phenomenon allows for modern applications (Wang, 2018). Some applications of nanotechnology in the food industry are categorized in Fig. 6.1. This technology is employed in food processing and agricultural production to improve texture, taste, flavor, stability, release bioactive compounds, detect pathogens, and identity preservation and tracing through nano-encapsulations and nano-emulsions, porous nanoparticles (NPs), and nano-sized salts and minerals. Food contact materials (e.g., antimicrobial metal/oxides, biodegradable nanocomposites, and nanosensors) are applied to produce active and intelligent packaging, improve thermal and mechanical strength, increase safety and food storage time, and monitor packaged food (Chaudhry et al., 2008; Kah, Beulke, Tiede, & Hofmann, 2013). Several applications are now commercialized, and most are being researched and developed. In 2015, the Woodrow Wilson International Center for Scholars reported a list of 117 products containing nanomaterials in the food and beverage category (Higashisaka, Yoshioka, & Tsutsumi, 2015).

Safety and Regulatory Issues of Nanoencapsulated Food Ingredients
http://dx.doi.org/10.1016/B978-0-12-815725-1.00006-9

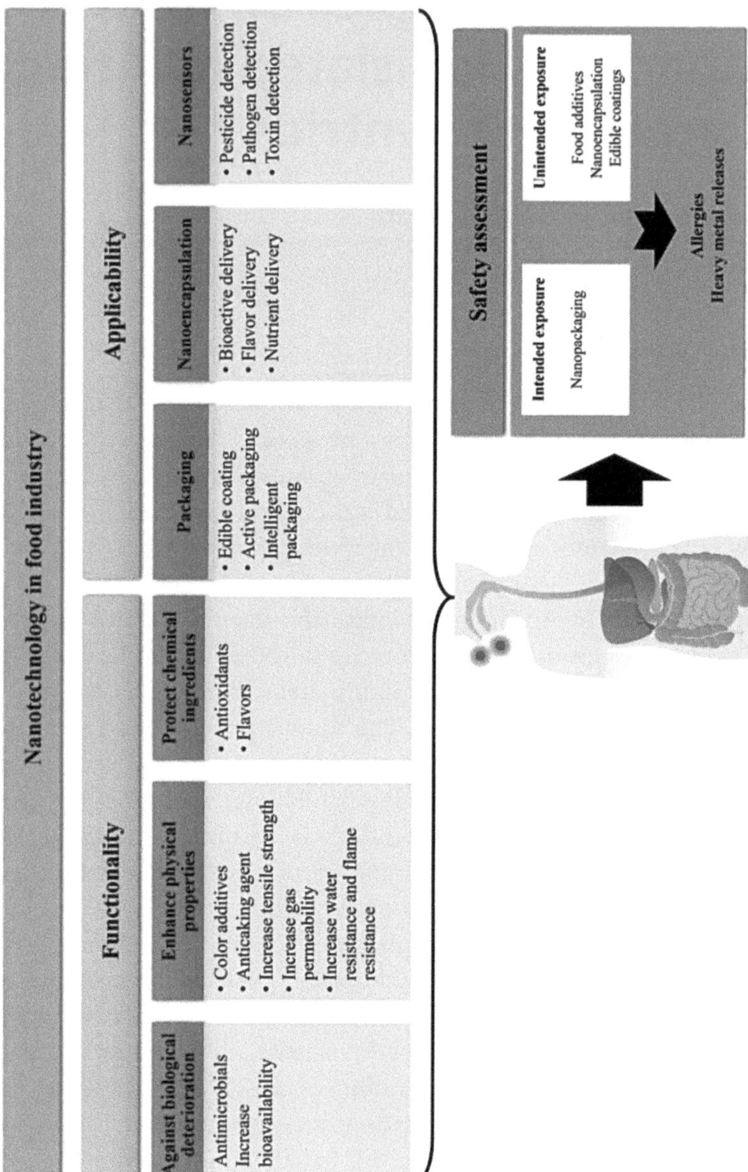

Figure 6.1 *Diagram revealing the development of nanotechnology in the food industry, functionality, applicability, and safety assessments.*

On the other hand, novel applications and products may have detrimental impacts on human health and the environment. On a nanoscale, material properties undergo many changes. As a result of these changes, it is also difficult to predict, identify, assess, and control the health, safety, and environmental risks posed by nanomaterials. For instance, 50 and 70 nm NPs can enter into cells and lungs, respectively, while 30 nm NPs can even pass through blood and brain barriers. These particles can reach many tissues and organs, including the brain, bloodstream, lymph nodes, and nervous system. They can cause toxicity, intestinal disease, and cancer in various parts of the body (des Rieux, Fievez, Garinot, Schneider, & Préat, 2006; ETC-Group, 2006). However, nanomaterials may be toxic and have inadvertent harmful effects.

Research has revealed that nanomaterial production and processing methods may not produce products with toxic influences (Zhang et al., 2003). For example, the toxicity of a particular substance such as selenium may be significantly reduced by decreasing its size to nano. It has also been reported that pure carbon nanotubes inserted into the rat trachea may kill them. Meanwhile, adding nitrogen to carbon nanotubes reduced toxicity, and death risk, paving the way to apply them in food packaging technology. Although most existing non-toxic studies typically focus on non-food or consumer products, these relevant research findings can still help identify potential food poisoning based on nanotechnology (Chau, Wu, & Yen, 2007). Hence, around the world, measures are being taken to control nanomaterials' safe production, use, and operation. These measures are taken by legislation and regulation or by providing non-binding recommendations and guidelines.

For the first time, the US Food and Drug Administration (FDA), as a public sector, described the concept of nanotechnology and its utilization to manufacture nanoproducts (Anandharamakrishnan, 2014). Numerous conferences and symposia have been held in Asia and Oceania to clarify and exchange information on nanotechnology risks in the food industry to date. On November 18, 2013, the National University of Singapore Academy for GxP Excellence (NUSAGE), Food Science & Technology Program (FST), and Agri-Food & Veterinary Authority of Singapore (AVA) jointly held a symposium on "Nanotechnology in Food – Past, Present, and Future." Part of the lecture covered some of the fundamental mechanisms by which nanostructured materials interact with the human body and the risk-benefit analysis to be undertaken when such engineering practices are presented. Due to the limited information on nanomaterials' perils, these ultra-small scale materials have received much attention regarding health hazards. Failure to enter nanotechnology-based food products into the food market may ultimately jeopardize the potential benefits of nanotechnology in the food industry in the absence of a clear definition, public debates on food safety assessments, and proper nutrition regulations (Peters et al., 2014).

6.2 Terminology

6.2.1 Nano

Nanometer (nm) means one-billionth of a meter or one-millionth of a millimeter. The term nano is now widely adopted commercially, and the business community often refers to it as something tiny and something that works on a small scale. One of the essential points to note here is that particles behave differently at the nanoscale compared to the mass state. This fact has prevented the authorities from reaching a comprehensive decision to control and monitor nanomaterials (Mansoori & Soelaiman, 2005).

6.2.2 Nanotechnology

A simple but accurate nanotechnology description is "molecular fabrication and production," which is the process of manipulating matter at the level of an atom or molecule to achieve the desired arrangement. An atom is the most miniature unit of any chemical. Richard Smalley, the winner of the Nobel Prize in Chemistry in 1996, has defined nanotechnology as the science and art of building nanometer-scale materials (Linkov & Steevens, 2008).

Presently, there are many definitions of nanotechnology. The two main definitions of this word are as follows: (1) A single-based, provided by various organizations such as the European Commission and has implications for policy in this area, and (2) a list-based definition developed by the European Union Patent Office (EPO) includes a list of nanotechnology applications such as nanobiotechnology, nanoelectronics, nanomaterials, etc. (Karim & Munir, 2015). A precise analysis of all existing single-based definitions indicates that most of them are derived from a description provided by the United States' National Nanotechnology Initiative (NNI). However, NNI places limitations on the application of nanotechnology to objects and uses the term only if it has all of the following characteristics (National Science, 2007):

1. Research and development of technology at the atomic, molecular, or macromolecular levels, with a scale of about 1–100 nm;

2. Construction and use of structures, tools, and systems with new features and applications due to their small or medium size; and
3. Ability to control or manipulate at the atomic scale.

All these features, that is, having a nanoscale (1–100 nm) and the construction and application of structures and tools and the ability to control, distinguish modern nanotechnology from natural, accidental, or environmental nanotechnology.

As there is no single, universally accepted description of the term, 18 different definitions of nanotechnology were investigated by researchers and organizations between 2001 and 2005, including the Allianz Centre for Technology and Organization for Economic Co-operation and Development (OECD), Environmental Law Institute, ETC Group, National Nanotechnology Initiative, United States Environmental Protection Agency, European Commission, etc. (Karim & Munir, 2015). Finally, nanotechnology is not a term or abbreviation, stating that nanotechnology is not a discipline or a set of techniques, but a widespread set of fields, including engineering, materials science, biotechnology, medicine, physics, biology, and information technology (Hodge, Bowman, & Ludlow, 2009).

6.2.3 Nanoparticles

NPs are particles that range in size from 1 to 100 nm found in gases such as particulate air, liquids such as colloids, or polymer matrices such as nanocomposites. The exact definition of a particle diameter depends on how it is measured (Boverhof et al., 2015).

6.2.4 Ultrafine particles

The ultrafine term has traditionally been attributed to particles in the air $<$ 100 nm by researchers at the National Institute for Occupational Safety and Health (NIOSH). There is no specific distinction between NPs and ultrafine particles. Ultrafine is usually used for particles that arise accidentally or unintentionally in combustion, welding, and diesel engines. The term nanoparticle refers to the particle size-dependent physicochemical properties and is mostly used for engineered nanomaterials. In some papers, the ultrafine term is also used for engineered materials such as titanium dioxide (Stone, Johnston, & Clift, 2007).

6.2.5 Engineered nanoparticles

Engineered NPs are designed and manufactured based on size, shape, surface, and chemical properties and exist in particulate air, colloids, and powders. Particles produced unintentionally in the nanoscale (ultrafine particles) are usually physically and chemically heterogeneous, while engineered NPs are present in homogeneous media (Jeevanandam, Barhoum, Chan, Dufresne, & Danquah, 2018).

6.2.6 Nanomaterials

Nanomaterials are materials that have at least one dimension on the nanometer scale and at least half of their particles are NPs, that is, their size is between 1 and 100 nm. In simple terms, nanomaterials are materials that are produced by applying NPs. European Commission (EC) in a report on "Considerations on a Definition of Nanomaterial for Regulatory Purposes" gathered all definitions provided by international organizations such as OECD, the European Union (EU) Scientific Committee on Emerging and Newly Identified Health Risks, EU Cosmetics Regulation, etc.; Plus, definitions in municipal legislation in various countries such as Australia, Canada, Denmark, the United Kingdom, and the United States of America, are considered. EC defines nanomaterials as materials whose internal or external structures are within the nanometer scale (often within 100 nm). The limit separates the nanoscale from the micro and macro scales (Lövestam et al., 2010). Though the EU attempt has been welcomed, the definition has been criticized. The European Chemical Industry Council has concluded that this definition is an extensive one and cannot be used in developing the rules since it imposes unnecessary restrictions and additional costs on companies and causes reduce the effective use of resources and materials (Karim & Munir, 2015). However, it is generally assumed that the definition of nanomaterials is required for any monitoring operation; this definition still does not solve all the problems. Andrew Maynard, one of the leading experts who believed that a definition of nanotechnology should be given, later realized that "one size fit for all" might not be appropriate for this technology, so it is better to define nanomaterials on a case-by-case basis (Maynard, 2011). It should be noted here that the EC also prefers the case-by-case approach in terms of the safety aspect of engineered nanomaterials (European Comm'n, Nanomaterials: Commission Proposes Case by Case Approach to Assessment (2012), *available at* http://europa.eu/rapid/press-release_MEMO-12-732_en.htm).

6.3 Nanotechnology market in the food and beverage industries

The current world population is > 7 billion, half of whom live in Asia. Many of these people live in developing countries. As a result of environmental problems and political instability, they face food shortages every day while in developed countries, there is excess food. The driving force behind developing countries is drought- and pest-resistant plants, which maximizes production yields. In developed countries, consumer demand, currently focused on fresher and healthier foods, is controlled by the food industry (FAO, 2007). Food and beverages include most of the daily expenses. There is a vast potential market for future applications and nanomaterials' development throughout the agricultural food chain from farm to fork. Agricultural research has always been associated with agricultural production, food processing, food health, and the environmental consequences of food production, storage, and distribution. Nanotechnology provides a tool to pursue these goals (Dasgupta et al., 2015).

The nanoscale is not new to the food and beverage sector; in recent years, in the formulation, production, and processing of supplements and functional foods, various phenomena have been observed and used on this scale. For example, the science of colloids has long been used to produce food. A group of foods and beverages contain nanoscale components or are fabricated using nano-processes (such as dairy products); natural nanoparticle manipulation is also used in some cases (Sanguansri & Augustin, 2006). However, research to achieve innovative applications has only been conducted in recent years to reach new functions and more effective release mechanisms for food and beverages. Brand-new instruments and processes allow researchers to develop targeted release mechanisms and lead to intelligent release to optimize human health and physical, appearance, and sensory effects. Some potential applications of this discussion include (Rossi et al., 2014):

- Foods can change their color, taste, and even nutrients depending on the consumer's preference and health.
- Filters can separate toxins from food or change the flavor by moving specific molecules based on shape (rather than size).
- Packages that notify the consumer by changing the color if their content is damaged.

6.3.1 Key applications and market opportunities

The nanotechnology market in the food and beverage industry in 2007 was about 105 million USD, the majority of which was the packaging. There were few nano-based products on the market in this industry. Most likely, the first technologies that entered the market were used in the processing stage. The value of nanotechnology-based products and processes in the food and beverage sector was estimated at 2.135 billion USD by 2015 (Nanoposts, 2008; Nile et al., 2020).

Fig. 6.2A presents the market share of various applications in the food and beverage sector using nanotechnology by 2015. The highest revenue of the nanotechnology market in this area is related to packaging. By 2015, the field has developed obstructive packaging, intelligent, and interactive packaging. The delivery of nutrients and flavors using this technology is still not very attractive due to consumers' vague perceptions of the application of nanotechnology in food. However, large companies are performing research and development in this area. These companies are also interested in utilizing nanotechnology in food processing equipment. 29% of activities in the food and beverage industry, from processing to final production, in 2015 adopted nanotechnology-related technologies. The launch of nanotechnology-based applications in 2012 has had an impact by reducing the cost of packaging materials. Fig. 6.2B displays the market development of nanotechnology applications in the food and beverage industry from 2006 to 2015 (Nanoposts, 2008).

6.3.2 The main role players

Table 6.1 presents the leading manufacturers of nanomaterials and end-users of nanotechnology in the food and beverage industries. The table also shows emerging domains of the food and beverage sector affected by nanotechnology in the short- and medium-term (Nanoposts, 2008).

6.3.3 Development of nano-based products and technologies in food industry

This section provides some developed/developing products and technologies in the three main areas of food and beverage applications, that is, processing and safety, packaging, and targeted delivery and release. A summary of this section is provided in Table 6.2.

6.3.3.1 Processing and safety

Detection of pathogens: AgroMicron has developed a nanobioluminescence spray that can be used to detect a wide range of pathogens in food and beverages. The product employs a nanoscale luminescence protein molecule that has been modified to bind to target bacteria's surface (Plexus Institute. (2006). New nanotechnology food research—if it glows don't eat it. http://www.plexusinstitute.org/news-events/show_news.cfm?id=164).

Separation and filtration: Aquamarijn Micro Filtration has provided advanced nano/microfiltration technology, producing well-defined membranes (micro/nanosieve) with well-defined pore sizes by silicone or poly-

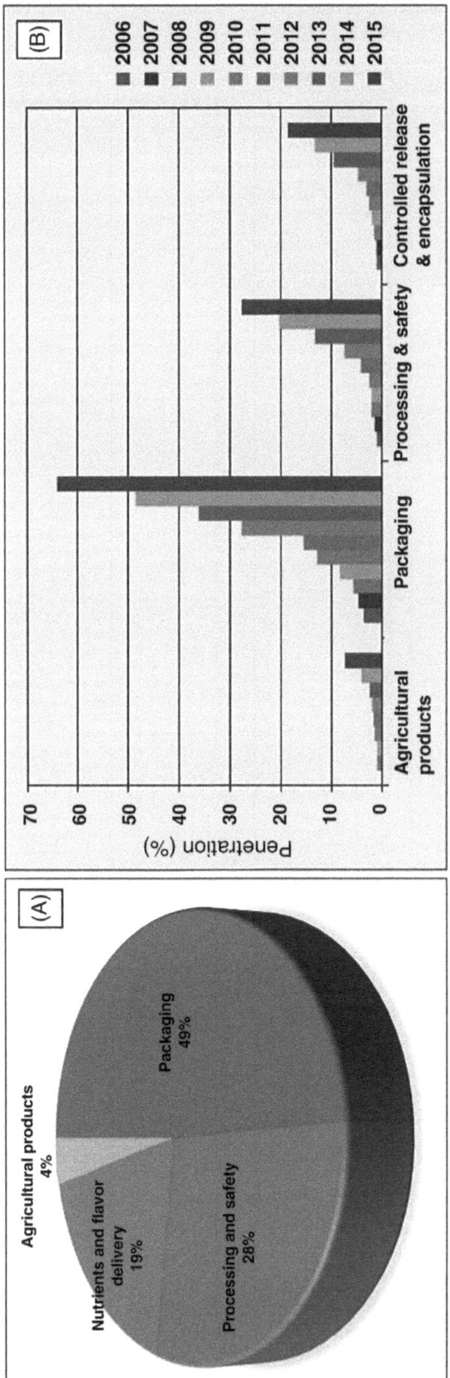

Figure 6.2 The market of nanotechnology in 2015 (A) and market development from 2006 to 2015 (B) in the food and beverage industry (Nanoposts, 2008).

Table 6.1 Global actors in the adoption of nanotechnology in the food and beverage sector by field of activity (Nanoposts, 2008).

Company	Agricultural products	Packaging	Processing and safety	Controlled release
AC Serendip Ltd.				√
Advanced Nanotechnology		√		
AgroMicron Ltd.			√	
Ambri Pty Ltd.			√	
Aquamarijn			√	
Aquanova		√		√
Bayer				√
Biodelivery Sciences				√
Bühler AG		√		
Chengdu Somo Nano-biology Co., Ltd.			√	
China Beijing Penta Nanotech Co., Ltd.			√	
Cornelius Group		√		
Crown Bio Technology Ltd.		√		
DGTec		√		
Heinz Honeywell		√		
InMat		√		
Iota NanoSolutions Limited		√		
Kodak		√		
Kraft			√	√
Landec Corporation	√			
Leatherhead Food International Ltd.			√	
Nanocor		√		
Nanomi				√
Nanosciences Inc.		√		
Nanova		√		
NGimat		√		
Nestle			√	√
Nutralease				√
PChem Associates		√		
Proctor & Gamble		√		√
Protista International AB			√	
Surface Innovations Ltd.		√		
Unilever			√	√

Table 6.2 Summary of nano-based products and technologies in the food industry.

	Company	Research field	Technology/Product	Website
Processing and safety	AgroMicron	Detection of pathogens	Nanobioluminescence spray	www.agromicron.com
	Aquamarijn Micro Filtration	Separation and filtration	Micro/nanosieve	www.microsieve.com
	Nanosens	Detection of biological molecules	Nanowire biosensors	www.nanosens.com
	Leatherhead Food International	Detection of viruses in real-time	Aluminum oxide/glass nanofiber composites	www.leatherheadfood.com
	Protista International AB	Separation of bio nanoparticles	Supermacroporous gels	—
Packaging	Cornelius Group	Beverage packaging	NanoZ	www.cornelius.co.uk
	Ingenia Technology	Barcoding of food packaging	Laser surface authentication	www.ingeniatechnologies.com
	Crown Bio Technology	Freshness indicator	Durable packaging	www.crownbio.com
	Nanova	Nanocomposites with a long shelf life	One-year PET nanocomposite	—
	InMat	Gas barrier coatings	Nanolok	www.inmat.com
	PChem Associates	Antibacterial plastic coatings	Packaging containing nanosilver	www.nanopchem.com
	ConverTec	Flexible blocking material	Corona beam technology	www.coronabeam.com/www.convertech.com
	Nanoplex Technologies	Anti-counterfeiting material	Nanobarcodes	—

(*Continued*)

Table 6.2 Summary of nano-based products and technologies in the food industry. (*Cont.*)

	Company	Research field	Technology/Product	Website
Targeted delivery and release	Vivamer	Smart material for encapsulation	Micro and nanoparticles, films, or gels	www.vivamer.com
	Aquanova	Nanocarrier	NovaSOL Sustain/ SoluE	www.aquanova.de
	Biodelivery Sciences International	Nutrient delivery	NanoSnails	www.bdsi.com
	Nanomi	Fat reduction nanoemulsions	Nano oil rods	www.nanomi.com
	AC Serendip Ltd	Nanoemulsions	High-pressure homogenization and emulsifying filters	www.ac-serendip.com

mer-based materials. Their properties (porosity, shape, size of pores, type of material, special coatings) can be adjusted for various applications. Current filtration applications focus on beer, milk, pharmaceuticals, and environmental filtration. These micro-sieves (with pores in the range of 100 nm to 10 μm) enable other research companies to enter the nanoscale and allow the patterning of structures with functional layers. The technology can also be used to make droplets for drug delivery, nanoencapsulation, or functional foods production (https://www.aquamarijn.nl/technology/).

Nanowire biosensors: Nanosens has used micro vibrators-based biosensors as promising tools for direct and highly accurate biomolecular interactions detection. After identifying biological molecules, these micro-vibrators react in the form of nanomechanical motion. Optical or compression sensor systems attached to these sensors show these movements. These sensing systems are portable and designed to distinguish biological and chemical species quickly. Systems have been developed to immediately detect hundreds of different specimens using various vibrators (http://nanosens.nl).

Detection of viruses in food: The main focus of Leatherhead Food International Ltd's research is on nanomaterials (such as NPs and sensor chips) and nanocomposites (e.g., nanofiltration mats). In collaboration with some other companies, the company has developed several technologies that benefit various food and beverage industries. One of the company's successful activities has been applying magnetic micro/nanoparticles to capture and decompose microorganisms (e.g., Listeria, Salmonella, and yeasts) in real-time. Another company's activity is aluminum oxide/glass nanofiber composites to detect food viruses in real-time (www.Leatherheadfood.com).

Separation of bio nanoparticles: Protista International AB has developed supermacroporous gels that can help separate biological NPs (such as viruses and cellular components). This technology can be employed in food processing (Lozinsky et al., 2003).

6.3.3.2 Packaging

Nanomaterials for beverage packaging: Cornelius Group from Advanced Nanotechnology has produced a type of zinc oxide nanoparticle called NanoZ that can be used as a barrier against ultraviolet light. NanoZ can be employed both inside the plastic and as a cover to prevent ultraviolet light from passing through. The advantages of NanoZ include the ability to block the passage of a wide range of ultraviolet light with high transparency (www.cornelius.co.uk).

Barcoding of food packaging: Ingenia Technology Ltd. has a technology called the Laser Surface Authentication (LSA) system that applies quantum dots to detect the inherent fingerprints in materials such as paper, plastic, metal, and ceramics. As provided in the relevant documentation, this technology can validate papers, plastics and metals, credit cards, and packaging (Cowburn, 2008).

Nanoscale freshness indicator: Crown Bio Technology Ltd. has unique ideas in applications close to the nano market, related to the release of various elements, compounds that create freshness (use in packaging), and durable packaging (recyclable) (www.crownbio.com).

Nanocomposites with a long shelf life: Nanova has developed a polyethylene terephthalate (PET) nanocomposite with one year's shelf life. The company was looking to take advantage of Nanotalc's unique features. Nanotalc composite is mineral-organic at the molecular level. In this nanocomposite, some nanocrystals of magnesium silicate are chemically attached to the organic polymer to form a huge molecule. As a result, the produced polymer composite is transparent and colorless, even with 20%. The addition of magnesium nanosilicate also improves the polymer's blocking properties against all gases and increases its thermal stability by about 100°C (DePolo & Baird, 2009).

Gas barrier coatings: InMat fabricates gas-blocking polymer coatings for packaging applications, which are environmental-friendly. The level of blockage created by these coatings is the same on different substrates. Production of Nanolok nanocoatings begins with an aqueous suspension of nano-dispersed silicates in a polymer substrate (www.inmat.com).

Antibacterial plastic coatings: PChem Associates develops antibacterial plastic coatings for food packaging applying silver NPs. The company claims that it is possible to use nanosilver as an antimicrobial coating and light reducer in packaging, films, and storage containers (www.nanopchem.com).

Flexible blocking material: ConverTec manufactures flexible blocking materials for pharmaceutical, medical and food packaging by corona beam technology. Identification of pores, holes, bumps, contaminants, pressure fractures, overlapping materials, formulation defects, and other irregularities < 0.5 nm in scale (www.coronabeam.com).

Anti-counterfeiting material: Nanoplex Technologies has developed nano barcode particles that contain several properties, making them ideal candidates for next-generation anti-counterfeiting and trademark protection applications. The company has expanded two nanoparticle-based technologies: nano barcode particles and sensor tags. Nano barcode particles have properties that make them suitable for use in the protection of documents and products. These barcodes are made of corrugated cylindrical metal NPs, each made of different metals such as gold, silver, platinum, or nickel. Typically, each of these cylindrical NPs has a diameter of 250 nm and a length of 6 μm (www.packagingdigest.com).

6.3.3.3 Targeted delivery and release

Smart material for encapsulation: The formulator Vivamer owns the intellectual property of the smart materials used for encapsulation. It is possible to release the desired products in response to specific environmental conditions such as temperature, acidic conditions, water activity, light, and exposure to specific chemical starters utilizing these materials. Besides, by changing these intelligent materials' chemistry, they can be adjusted to respond to particular environmental conditions within a well-defined range. These materials are stable in aqueous solutions, non-toxic, and biodegradable. These compounds are odorless and can be converted into micro and NPs, films, or gels. Along with multinational companies, the company is employing smart materials in home products and the food industry (www.vivamer.com).

Food nanocarriers: Aquanova has developed a new technology that combines two active ingredients, one to reduce fat and the other to create satiety in a single nanocarrier (micelles with an average diameter of 30 nm) called NovaSOL Sustain. NovaSOL technology has also been applied to generate a new formulation of vitamin E called SoluE. Since NovaSOL can protect its contents against stomach acids, it can deliver other nutrients to the body (www.aquanova.de).

NanoSnails in nutrient delivery: Biodelivery Sciences International has developed structures called NanoSnails. NanoSnails are 50-nm-diameter helical NPs that can be utilized to deliver nutrients such as vitamins, lycopene, and omega-3 fatty acids without affecting the color or taste of food (www.bdsi.com).

Fat reduction nanoemulsions: Nanomi has developed a tool that can be employed to produce droplets in fluids continuously. Pressing the oil and passing it through a filter generates oil rods. Passing water with high pressure along this filter causes the upper part of this rod to separate and form drops. The company has patented the mentioned method, which can control droplets' shape and composition. Customers decide on droplet size and order dual emulsions. The inside of the oil droplets can be filled with water to reduce the food's oil content or fill the drops with healthy ingredients that are not tasty so that these substances are released only in the stomach (www.nanomi.com).

Nanoemulsions: AC Serendip Ltd. has manufactured nanoemulsions for milk, mayonnaise, sauces, and soups. The company's technology is based on high-pressure homogenization and emulsifying filters that result in finer particles, faster AI transfer, or direct particle size control (www.ac-serendip.com).

6.4 Potential risks and toxicity of food nanoparticles

NPs increase the bacterial mutation rate and pose a potential threat to the environment and human health. The size of these particles, which are crucial properties, can threaten health and environment. These particles are also smaller than the pollen of plants and common allergens and can cause allergies. The transfer of NPs to the human body and their behavior in biological systems is challenging to predict due to their tiny size, the surface to mass ratio, more stability, and differences in biocompatibility and excretion from the body. Contact with NPs occurs through the respiratory tract, gastrointestinal tract, dermal, or venous uptake. The accumulation rate of NPs, removal from tissues, and chemical activity of the target organ/tissue affect toxicity. Thus, more stable NPs generate more toxicity over a more extended period.

When nanomaterials contact biological fluids, the liquids penetrate the pores and are covered by proteins and biomolecules (Buzea, Pacheco, & Robbie, 2007). This cover may affect the biological response to NPs. The accumulation and separation of proteins from nanomaterials depend on the particles' hydrophobicity and their size and radius of curvature. The binding or non-binding and formation of transient complexes with nanomaterials depend on the identity of the protein. For example, albumin and fibrinogen play a significant role in community and differentiation compared to apolipoproteins (Wijnhoven et al., 2009). One of the NPs' concerns is their entry into cells and intracellular structures such as mitochondria and nuclei, which direct and indirect mechanisms cause oxidative damage to the cell and genome (Unfried et al., 2007). Zinc oxide and titanium dioxide particles produce reactive oxygen species (ROS) in skin cells and damage DNA. This destruction causes mutations and changes in the structure of the protein that may cause cancer and tumors. NPs induce cell death by damaging cell membranes and producing ROS (Fu, Xia, Hwang, Ray, & Yu, 2014), as shown in Fig. 6.3.

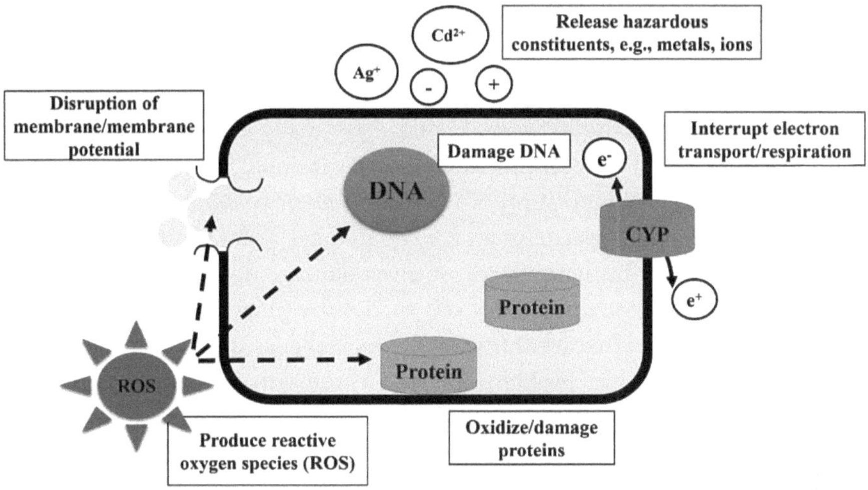

Figure 6.3 *Mechanism of cell death by nanoparticles.*

6.4.1 Interaction of nanoparticles and proteins

In biological fluids, proteins attach onto the surface of NPs and form a coating called "Corona," affecting NPs' interaction with living systems. Corona means that proteins are placed like a halo around NPs. The binding of the protein to the NPs' surface is due to the NPs' very high surface energy. Interaction of proteins with NPs causes changes in the protein structure, altering enzymatic activities or information pathways. It can also alter protein function or protein aggregation (Saptarshi, Duschl, & Lopata, 2013). The corona is divided into hard corona and soft corona. In the hard corona, proteins bind directly to the surface of NPs. In the soft corona, proteins attach to the hard corona surface (protein–protein binding). There is a competition between all plasma proteins for binding to the nanoparticle surface. Proteins found in almost all hard corona cases include albumin, immunoglobulin G, fibrinogen, and apolipoprotein (Milani, Baldelli Bombelli, Pitek, Dawson, & Radler, 2012). There are different coronas due to the tendency of proteins to bind to the surface of NPs. Initially, more abundant proteins in plasma bind to particles. More tending proteins then replace these proteins. Corona diversity depends on several factors, including size, shape, surface charge, functional groups, hydrophobicity, or hydrophilicity. The corona's thickness also depends on the size and concentration of the protein (Rahman, Laurent, Tawil, & Mahmoudi, 2013).

6.4.2 Passage of nanoparticles in the body

6.4.2.1 Inhalation

The lungs are more defenseless than other organs due to their large surface area. This property of the lungs makes the NPs easily enter the tissue compared to larger materials, leading to dangerous lesions. The three main factors, size, shape, and composition of NPs, affect their toxicity and a potential risk to the lungs. These particles attack the defense and immune systems of living organisms and humans. Some of them damage the lungs' air sacs after inhalation, in which macrophages try to destroy them and prevent passing through the bloodstream. However, macrophages detect particles with diameters < 70 nm are problematic, and these particles easily penetrate the blood. NPs such as carbon black and titanium dioxide, widely used in industrial processes and contribute to air pollution, have been reported to cause inflammation and skin injuries and accumulate in the lungs (Cho et al., 2010; Kim, Cho, & Park, 2016).

Studies have revealed that microclusters of NPs are virtually inactive, and particles smaller than this size are absorbed and found in the blood after inhalation. Generally, by inhaling through the nose, with the deposition system, the entry of particles with a diameter > 6 μm is prevented by the nasal hairs. Particles 1–5 μm in size settle in smaller bronchioles. Terminal bronchioles are very common in coal miners due to sedimentary dust particles. Particles < 100 nm in diameter (NPs) penetrate air bubbles and settle there (Garcia, Schroeter, & Kimbell, 2015).

6.4.2.2 Gastrointestinal system

As many NPs and nanomaterials are produced and prepared in liquid, one way to exposure can be through skin uptake or oral ingestion. The kinetics of transmission and displacement of soluble solid particles in the intestine depends on three parameters: (1) the rate of particle diffusion, (2) the ease of access of the intestinal mucosa to them, and (3) the surface properties of the NPs (Schneider, Stracke, Hansen, & Schaefer, 2009). In research, the diffusion rate of NPs (given their size and charge) was investigated through the mucosal layer to the enterocyte surface. Cationic latex NPs (positive charge) and anionic latex ones (negative charge) were utilized in this regard. The results showed that cationic latex NPs were trapped in large quantities in the mucosal layer with a negative charge and entered the mucosa. Nevertheless, anionic latex NPs are resisted. According to this study, the smaller cationic NPs can enter the mucosa faster. For example, particles with a diameter of 14 nm in < 2 min and particles with a diameter of 415 nm in 30 min reach enterocytes. Particles with a diameter of 1000 nm will never pass through this layer (Szentkuti, 1997).

Uptake takes place mainly in the small intestine. NPs can enter both lymph and capillaries when they enter the subcutaneous tissue of the small intestine. The entry of NPs into the lymph causes the secretion of immune responses (antibodies). In contrast, their entry into the capillaries causes the transfer to the bloodstream (Ashammakhi et al., 2020). In one study, the transfer of NPs through the small intestine to other organs in the body was investigated. The transfer of spherical polystyrene particles in the sizes between 50 nm and 3 μm (3000 nm) was examined in rats at a dose of 1 mg/kg body weight. Approximately 7% of 50 nm particles and 4% of 100 nm particles were observed in liver, spleen, blood, and bone marrow. Particles larger than 100 nm did not reach the bone marrow, and particles > 300 nm were not found in the blood. No particles were found in the lung and heart tissue (Jani, Halbert, LANGRIDGE, & Florence, 1990).

6.4.2.3 Nanoparticles and blood–brain barrier

One of the most promising gateways in nanotechnology is the drug delivery to specific cells by NPs. It is expected that NPs may cross the blood–brain barrier (BBB) through passive diffusion or mediator-carrying endocytosis. Impregnation of particles with polysorbate (e.g., polysorbate 80) involves inhibition of apolipoprotein E or other blood elements. Surface-modified particles mimic LDL particles and have internal interactions with LDL receptors, leading to endothelial cells' elevation. The drugs (loaded on the particles) may spread into these cells and diffuse inside the brain, or the particles may turn into cytoses (Kreuter, 2004). Drug delivery systems that pass through the BBB are quite useful. Yet, these systems create the concept that unwanted passage through the BBB is also possible. Therefore, proper safety assessments are required.

6.5 Nano regulations in Asian countries

Asia is the world's largest and most populous continent, and having cheaper labor can be highly profitable for multinationals. China and India can be considered global producers of nanotechnology-based products. Japan, South Korea, Iran, Singapore, and Taiwan have always played a key role in advancing nanotechnology research (Karim, 2013). Hence, these countries have made remarkable advances in nanotechnology development and research and have pioneered global nanotechnology competition in investment, patents, and scientific publications in reputable scientific journals. In addition to national laws, several countries have established standards and certification systems for nano-products (Sabatier & Chollet, 2017).

Asian countries have different legal systems. For example, in China, Indonesia, Japan, South Korea, Taiwan, and Vietnam, the civil law system prevails; India, Singapore, and Malaysia have often adapted from the common law system of England; in Saudi Arabia, the legal structure of Islamic law commands; in Iran, there is a legal system of Shia Islam, and in the Philippines, there is a combination of civil law and common law. Although there are differences between these legal systems, all of these countries have written constitutions. Most of them also have legal regulations in the cases mentioned above, such as nanomaterials' life cycle in the form of codified law, common law, and secondary law. There are particular regulatory organizations to enforce these rules, and in addition to them, there are guidelines and standards set by these bodies. Most of these countries are also members of various international organizations such as the United Nations, the OECD, ISO, the Association of Southeast Asian Nations (ASEAN), and the Asia-Pacific Economic Cooperation (APEC). They are members of many regional and international human rights treaties. Some other treaties and agreements are related to many of the above (Karim & Munir, 2015; Karim, 2013). Many of these Asian countries, such as Japan, China, Singapore, India, etc., have joint programs with universities and research institutes in Europe and the United States. In these programs, it is always assumed that nanomaterials' rules, standards, and guidelines are strictly observed.

Here, only eight countries are covered: China, Taiwan, Japan, South Korea, India, Iran, Malaysia, and Thailand. However, other countries, including Indonesia, Saudi Arabia, Sri Lanka, Bangladesh, and others, have taken actions. A summary of the regulatory rules for the application of nanomaterials in the food industry is given in Table 6.3.

6.5.1 China

In China, food safety is regulated by the "Food Safety Law," which has not specified nanomaterials. In March 2003, the Chinese Academy of Sciences (CAS) and the Ministry of Education jointly established the National Center for Nanoscience and Technology (NCNST). This center consists of numerous divisions: (1) nano-processing and nano-device laboratory, (2) nano-materials and nano-structure laboratory, (3) nano-medicine and nano-biotech laboratory, (4) nano-structure characterization and test

Table 6.3 Asia and Oceania status for regulation of nanoproducts in agri-food systems.

Country	Regulatory Body/Responsible Organization	Legislation/ Concerned Documents	Available Provisions and Remark
China	Ministry of Agriculture, Ministry of Health	Food Safety Law of China	National Centre for Nanoscience and Technology (NCNST) and the Commission on Nanotechnology Standardization. Nano-products in agri-food systems is not yet approved by the authority
Taiwan	Taiwan Nanotechnology Industry Development Association (TANIDA)	Nano-Mark system to certify the nanoproducts	Nanoproducts are certified
Japan	Ministry of Health, Labor, and Welfare	Food Sanitation Law	No NM specific regulation
South Korea	Ministry of Food and Drug Safety (MFDS)	National Nano-Safety Strategies Plan	No NM specification
	Korean Food and Drug Administration (KFDA) Korean Agency for Technology and Science (KATS)	(2012/2016) and Food Sanitation Act	No NM specification
India	Government of India	Food Standards and Safety Act (2006)/ Nano Mission Program	No NM specification
Iran	Iran Nanotechnology Initiative Council (INIC), Nanotechnology Committee of Food and Drug Organization (FDO)		FDO guidelines for nanoproducts in food, beverages, pharmaceutical, medical equipment. But agriculture is not yet included.

(Continued)

Table 6.3 Asia and Oceania status for regulation of nanoproducts in agri-food systems. (*Cont.*)

Country	Regulatory Body/Responsible Organization	Legislation/ Concerned Documents	Available Provisions and Remark
Malaysia	Ministry of Health	National Food Safety Policy (NFSP), the Food Act 1983 (Act 281), and Food Regulations 1985	No NM specification
Thailand	Food and Drug Administration of the Ministry of Public Health	Certification by Nanotechnology Association of Thailand	NanoQ label has been introduced for nanoproducts (but not for agri/food) that are certified by the Nanotechnology Association of Thailand
Australia and New Zealand	Food Standards Australia New Zealand (FSANZ)	Australian New Zealand Food Standards Code	Amended FSANZ Application Handbook to support new food regulations to manage risks from nanoproducts

laboratory, and (5) coordination laboratory. In 2005, the NCNST-affiliated Nanotechnology Standardization Commission was established to develop national standards, including terminology, protocols, and safety requirements for particle size measurement, nano processing, nanomaterials, and nanotechnology. Under China's existing national standard, the Commission enables the control and evaluation of nanotechnology products facilitating nanotechnology companies to improve product quality, decrease health risks, keep safety, and enhance the development of new products (Magnuson et al., 2013).

In September 2010, the Food and Environmental Hygiene Department of Center for Food Safety presented standards legislated by the World Health Organization (WHO), Organization of the United Nations (FAO), European Food Safety Authority (EFSA), Health Canada, FDA, and Food Standards Australia New Zealand (FSANZ) in the Risk Assessment Studies Report 41. Accordingly, in this document, nanomaterials' behavior and evolution in the gastrointestinal tract, including uptake, distribution, metabolism, and excretion/elimination, were reviewed (Takeuchi, Kojima, & Luetzow, 2014). Until 2013, the use of nano minerals or nanomaterials for use as food ingredients has been rejected by Chinese regulators (Amenta et al., 2015). In March 2017, the Risk Assessment Section of the Food Safety Center expressed concerns about the use of nanomaterials in food in four:

1. NPs turn to an indirect source of food contaminant,
2. Changes in the body's uptake and metabolism due to the uptake of nanomaterials,
3. The toxicity of NPs remains unknown, and
4. Lack of effective analytical method and predictive model for the safety assessment of NPs.

There are also two recommendations for trading nanomaterials-based products: (1) Traders should guarantee that products are safe for humans, and (2) nanomaterials that have not been evaluated for safety should not be exchanged. Furthermore, the public has been advised to purchase their products only from reliable suppliers, as nanofoods are mainly available in the global market through internet trading (Hardy et al., 2018).

6.5.2 Taiwan

In 2004, the Industrial Technology Research Institute implemented a nanotechnology-based certification system for consumer products. The world's first "Nano Mark" certification for verified nanoproducts was launched worldwide to validate these products. This certification system is issued based on promoting the industrialization of nanotechnology, encouraging the development of nanoproducts with a high standard of consumer protection, and enhancing the acceptance of nanoproducts by consumers. Products that meet at least the following two main requirements can be certified as Nano Mark: First, one of the dimensions of the product's functional ingredient is approximately 1–100 nm. Second, the nano product has innovative characteristics or improved usability because of its ultra-small size (Chau et al., 2007).

In addition to the government certification system, in 2006, a non-governmental organization called the "Nano Manufacture and Inspection Union" was established in Taiwan. The union has provided an approach to developing analytical methods, evaluation standards, inspection protocols, standard performance solutions, and services such as nanomaterials processing, analysis, product inspection, information exchange, and technical consulting. In 2007, the union initiated a plan to develop an efficient process and official inspection protocol for pearl powder production on a scale of 100–500 nm, the bulk form of traditional edible material in Chinese medicine. Pearl powder evaluation criteria include particle size, heavy metal, nutrition label, amino acids, functional properties, and origin (Amenta et al., 2015). An article examined Taiwan's regulations and strategies for nanotechnology-based pharmaceuticals that align with international legislation and considerations. Taiwan's nanomedicine regulations have also been compared to those of the United States, the European Union, and Japan. This is a list of nanomedicine approved by the Taiwan FDA.

Accordingly, until 2014, there were no international regulations for nanomedicine, but it was agreed worldwide that more attention should be paid to the quality, safety, and effectiveness of nanotechnology-based drugs. There has been a lack of adequate evaluation techniques to deal with the adverse effects of nanotechnology. Pharmaceutical products that required critical consideration included liposomes, followed by metal colloids and block copolymer micelles (Guo et al., 2014). In 2017, new guidelines were presented on the usage of nanomaterials in food packaging by Taiwan FDA. Therefore, any food packaging containing nanomaterials must be evaluated for safety and approved before entering the market (He, Deng, & Hwang, 2019b).

6.5.3 Japan

According to estimates, the Japanese market for food containing nanomaterials could reach 250 billion yen by 2030 (Tachikawa, 2012). In Japan, the development of the nanotechnology industry is strongly supported by Japanese government agencies, including the Ministry of Economy, Trade and Industry (METI), the Ministry of Environment (MOE), the Ministry of Health, Labor and Welfare (MHLW), the Ministry of Agriculture, Forestry and Fisheries, the Ministry of Education, Culture and Sports Science and Technology (MEXT). MEXT is responsible for research and development and conducts collaborative research between industry, the scientific community, and the government to promote research and development in nanotechnology and materials science. MEIL works to standardize safety assessment test methods of nanomaterial, and MHLW develops methods for assessing nanomaterials' health effects. Besides, the Nanotechnology Researchers Network Center (Nanonet), set up by the National Institute for Materials Science (NIMS), develops informational support for the improvement of the nanotechnology network (Chau et al., 2007).

In March 2010, a document entitled "A Basic Survey Report on Safety Assessment Information on the Use of Nanotechnology in the Food Sector" was published by the Food Safety Commission of Japan (FSCJ), which provided viewpoints on the state of application of nanotechnology in the Japanese food sector (Chowdhury, Gogoi, Borchetia, & Bandyopadhyay, 2017). The document reports nanomaterials in food products in those years, while the definition of nano was ambiguous. In this document, a report is presented in which the possibility of the transfer of artificial NPs from packaging materials to food products by a physicochemical method has been investigated. The results revealed that when a polymer matrix with relatively low dynamic viscosity does not react with artificial NPs and tiny synthetic NPs with a diameter of 2 nm are used in packaging materials, manufactured NPs can be transferred from the packaging to food. These observations can be applied to nanocomposites consisting of silver and polyolefins (LDPE, HDPE, and PP). However, in larger artificial NPs and polymeric matrices with relatively high dynamic viscosities, such as polystyrene and polyethylene terephthalate, a large amount of transfer is probably not considered (ŠIMON, Chaudhry, & BAKOŠ, 2008).

The "Committee on Safety Management for Nanomaterials" established by the METI also issued a new notice on precautionary measures to prevent exposure to nanomaterials replacing the previous one in 2013 (Karim, 2013). Furthermore, in January 2016, the MHLW issued new safety standards for agriculture and food in the 190th Conference for the Promotion of Food Import Facilitation report which did not include any specifications of nanomaterials (He, Deng, Aker, & Hwang, 2019a).

6.5.4 South Korea

The Republic of South Korea is the first Asian country to adopt a special law on nanotechnology entitled the Nanotechnology Development Promotion Law 2008 (No. 8852) to organize a national and systematic effort to promote research in nano and its application in industry. This law was implemented on February 29, 2008, along with the announcement of implementing the Nanotechnology Development Promotion Law 2008 (No. 20740) (MINISTRY OF GOVERNMENT LEGISLATION, http://www.moleg.go.kr/english/korLawEng?pstSeq=52226). Moreover, before this law, in 2002, South Korea legislated another rule with a similar title. The 2002 law, which provided a sufficient impetus for nanotechnology's strategic development, was adopted at the 234th General Assembly of the Korean National Assembly on November 12, 2002. It was enacted as Law No. 6812 in December 2002. Subsequently, this law was implemented in 2003 with the Technology Development Implementation Order (Executive Order 18011) (So Young Kim, *National Nanotechnology Development Plan (South Korea), in* ENCYCLOPEDIA OF NANOSCIENCE AND SOCIETY 556-557 (David H. Guston ed. 2010).).

The "Food Sanitation Act" is the principal organization responsible for food, food additives, and food packaging legislation. Moreover, some food standards such as the Food Regulations, the Food Additives Regulations, and the Food Labeling Standards are listed on the Korean FDA website. A "National NanoSafety Strategic Plan (2012/2016)" has been acting in this country. A research and development platform for nanomaterial risk assessment is presented in Fig. 6.4. Critical steps for risk assessment must be performed first, such as collecting information on appropriate measurement methods, monitoring marketable products, and valid test methods (Hwang et al., 2012). In 2011, the Ministry of Science and Economy and the Korea Standards and Technology Organization published a standard entitled "Guidelines for the safety management of nano-based products." In this standard, nanomaterials are defined as nanostructured materials in solid forms that are smaller than 100 nm (Mantovani, Porcari, Morrison, & Geertsma, 2012; Park, 2012).

The definitions of nanomaterials and nanoingredients in the food industry provided by the Korean Ministry of Food and Drug Administration (MFDS) are as follows (Kim, Lee, & Lee, 2014):

Nanomaterials are substances within the range size of 1–100 nm and can be observed by electron microscopy.

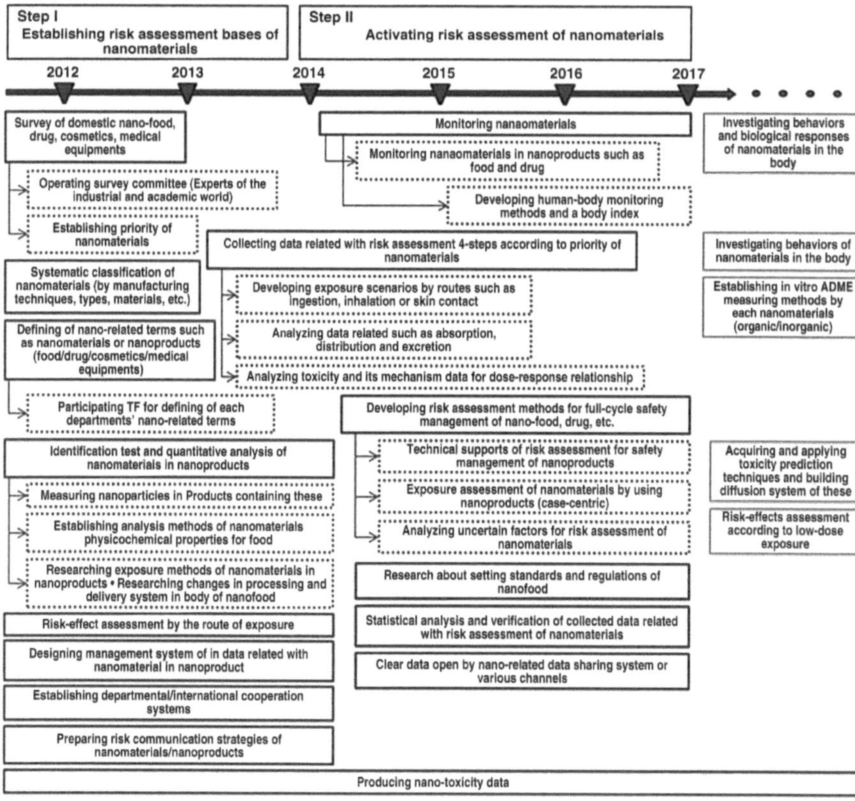

Figure 6.4 ***Research and development roadmap for the risk assessment of nanomaterials (Hwang et al., 2012).***

Nanoingredients are also referred to:

- manufactured materials with a size of 1–100 nm;
- manufactured materials containing > 50% particles with a size distribution between 1 and 100 nm;
- manufactured materials exhibiting properties at the nanoscale, if the average size does not exceed 1000 nm considering minimum one dimension; and
- the phenomenon of aggregation in these materials may occur.

6.5.5 India

In India, the central body of food safety regulations is the Food Standards and Safety Act (2006). In October 2001, the Government of India originated a Pioneering Nanoscience and Technology program, which was followed in 2007 by the Nano Mission Program. Several research activities have been started under this program, and by 2015, some of the leading projects addressed hazard and safety issues (Barpujari, 2011; Sharma & Chugh, 2009). Further, in March 2020, the Government of India has prepared a preliminary draft guideline for monitoring agricultural nano products to utilize it to monitor the safety of these products. Nanotechnology is supporting the Indian agricultural market to offer products with higher efficiency and lower cost. As the commercialization of this technology increases, the Government of India intends to establish regulations for the rapid management of safety issues in applying this technology. The laws aim to regulate a range of nanoproducts used in the agricultural sector to prevent environmental and health problems. These nanoproducts contain compounds employed in agriculture to pursue specific goals. For example, they kill germs and pests, promote better plant growth, feed crops, and increase packaging and storage efficiency.

The Department of Biotechnology, under the auspices of the Indian Ministry of Science and Technology, announced that the initial draft of the guidelines for evaluating nanoproducts used in the agricultural sector had been prepared. All sectors, such as government agencies, institutions, industries, and researchers, are called upon to promote safety. This directive of the Department of Biotechnology has been developed to improve the quantity and quality of commercialization and innovation of technologies based on nanotechnology. This guideline also attempts to ensure the safety of products. Part of this guideline states: The unique properties of NPs cause these substances' toxic behavior to humans and the environment. Previously, the department had developed guidelines for pharmaceutical nanoproducts (https://static.pib.gov.in/WriteReadData/userfiles/NanoAgri_15.6.2020.pdf).

6.5.6 Iran

The Iran Nanotechnology Initiative Council (INIC) was established in 2003 to determine general advancement and market policies for nanotechnology products. The Food and Drug Organization (FDO) is the main body responsible for certifying and licensing food products in the country, in which a nanotechnology committee was organized to perform specific guidelines for the evaluation and approval of drugs, medical equipment, cosmetics, food, supplements, and beverages (http://nanohealth.ir/pages/static_page.php?id%C2%BC14&site%C2%BC1&lang%C2%BC2). A flowchart of nano-health product approval guidelines is available on the committee's website (http://nanohealth.ir/contents/content_Nanohealth%20product%20list.pdf). The website lists safety aspects and specifications that should be considered for approval of food packaging, food additives, supplements' raw materials, and complementary products based on nanotechnology. A list of approved nano health products and product applications such as supplements, fruit packaging, and disinfectants is also available. Iran has introduced a nano-label as an industry-standard certificate used for "products related to medicine, medical equipment, beauty and healthcare, food and beverages and dietary supplements."The credit of this mark is validated through the "Nanotechnology Committee" affiliated with the FDO. Until 2015, > 12 products have received this certificate (Amenta et al., 2015).

6.5.7 Malaysia

The legal framework for food safety in Malaysia is governed by the National Food Safety Policy (NFSP), the Food Act 1983 (Act 281), and Food Regulations 1985. NFSP is responsible for ensuring food safety at all levels of the food chain from farm to storage. The purpose of Act 281 is to protect people against food-related risks and frauds and to promote and motivate the preparation, management, distribution, sale, and consumption of healthy, high-quality food. Food regulations include food-labeling regulations, the list of permitted food additives and food additives, the packaging requirements, and the list of harmful substances banned in food. The primary regulators and implementers of food safety are the Ministry of Health, the Ministry of Agriculture, the Ministry of Taxation and Royal Taxation, and local municipal councils. The Ministry of Health reviews all relevant laws and determines that, if possible, the legal principles of food safety to be extended to nano-based foods (Awang, 2016). According to these rules, the definition of nanomaterials for regulatory purposes is not only based on size but also includes nanoscale-related phenomena such as physicochemical properties (Chaudhry, Watkins, & Castle, 2010).

The current threshold does not define the particle size range or features of a particulate material used as a food additive or preservative (Awang, 2016). The maximum allowance guidelines include any scale of food additives, including those produced at the nanoscale. Food Regulations 1985 do not concretely address the application of nanotechnology in food. However, nanoscale additives must be evaluated for safety before use in food. According to the approaches to risk assessment of food products produced through novel technologies, the Malaysian Ministry of Health follows other standards and safety regulations prepared by Codex member countries, in the absence of specific international standards on the safety of nano-based food products (Salleh, Lani, Abdullah, Chilek, & Hassan, 2017).

6.5.8 Thailand

The Food and Drug Administration of the Public Health Organization is responsible for controlling food products' safety and authority in Thailand. It has attempted to develop a strategic safety plan and ethical principles for nanotechnology. The five-year project began in 2011 and lasted until 2016, which includes industry standards for nanotechnology products. The National Nanotechnology Centre (NANOTEC) is one of the main elements in implementing this national program (www.nanotec.or.th/en/?p=4014). While the NanoQ label is intended for nano-based products in Thailand, such as in the textile, plastics, and other industries, it is not yet used for foods and beverages. A similar program is being implemented by this center for the years 2017–21 (Karim et al., 2015).

6.6 Nano regulations in Australia and New Zealand

All food products commercialized in Australia and New Zealand must meet the Australian-New Zealand Food Standards (FSANZ) and be deemed safe for human consumption. FSANZ is responsible for regulating food rules in both countries. The organization was established by the Australian New Zealand Food Standard 12 Act 1991 and is responsible for maintaining the Australian New Zealand Food Standard Code. The Act sets standards for the composition, labeling, safe use, and initial production of foods traded in Australia and New Zealand. Producing food that does not comply with relevant food standards is a crime in both countries. The three main goals pursued in this Act are (1) protection of public health and safety, (2) providing adequate food information to users, and (3) preventing misleading or deceptive behaviors (Fletcher & Bartholomaeus, 2011). The organization has also developed strategies to manage the potential risks of adopting nanotechnology in food for human health and believes that functional nano foods should undergo safety and toxicity tests before wholesale. Authorities are optimistic about nano-foods' current regulatory tools (Jafari, Katouzian, & Akhavan, 2017).

In 2009, the National Industrial Chemicals Notification and Assessment Scheme (NICNAS), the Australian Government's regulator of industrial chemicals, specified industrial nanomaterials' regulatory principles as its approach. Accordingly, NICNAS has developed an empirical definition for investigating industrial nanomaterials (usable only for this field), extended through consultation with its stakeholders and nanotechnology advisory group. The definition of industrial nanomaterials is as follows (Boverhof et al., 2015):

Industrial materials that are intentionally manufactured, produced, or engineered to have unique properties or a specific composition at the nanoscale (in the range of 1–100 nm) and be nanoobject (i.e., in one, two, or three dimensions at the nanoscale) or nanostructure (i.e., have an internal or surface structure at the nanoscale). It is necessary to explain that:

1. Intentionally manufactured, produced, or engineered materials are distinct from randomly produced ones.
2. Unique properties refer to physical or chemical properties that change due to nanoscale (compared to the same non-nanoscale material) and cause unique phenomena (e.g., improved strength, chemical reactivity, or conductivity) that makes possible new applications.
3. A substance is composed of 10% or more of the materials with features mentioned above (size, unique properties, intentionally produced); NICNAS will consider it a nanomaterial.

In 2011, NICNAS announced new administrative regulations for new nanoscale chemicals not on the national commodity list. NICNAS controls the extent to which the industry complies with these administrative regulations by monitoring measures. The benefit of this change in regulation in the industry is that new nano chemicals cannot enter Australia except human and environmental exposure can be predicted. Also, self-assessment by industry cannot be the basis for the evaluation of new chemical nanomaterials. This change of regulation ensures that for all new nanomaterials, pre-market evaluation is performed by NICNAS. NICNAS guides these regulatory actions (Dalton-Brown, 2016). The NICNAS Industrial Chemicals Division has made three requests for risk assessments for materials that, by definition, are nanomaterials. No application in pesticides and food has been received by this center (OECD, 2008).

Another step taken by NICNAS was in 2013 when an electronic version of its Handbook was published to guide importers and manufacturers of industrial chemicals in Australia. One of the appendices to this document contains guidelines and requirements for notifying chemicals considered industrial nanomaterials (Amenta et al., 2015). Department of health and aging (DoHA) recommended a regulation that new industrial nanomaterials must be tested for safety before entering the market. Otherwise, nano-foods and food additives are not included in the industrial chemical groups, so the food industry is not covered by this rule (Jafari et al., 2017). Further, the approval process for using new food packaging elements is generally optional in these two countries if it has the US or the EU's approval (Tager, 2014).

6.7 Nano-labeling systems

At least three Asian countries, Iran, Taiwan, and Thailand, have used the nano-marking system, a product labeling system. It is interesting to note that all three countries' marking systems are different from each other, although applied for the same purposes and are standard and trusted by the consumer (Karim, 2014).

Taiwan has employed the Nano-Mark design, a product certification system with nano-labeling. The project is a voluntary program for manufacturers that have increased competition between companies since 2003. The Industrial Development Bureau (IDB) of the Ministry of Economic Affairs (MOEA) approved the Industrial Technology Research Institute (ITRI) to improve the nano-product certification system to implement the nano-product certification specifications. ITRI formed the world's first technology certification system. Between 2004 and 2014, 38 manufacturers and 1488 products were certified. It has been observed that with the increase in consumer faith and the improvement of the company's image, the Nano-Mark certification has increased the trading price of products by 20%. This certificate is valid for 3 years, and the renewal application must be submitted 2–6 months before the expiration date of the previous one (Snir, 2014). According to the certification, the main focus is on nano size and product performance. However, it is clear that for further success, issues such as the health and environmental implications of these nanoproducts should be confirmed by other quality control systems (Karim & Munir, 2015).

In 2010, Iran published the second national standard for nanotechnology, entitled "Nanotechnology, Occupational Health and Safety - Operation Technologies" (ALI BEITOLLAHI, NANO MARK IN IRAN, *available at* http://irannano.org/filereader.php?p1=main_cb0df69b4d3e9327f05075052c9ea583.pdf&p2=news&p3=3&p4=1). This standard is based on ISO, ASTM National and International Organization for Occupational Safety and Health (NIOSH), USA. The country has started utilizing the Nano-Mark certification scheme. The purpose of this certificate is to verify the competence of the product in various aspects related to engineered nanomaterials with dimensions less than 100 nm under ISO/TC 299. The brand will also represent health, safety, and environmental information for future consumers and regulatory bodies. This is a creative initiative since, on the one hand, it is a voluntary project. Still, on the other hand, to encourage start-ups, small and medium-sized companies, and large industrial companies active in the field of nanotechnology, it has been stated that commercial organizations with Nano-Mark have priority in using the facilities and support of the INIC. More than 294 applications for

Nano-Mark have been registered, and the INIC has approved 98 products (Karim, 2014; Karim & Munir, 2015).

Thailand is enthusiastic about being a pioneer in the field of nanotechnology. To this aim, it has taken adequate steps so far compared to other developing economic systems in Asia. The country has applied the Nano-Mark system to assess the safety of nano. This brand is known as "Nano Q," including a safety label on products containing NPs. This mark is printed by the Thai Nanotechnology Association and this association has approved the products. Nano Q is expected to enhance public reliance, support consumers and business organizations, and facilitate trade (SIRIRURG SONGSIVILAI, NATIONAL NANOTECHNOLOGY CENTER (NANOTEC) THAILAND (2011), http://nanotech.apctt.org/countryreports/Thailand%20Country%20Report.pdf.).

6.8 Asian Nanotechnology Association

Asian Nanoscience and Nanotechnology Association (ANNA) is a significant organization consisting of 15 economic powers in the Asia-Pacific area based in Singapore. All principal Asian countries in nanotechnology, including China, Japan, South Korea, and Singapore, are ANNA members. This Asian Forum creates sufficient space for communication between governments and industries of several countries. It develops discussions and exchanges to increase government, the people, and the private sector's awareness (Maclurcan, 2012). The organization is active in the field of nanotechnology standards under ISO/TS 229. The forum has established working groups, including the Standards and Risk Management Group, the Education Group, the Resources and Infrastructure Group, the NanoSafety Group, and the Water and Energy Group. Despite all these actions of the forum, we see that the organization's primary focus is nanotechnology's technical and commercial aspects, not on the legal issues. This organization can act as a leading association and connect the stakeholders in this field with programs in the area of nanotechnology's legal aspects (Karim & Munir, 2015; Karim et al., 2015).

6.9 Conclusion

Nanotechnology has opened up a world of new possibilities for the food industry. Nevertheless, this technology is commonly referred to as a double-edged sword. There is no doubt that the entry of fabricated NPs into the food chain can lead to the accumulation of toxic contaminants in foods and their harmful effects on health. On the other hand, not all nanomaterials mentioned in scientific journals are dangerous. Some nanomaterials such as silver NPs, titanium dioxide, carbon nanotubes, zinc NPs, etc., used in many marketable products are considered hazardous. At the nanoscale, particles may exhibit different behaviors from ordinary particles in contact with biological environments, indicating the necessity to investigate these particles' toxicity from different perspectives. The presentation of products that have not been evaluated for toxicity can cause critical, chronic, or long-term side effects. Therefore, legislation needs to intervene directly in the field of these NPs and nanomaterials. To fill the information gaps, continue the food industry's growth, and avoid any unexpected health risks, further studies are needed on nanotechnology applications in food processing and packaging, enacting relevant rules and regulations, and risk analysis.

Numerous nanotechnology initiatives, commissions, and centers have been set up by governments, academia, and the private sector in Asian and Oceanian countries to ensure rapid development and improved innovation. Some of them have been involved in proposing legislation to increase the protection of human health and the environment. These organizations or research centers, mainly funded by government resources, play a crucial role in conducting or supporting nanotechnology research, including basic research on nanotechnology, its applications, nanomaterial safety assessment, and regulatory control development. Yet, it appears that Asian countries, despite notable activities in the production of nano-based food compared to their American and European counterparts, have not planned a specific regulatory action for nanotechnologies and have only observed legislation in developed American and European countries to be able to employ these rules for themselves. Several European countries (UK, Sweden, Denmark), Oceania (Australia and New Zealand), Canada, and the United States have completed their assessments of the adequacy or inadequacy of the existing legal and regulatory framework for managing various aspects of nanotechnology.

In contrast, Asian countries have not addressed such an issue. Now is the time for Asian countries to consider the importance of evaluating their regulatory and legal systems in dealing with nanomaterials and reassuring consumers, the business community, and all those involved. This chapter attempts to review nanotechnology's regulatory aspects and the use of food NPs in some Asian and Oceanian countries. The effort made is only the prelude to this discussion and is by no means a comprehensive study. In the end, precise coordination of knowledge and policy between different government agencies is essential for the development of harmonized regulations on the application of nanotechnology. The labeling of nanoproducts is one of the priorities in developing standards, leading to consumer reliance and nanoproducts development by providing appropriate information to consumers about these products' safety.

References

Amenta, V., Aschberger, K., Arena, M., Bouwmeester, H., Moniz, F. B., Brandhoff, P., et al. (2015). Regulatory aspects of nanotechnology in the agri/feed/food sector in EU and non-EU countries. *Regulatory Toxicology and Pharmacology*, *73*(1), 463–476.

Anandharamakrishnan, C. (2014). Techniques for formation of nanoemulsions. In *Techniques for nanoencapsulation of food ingredients* (pp. 7–16). Springer.

Ashammakhi, N., Darabi, M. A., Çelebi-Saltik, B., Tutar, R., Hartel, M. C., Lee, J., et al. (2020). Microphysiological systems: next generation systems for assessing toxicity and therapeutic effects of nanomaterials. *Small Methods*, *4*(1), 1900589.

Awang, M. N. (2016). Regulating safety of nanotechnology in food: the legal case for Malaysia. In Bowman et al. (Ed.), *Responsibility and emerging technologies. experiences, education and beyond*, IOS Press, Forthcoming.

Barpujari, I. (2011). Attenuating risks through regulation: issues for nanotechnology in India. *Journal of Biomedical Nanotechnology*, 7(1), 85–86.

Boverhof, D. R., Bramante, C. M., Butala, J. H., Clancy, S. F., Lafranconi, M., West, J., et al. (2015). Comparative assessment of nanomaterial definitions and safety evaluation considerations. *Regulatory Toxicology and Pharmacology*, *73*(1), 137–150.

Buzea, C., Pacheco, I. I., & Robbie, K. (2007). Nanomaterials and nanoparticles: sources and toxicity. *Biointerphases*, *2*(4), MR17–MR71.

Chau, C. -F., Wu, S. -H., & Yen, G. -C. (2007). The development of regulations for food nanotechnology. *Trends in Food Science & Technology*, *18*(5), 269–280.

Chaudhry, Q., Scotter, M., Blackburn, J., Ross, B., Boxall, A., Castle, L., et al. (2008). Applications and implications of nanotechnologies for the food sector. *Food Additives and Contaminants*, *25*(3), 241–258.

Chaudhry, Q., Watkins, R., & Castle, L. (2010). Nanotechnologies in the food arena: new opportunities, new questions, new concerns. *Nanotechnologies in Food*(14), 1–17.

Cho, W. -S., Duffin, R., Poland, C. A., Howie, S. E., MacNee, W., Bradley, M., et al. (2010). Metal oxide nanoparticles induce unique inflammatory footprints in the lung: important implications for nanoparticle testing. *Environmental Health Perspectives*, *118*(12), 1699–1706.

Chowdhury, P., Gogoi, M., Borchetia, S., & Bandyopadhyay, T. (2017). Nanotechnology applications and intellectual property rights in agriculture. *Environmental Chemistry Letters*, *15*(3), 413–419.

Cowburn, R. (2008). Laser surface authentication–reading Nature's own security code. *Contemporary Physics*, *49*(5), 331–342.

Dalton-Brown, S. (2016). Public engagement and nanotechnology in Australia. *Cambridge Q. Healthcare Ethics*, *25*, 518.

Dasgupta, N., Ranjan, S., Mundekkad, D., Ramalingam, C., Shanker, R., & Kumar, A. (2015). Nanotechnology in agro-food: from field to plate. *Food Research International*, *69*, 381–400.

DePolo, W., & Baird, D. (2009). Particulate reinforced PC/PBT composites. I. Effect of particle size (nanotalc versus fine talc particles) on dimensional stability and properties. *Polymer Composites*, *30*(2), 188–199.

des Rieux, A., Fievez, V., Garinot, M., Schneider, Y. -J., & Préat, V. (2006). Nanoparticles as potential oral delivery systems of proteins and vaccines: a mechanistic approach. *Journal of Controlled Release, 116*(1), 1–27.

ETCGroup. (2006). *Nanotech product recall underscores need for nanotech moratorium: Is the magic gone?* News release, Ottawa, Available from: www.etcgroup.org/en/materials/publications.html.

FAO, F. (2017). *The future of food and agriculture–Trends and challenges.* Annual Report.

Fletcher, N., & Bartholomaeus, A. (2011). Regulation of nanotechnologies in food in Australia and New Zealand. *International Food Risk Analysis Journal, 1*(2), 33–40.

Fu, P. P., Xia, Q., Hwang, H. -M., Ray, P. C., & Yu, H. (2014). Mechanisms of nanotoxicity: generation of reactive oxygen species. *Journal of Food and Drug Analysis, 22*(1), 64–75.

Garcia, G. J., Schroeter, J. D., & Kimbell, J. S. (2015). Olfactory deposition of inhaled nanoparticles in humans. *Inhalation Toxicology, 27*(8), 394–403.

Guo, J. -W., Lee, Y. -H., Huang, H. -W., Tzou, M. -C., Wang, Y. -J., & Tsai, J. -C. (2014). Development of Taiwan's strategies for regulating nanotechnology-based pharmaceuticals harmonized with international considerations. *International Journal of Nanomedicine, 9*, 4773.

Hardy, A., Benford, D., Halldorsson, T., Jeger, M. J., Knutsen, H. K., More, S., et al. (2018). Guidance on risk assessment of the application of nanoscience and nanotechnologies in the food and feed chain: Part 1, human and animal health. *EFSA Journal, 16*(7), 5327.

He, X., Deng, H., Aker, W. G., & Hwang, H. -m. (2019a). Regulation and safety of nanotechnology in the food and agriculture industry. In *Food applications of nanotechnology* (pp. 12). CRC Press.

He, X., Deng, H., & Hwang, H. -m. (2019b). The current application of nanotechnology in food and agriculture. *Journal of Food and Drug Analysis, 27*(1), 1–21.

Higashisaka, K., Yoshioka, Y., & Tsutsumi, Y. (2015). Applications and safety of nanomaterials used in the food industry. *Food Safety, 3*(2), 39–47.

Hodge, G. A., Bowman, D., & Ludlow, K. (2009). *New global frontiers in regulation: The age of nanotechnology*. Edward Elgar Publishing.

Hwang, M., Lee, E. J., Kweon, S. Y., Park, M. S., Jeong, J. Y., Um, J. H., et al. (2012). Risk assessment principle for engineered nanotechnology in food and drug. *Toxicological Research, 28*(2), 73–79.

Jafari, S. M., Katouzian, I., & Akhavan, S. (2017). Safety and regulatory issues of nanocapsules. In *Nanoencapsulation technologies for the food and nutraceutical industries* (pp. 545–590). Elsevier.

Jani, P., Halbert, G. W., Langridge, J., & Florence, A. T. (1990). Nanoparticle uptake by the rat gastrointestinal mucosa: quantitation and particle size dependency. *Journal of Pharmacy and Pharmacology, 42*(12), 821–826.

Jeevanandam, J., Barhoum, A., Chan, Y. S., Dufresne, A., & Danquah, M. K. (2018). Review on nanoparticles and nanostructured materials: history, sources, toxicity and regulations. *Beilstein Journal of Nanotechnology, 9*(1), 1050–1074.

Kah, M., Beulke, S., Tiede, K., & Hofmann, T. (2013). Nanopesticides: state of knowledge, environmental fate, and exposure modeling. *Critical Reviews in Environmental Science and Technology, 43*(16), 1823–1867.

Karim, M. (2014). *Asia: No law for nano?* Eleventh Asian Law Institute Conference, University of Malaya, Kuala Lumpur, Malaysia.

Karim, M., & Munir, A. B. (2015). Nanotechnology in Asia: a preliminary assessment of the existing legal framework. Abu Bakar, *Nanotechnology in Asia: a preliminary assessment of the existing legal framework* (April 4, 2015).

Karim, M.E. (2013). Nanotechnology safety in Asian perspective: A legal analysis. In: Asia Pacific Symposium on Safety (APSS) 2013, 17 October 2013, Singapore.

Karim, M. E., Munir, A. B., Reza, A. W., Muhammad-Sukki, F., Yasin, S. H. M., Abu-Bakar, S. H., et al. (2015). Too enthusiastic to care for safety: Present status and recent developments of nanosafety in ASEAN countries. *Technological Forecasting and Social Change, 92*, 168–181.

Kim, H.-K., Lee, J.-G., & Lee, S.-Y. (2014). Current regulation of nanomaterials used as food ingredients. In *Nano-and microencapsulation for foods* (pp. 383). John Wiley & Sons.

Kim, K. E., Cho, D., & Park, H. J. (2016). Air pollution and skin diseases: Adverse effects of airborne particulate matter on various skin diseases. *Life Sciences, 152*, 126–134.

Kreuter, J. (2004). Influence of the surface properties on nanoparticle-mediated transport of drugs to the brain. *Journal of Nanoscience and Nanotechnology, 4*(5), 484–488.

Linkov, I., & Steevens, J. A. (2008). *Nanomaterials: risks and benefits:*. Springer Science & Business Media.

Lövestam, G., Rauscher, H., Roebben, G., Klüttgen, B. S., Gibson, N., Putaud, J.-P., et al. (2010). Considerations on a definition of nanomaterial for regulatory purposes. *Joint Research Centre (JRC) Reference Reports, 80*, 00–41.

Lozinsky, V. I., Galaev, I. Y., Plieva, F. M., Savina, I. N., Jungvid, H., & Mattiasson, B. (2003). Polymeric cryogels as promising materials of biotechnological interest. *Trends in Biotechnology, 21*(10), 445–451.

Maclurcan, D. (2012). *Nanotechnology and global equality*. CRC Press.

Magnuson, B., Munro, I., Abbot, P., Baldwin, N., Lopez-Garcia, R., Ly, K., et al. (2013). Review of the regulation and safety assessment of food substances in various countries and jurisdictions. *Food Additives & Contaminants: Part A, 30*(7), 1147–1220.

Mansoori, G. A., & Soelaiman, T. F. (2005). Nanotechnology—an introduction for the standards community. *Journal of ASTM International, 2*(6), 1–22.

Mantovani, E., Porcari, A., Morrison, M., & Geertsma, R. (2012). Developments in nanotechnologies regulation and standards 2011-Report of the observatory nano. Brussels, Observatory Nano. Available from: http://www.observatorynano.eu/project/filesystem/files/ObservatoryNano_Nanotechnologies_RegulationAndStandards_2012.pdf.

Maynard, A. D. (2011). Don't define nanomaterials. *Nature, 475*(7354), 31–131.

Milani, S., Baldelli Bombelli, F., Pitek, A. S., Dawson, K. A., & Radler, J. (2012). Reversible versus irreversible binding of transferrin to polystyrene nanoparticles: soft and hard corona. *ACS Nano, 6*(3), 2532–2541.

Momin, J. K., & Joshi, B. (2015). Nanotechnology in foods. In *Nanotechnologies in Food and Agriculture* (pp. 3–24). Springer.

Nanoposts, R. (2008). The Market for Nanotechnology in Food and Drink to 2015, 2008. Nanoposts.com.

Nile, S. H., Baskar, V., Selvaraj, D., Nile, A., Xiao, J., & Kai, G. (2020). Nanotechnologies in food science: applications, recent trends, and future perspectives. *Nano-Micro Letters, 12*(1), 45.

OECD. (2008). Environment Directorate Joint Meeting of the Chemical Committee and the Working Party on Chemicals, Pesticides and Biotechnology. Series on Testing and Assessment, Number 43. Guidance document on mammalian reproductive toxicity testing and assessment. ENV/JM/Mono(2008)16. Adopted on 24 July 2008.

Park, Y. (2012). http://www.egg2012.de/tl_files/pdf/Paper/EGG2012_C6_1_Park_presentation_slides.pdf.

Peters, R., Brandhoff, P., Weigel, S., Marvin, H., Bouwmeester, H., Aschberger, K., et al. (2014). Inventory of nanotechnology applications in the agricultural, feed and food sector. *EFSA Supporting Publications, 11*(7), 621E.

National Science and Technology Council Committee on Technology Subcommittee on Nanoscale Science, Engineering, and Technology. The National Nanotechnology Initiative Strategy Plan December (2007). Washington D.C. 20502. Office of Science and Technology Policy.

Rahman, M., Laurent, S., Tawil, N., & Mahmoudi, M. (2013). Nanoparticle and protein corona. In *Protein-nanoparticle interactions* (pp. 21–44). Springer.

Rossi, M., Cubadda, F., Dini, L., Terranova, M., Aureli, F., Sorbo, A., et al. (2014). Scientific basis of nanotechnology, implications for the food sector and future trends. *Trends in Food Science & Technology*, *40*(2), 127–148.

Sabatier, M., & Chollet, B. (2017). Is there a first mover advantage in science? Pioneering behavior and scientific production in nanotechnology. *Research Policy*, *46*(2), 522–533.

Salleh, W., Lani, M. N., Abdullah, W. W., Chilek, T. Z. T., & Hassan, Z. (2017). A review on incidences of foodborne diseases and interventions for a better national food safety system in Malaysia. *Malaysian Applied Biology*, *46*(3), 1–7.

Sanguansri, P., & Augustin, M. A. (2006). Nanoscale materials development–a food industry perspective. *Trends in Food Science & Technology*, *17*(10), 547–556.

Saptarshi, S. R., Duschl, A., & Lopata, A. L. (2013). Interaction of nanoparticles with proteins: relation to bio-reactivity of the nanoparticle. *Journal of Nanobiotechnology*, *11*(1), 26.

Schneider, M., Stracke, F., Hansen, S., & Schaefer, U. F. (2009). Nanoparticles and their interactions with the dermal barrier. *Dermato-endocrinology*, *1*(4), 197–206.

Sharma, K., & Chugh, A. (2009). Legal aspects of nanobiotehnology inventions: An Indian perspective. *SCRIPTed*, *6*, 433.

ŠIMON, P., Chaudhry, Q., & BAKOŠ, D. (2008). Migration of engineered nanoparticles from polymer packaging to food--a physicochemical view. *Journal of Food & Nutrition Research*, *47*(3).

Snir, R. (2014). Trends in global nanotechnology regulation: the public-private interplay. *Vanderbilt Journal of Entertainment & Technology Law*, *17*, 107.

Stone, V., Johnston, H., & Clift, M. J. (2007). Air pollution, ultrafine and nanoparticle toxicology: cellular and molecular interactions. *IEEE Transactions on Nanobioscience*, *6*(4), 331–340.

Szentkuti, L. (1997). Light microscopical observations on luminally administered dyes, dextrans, nanospheres and microspheres in the pre-epithelial mucus gel layer of the rat distal colon. *Journal of Controlled Release*, *46*(3), 233–242.

Tachikawa, M. (2012). Uncertainty of, and stakeholder response to, emerging technologies: food nanotechnology in Japan. *Ethics in Science and Environmental Politics*, *12*(2), 113–122.

Tager, J. (2014). Nanomaterials in food packaging: FSANZ fails consumers again. *Chain Reaction*, *122*, 16–17.

Takeuchi, M. T., Kojima, M., & Luetzow, M. (2014). State of the art on the initiatives and activities relevant to risk assessment and risk management of nanotechnologies in the food and agriculture sectors. *Food Research International*, *64*, 976–981.

Unfried, K., Albrecht, C., Klotz, L. -O., Von Mikecz, A., Grether-Beck, S., & Schins, R. P. (2007). Cellular responses to nanoparticles: target structures and mechanisms. *Nanotoxicology*, *1*(1), 52–71.

Wang, G. (2018). Nanotechnology: The new features. *arXiv preprint arXiv:1812.04939*.

Wijnhoven, S. W., Peijnenburg, W. J., Herberts, C. A., Hagens, W. I., Oomen, A. G., Heugens, E. H., et al. (2009). Nano-silver–a review of available data and knowledge gaps in human and environmental risk assessment. *Nanotoxicology*, *3*(2), 109–138.

Zhang, Q., Kusaka, Y., Zhu, X., Sato, K., Mo, Y., Kluz, T., et al. (2003). Comparative toxicity of standard nickel and ultrafine nickel in lung after intratracheal instillation. Journal of Occupational Health, 45(1), 23–30.

CHAPTER 7

Regulatory principles on food nano-particles legislated by North and South American countries

Masoud Faghih Akhlaghi[a], Marjan Daeihamed[b], Seid Mahdi Jafari[c]
[a]Department of Medicinal Chemistry, School of Pharmacy, Guilan University of Medical Sciences, Rasht, Iran
[b]Department of Pharmaceutics, School of Pharmacy, Guilan University of Medical Sciences, Rasht, Iran
[c]Department of Food Materials and Process Design Engineering, Gorgan University of Agricultural Science and Natural Resources, Gorgan, Iran

7.1 Introduction

Since the entrance of nanotechnology to food industries in 2003, some serious questions were raised about their safety and regulation. Although the scientific data about this new technology was not adequate to set a regulatory framework but some regulating agencies all over the world started to consider regulatory issues for nanotechnology-based food products (Scott, Chen, & Rutzke, 2003). This rapidly growing innovation made significant changes in different sections of food industry and brought a lot of novelties and economic potential. Nanomaterials resulted in manufacturing of high efficiency food products, which was mainly attributed to their small size and extremely high surface area (He & Hwang, 2016; Amenta et al., 2015). However, studies confirmed higher reactivity and interaction of nanomaterials in all aspects including human exposure, which can completely change their toxicity profile, calling for detailed considerations in regulatory frameworks (Wacker, 2014).

Although the scientific data and methods regarding safety of nanotechnology-based food products are not clear in all aspects but different countries try to set regulatory frameworks in a way that ensure the safety of products and also help the manufacturers to benefit from this emerging technology. Most of the published regulatory data on food-related nanomaterials share the common basic concepts but there is still a long way for harmonized and clear legislations between countries.

The Food and Drug Administration (FDA) of the United States of America (USA) was among the first governing agencies, which was aware

Safety and Regulatory Issues of Nanoencapsulated Food Ingredients
http://dx.doi.org/10.1016/B978-0-12-815725-1.00007-0

of the challenges related to the regulation and safety of nanotechnology and started considering nanotechnology-related issues for all products including food. Initially, there were only a few nanotechnology-based food products (mostly as dietary supplements) in the United States, but the implications of nanotechnology expanded rapidly in this country and it was in the top list of countries using nanotechnology-based products. The Nanotechnology Task Force of the FDA mentioned the regulatory challenges regarding nanotechnology-based products in a report released in 2007 (FDA, 2007) and some years later released some documents and guidance about nanotechnology-based products for manufacturers.

Although United States was the first American country that mentioned nanotechnology issues in regulation of food products, some other North and South American countries were also interested in nanotechnology-based food products and some activities were conducted regarding their regulation. In this chapter, a general view on the legislations in the United States and other North and South American countries is given.

7.2 Current and future applications of nanotechnology in food industries in North and South American countries

As described before, nanotechnology has been applied in different sectors of food chain including agriculture, feed and food, like food additives or food contact materials. Preparing an exact and complete inventory for nanotechnology-related food products in each country is very challenging, because harmonized and transparent regulatory items are not available to exactly define whether a product is considered "nano" or contains nanomaterials. Although some products are "claimed to contain nano", but some applications of nanotechnology may be missed or not claimed. Nanomaterials may be passed into food products from previous steps of food preparation chain (i.e., applications of nanotechnology in agriculture or animal feed) or through contaminations. The Center for Food Safety (CFS) (Center for Food Safety CSF, 2018) provides an inventory of food products containing nanotechnology. According to this database, nanotechnology-related products are widely used in American countries. The United States have the most number of records in this list with more than 300 records in 2018. FDA estimated the market growth of nanofood to reach up to more than 20 billion dollar in 2020, from 4 million in 2006 and 6 billion in 2012 (Chaturvedi & Dave, 2020). The potential applications of nanotechnology in North and South American countries are the same as other countries (Chapter 5) and some important applications include:

- Direct food additives at nanoscale including flavors, preservatives, thickening agents, and etc. which have higher efficacies,
- Increasing the quality of food by nanoencapsulation and nanoemulsions like taste masking of tuna fish oil in bread,
- Increasing the nutritional value of products providing higher availability of vitamins, proteins, minerals and phytochemicals, or lower carbohydrates and sugar,
- Edible nano-coating on food materials to provide a barrier against moisture and gas and increase the shelf life of food,
- Filtration and water purification systems,
- Fertilizers used for plant growth increasing nutrient and oxygen uptake,
- Different food packaging and food contact materials consisting of nanoclay or nano-titanium dioxide which extend the product freshness and shelf life,
- Nanomaterials exhibiting antimicrobial properties in food packaging, mostly based on of silver nanoparticles,
- Beverage bottles consisting of nanomaterials to prevent gas exchange,
- Nanosensors to detect trace amounts of contamination, and
- Nanoscale sensors for tracking of food (Duvall, 2012; Center for Food Safety CSF, 2018; He & Hwang, 2016; He, Deng, & Hwang, 2019).

The increasing number of records in inventories and cutting-edge novelties of nanotechnology shows that nanomaterials would be present in a wide variety of food products. Hence, clear consumer awareness and proper regulatory and risk assessment protocols are required to ensure consumer safety.

7.3 Regulations for food nanotechnology in North and South American countries

Although nanomaterials bring a lot of advantages in food technology, some new risks and safety issues raised for these small-sized particles regarding environmental and human health. These toxicity issues required new methods for risk evaluation and legislations to ensure their safe consumption and reduce the potential risks (Dasgupta & Ranjan, 2018; Park & Yeo, 2016). Numerous concerns regarding the safety of food-related nanomaterials lead to publication of legislations and guidance considering possible risks. The existing legislations and legal issues in North and South American countries are described further.

7.3.1 Regulations for food nanotechnology in North American countries

In the United States, the FDA deals with the safety and regulations of commercialized food products, typically food and feed additives and food contact materials, under the authority of the Federal Food, Drug, and Cosmetic Act (FFDCA). FDA tried to use the existing legislations for nanotechnology-based products as well and considered them adequate for these types of products. Though FFDCA do not include specific issues for nanomaterials and a clear definition of nanomaterials was not involved in FDA regulations (Hamburg, 2012; Amenta et al., 2015). There are no obligatory regulations for nanotechnology-based food products and FDA recommends the manufacturers to consult with them before their products are released in the market or considered for review. It is also notified that nanotechnology-based products may be subjected to additional safety reviews on a case-by-case approach (FDA, 2014a; Ravichandran, 2010).

However, some useful guidance about nanotechnology-based products was provided by FDA for the industry. A guidance titled "Considering Whether an FDA-Regulated Product Involves the Application of Nanotechnology" was published in 2014 (FDA, 2014b), which defines the characteristics needed to consider any FDA-regulated product, including food, as a nanotechnology-based product. The "Guidance for Industry: Assessing the Effects of Significant Manufacturing Process Changes, Including Emerging Technologies, on the Safety and Regulatory Status of Food Ingredients and Food Contact Substances, including Food Ingredients that are Color Additives." (FDA, 2014a), was another food-specific guidance published in 2014, which considers the use of nanotechnology as a significant change in the manufacturing process of food additives and food contact materials. Although FDA does not consider nanotechnology-related products as intrinsically harmful, but according to this guidance if food or any food-related materials are prepared based on nanotechnology techniques, it should be proved that the final products are safe using proper and validated methods for nanomaterials. The guidance states that:

> *"As with any other manufacturing technologies applied to food, nanotechnology and other emerging technologies may introduce issues that warrant additional or different evaluation during a safety assessment of a food substance. For example, so-called nano-engineered food substances can have substantially altered bioavailability and may, therefore, raise new safety issues that have not been seen in their traditionally manufactured counterparts. There is debate regarding the appropriate testing to judge the safety of food substances where physical and chemical properties are manipulated by engineering particle size distributions in the*

nanometer range, in some instances to intentionally manipulate the biological properties, as also noted by some scientific and regulatory authorities. Guidance and extrapolation from data on traditionally manufactured food substances can generally be accomplished only on a case-by-case basis. When a food substance is manufactured to include a particle size distribution shifted more fully into the nanometer range, safety assessments should be based on data relevant to the nanometer version of the food substance. Where nano-engineered food substances have new properties, additional or different testing methods may be necessary to determine the safety of the food substance."

Generally, food and color additives need premarket authorization by the FDA, while for food ingredients generally recognized as safe (GRAS), premarket authorization is not required (FDA, 2017). According to this guidance nanotechnology-based products are unlikely to be considered as GRAS, and a formal premarket authorization may be required (FDA, 2014a).

However, it should be noted that these guidance are not sufficient for controlling the risk of nanotechnology-related products and manufacturers are recommended to conduct the following actions when there is a significant change to the manufacturing process, including the utilization of nanotechnology:

- Determine what changes have been made to the food substance, including its physicochemical structure and properties, purity and impurities
- Conduct a safety assessment for the food substance, including the factors set forth above, bioavailability, and toxicity
- Consider whether the substance is within a relevant regulation, GRAS determination, or food contact notification
- Consult with FDA about the conclusions
- Make a regulatory submission, if warranted

In conclusion, manufacturers of any food product or food-related materials, including food additives and food contact substances, who want to use nanotechnology approaches should conduct safety testing of product through tests that has been validated for nanomaterials, and try to obtain premarket approvals from FDA (FDA, 2014a).

In Canada, the "Health Canada" department is responsible for regulating all consumed products including food and food packaging substances. Currently there is no specific regulation regarding the applications of nanotechnology in food products and Health Canada evaluates nanotechnology-based food products under the existing legislation and regulatory frameworks. Actually, a case-by-case approach has been recommended to exactly assess the safety of nanotechnology-related products and ensure consumer safety while profit from new technological developments. They also encourage the manufacturers to communicate the related regulatory departments in early steps of product development, if they are planning to use nanotechnology-based approaches for food products (Amenta et al., 2015).

Health Canada has provided a working definition of nanomaterials in 2011 published under "Policy Statement on Health Canada's Working Definition for Nanomaterials" based on nanoscale dimensions (1–100 nm) or special properties that are different from conventional materials (Health Canada, 2011). Health Canada has stated that in cases where size range is above the 100 nm, up to the upper limit of 1000 nm, but nanoscale properties and phenomena are suspected, individual regulatory programs can request information for risk assessment. Health Canada suggests the manufacturers to request a meeting, prior to entering of the product to the market, with the corresponding regulatory authority to discuss the safety assessment. Some information about the intended use of the nanomaterial, manufacturing methods, physicochemical properties of nanomaterial, toxicological and environmental fate of nanomaterial and risk assessment and management strategies may also help in this regard (Environment Canada, 2014). Canada and United States are trying to set up similar regulatory approaches for nanomaterials under the US-Canada Regulatory Cooperation Council (Canada-United States Regulatory Cooperation Council RCC, 2014).

Mexico, which is a Latin American country in North America, was one of the first countries working and investing on nanotechnology research and industry. In 2011, the Federal Government started to work on nanotechnology-related regulations by contribution of academics, policy makers, and manufacturers. They concluded that the present regulatory framework includes useful regulations for nanotechnologies, but it was necessary to prepare a set of guidelines for some specific issues (Delgado-Ramos, 2014). Due to the commercial relationship between Mexico and United States, the regulation of nanotechnology had most parts in common with US regulatory. In 2012, a nanotechnology-related guideline titled "Regulatory guidelines on nanotechnologies for the achievement of competitiveness and protection of the environment, health and consumer safety" was published by the Secretary of Economy (Foladori & Lau, 2014). However, the guidelines are nonbonding and there is no mandatory regulation for risk assessment of nanotechnology at present in Mexico, while the legislation lack specific issues regarding the applications of nanotechnology in food products.

7.3.2 Regulations for food nanotechnology in South American countries

Brazil and Argentina are the leading countries in developing nanotechnology-related applications and research among other South American countries (Kay & Shapira, 2009). In Brazil, significant efforts have been done on research and investments in nanotechnology applications but there is no specific regulation in this country. Since 2011 some efforts have been made by Brazilian nanotechnology experts to prepare legislations and standards in a meeting in São Paulo. In 2013, a proposal was introduced about proper labeling of food, drugs, and cosmetics manufactured by nanotechnology techniques. This proposal was rejected by the Brazilian congress because of possible negative economic effects on companies, which had invested significant amounts of money on nanotechnology (Bumbudsanpharoke & Ko, 2015; Amenta et al., 2015). However, there are a lot of regulatory debates regarding the toxicological impacts of nanotechnology and there is no legislation in this regard until present.

In Argentina, nanotechnology was introduced in the public agendas from 2003 by the Ministry of Science and Technology Secretariat and it was considered as a priority for funding and research but there are no current regulations regarding nanotechnology in this country (Foladori, 2013).

Although other South American countries including Colombia, Equator, Chile, Peru, Uruguay, and etc. have participated in nanotechnology research and some institutions have been active in this regards but there are no official and formal legislations for nanotechnology-based products especially food products (Kay & Shapira, 2009; Foladori, 2013).

7.4 Regulatory aspects considered in food nanotechnology in North and South American countries

Although food-related nanotechnology guidelines in North and South American countries are not as comprehensive as European countries but the most important and highlighted parts in guidelines are the properties that result in consideration of a product as nano-based product.

In the initial years, FDA did not define standard and regulatory definitions of nanotechnology-related terms because they believed that the scientific base was insufficient to provide a definition suitable for the purposes of regulation (Amenta et al., 2015). However, in the guidance published in 2014 for industry (FDA, 2014a, 2014b), an FDA-regulated product including food is considered to exhibit utilization of nanotechnology, if the final product or any material in the product is "engineered" to have these properties:

1. At least one external dimension, or an internal or surface structure, in the nanometer range (1–100 nm),
2. Exhibit properties or phenomena, including physical or chemical properties or biological effects that are attributable to its dimension(s), even if these dimensions fall outside the nanoscale range, up to 1000 nm.

Briefly, FDA looks at the material's size or properties/phenomena related to its dimension even if the dimensions are not in the nanoscale range. These size-dependent properties are distinguishable from the chemical or physical properties of individual atoms, molecules or bulk material and are not normally observed or expected in larger particles with the same chemical composition. These may include the functional properties like improved packaging properties or better delivery of nutrients (FDA, 2014b).

The guidance refers to both primary particles having 1–100 nm external dimensions, and any aggregates or agglomerates formed by these primary particles, as well as coated or assembled structures that include internal or surface nanoscale entities, which may have modified properties that can affect the safety or effectiveness of product. It covers both the finished products, as well as materials and ingredients that are intended for use in final products, like food additives. The term "engineered" refers to deliberate and purposeful manipulation of particle size by the application of nanotechnology and shows that the process is engineered and controlled. If some products naturally or unintentionally contain particles in nanometer range are not covered under the scope of this guidance. Therefore, conventionally manufactured food products that may contain some nanometer-sized particles are not considered in this guidance (FDA, 2014b).

However, if a product encompass within the above-mentioned criteria, any unique characteristics or biological effects of the product that may influence its safety, effectiveness, public health impact, or regulatory status should be considered and evaluated by FDA and industry. Generally, in the food sector, a case-by-case risk assessment of nano-based products is recommended by FDA.

FDA does not consider all nanotechnology-based products as a separate category but considers the potential impact of nanotechnology as a significant manufacturing process on the safety and regulation of food products. However, like other conventional food products, for nanotechnology-based products FDA considers the properties of finished product and the safety of its intended use. Generally, if nanotechnology is used in the manufacturing process of any food products including food or color additives that have been approved by regulation, foods identified as GRAS, or food contact substances that have been approved previously, safety evaluations should be conducted by manufacturers using tests that are validated for nanomaterials. The manufacturers should also consult to FDA and obtain the approval prior to marketing of products, or accept possible FDA's enforcement action (FDA, 2014a).

A working definition with the same concept is given by Health Canada based on nanoscale dimensions (1–100 nm) or special properties that are different from conventional materials. In cases where size range is above the 100 nm but nanoscale properties and phenomena are suspected (up to the upper limit of 1000 nm), individual regulatory programs can request information for risk assessment (Health Canada, 2011).

7.5 Labeling

The expanding market of nanotechnology-based food products calls for regulated and mandatory labeling of these products to help consumers make informed choices. However, there are no specific labeling requirements for nanomaterials in most countries and labeling of nanotechnology-based food products is a global controversial subject.

In the United States, FDA had no plans for labeling of nanotechnology-based products. Labeling in the United States is based on the materials existing in the product in order to use the product safely. Labeling does not cover any information about the manufacturing process of the product. Unlike EU, FDA labeling rules are not based on the "right to be informed" and FDA is concerned about overcrowded labels having detailed information, which are not practically useful for consumers. For instance, genetically modified foods are labeled in the EU but not in the United States. However, FDA suggests product labeling on a case-by-case basis and individual products may have further information, if it is needed (Falkner et al., 2009).

Among other American countries, some guidelines were published in Mexico considering regulation of nanotechnologies by the Mexican Secretary of Economy. This guideline entitled the "Regulatory guidelines on nanotechnologies for the achievement of competitiveness and protection of the environment, health and consumer safety," is divided into four chapters and some specific principles are discussed in these chapters to guarantee competitiveness while trying to justify consumer and environmental protection. The 8th principle in the second chapter calls for "Informing society, particularly consumers, about the products containing nanomaterials and the possible short, medium and long term effects," but this principle does not mention mandatory or voluntary labeling (Foladori & Lau, 2014). Moreover, ISO has also published a voluntary labeling guide for consumer products containing manufactured nanomaterials, which may help for labeling of nanotechnology-related food products (ISO, 2013).

7.6 Conclusions, challenges, and future remarks

With the rapid development of nanotechnology-based products in different industries including food, various regulatory approaches were considered in different countries. These policies are basically similar but are customized in each country according to its existing regulatory framework and are updated thoroughly to ensure consumer safety and help the industry in implications of this new technology. Among different North and South American countries the United States was the first country having various nanotechnology-based products in the market. It was also the first country in this region that released some guidance and definitions for industries and further considered nanotechnology as a significant change in the manufacturing process of food products.

However, FDA suggests a case-by-case and early consultation approach for safety evaluations of nanotechnology-based food products and the guidelines are not as comprehensive as European guidelines. FDA admires the manufacturers to provide safety information for their products but there is a gap of clear information about validated safety tests for nanotechnology-based food products. This may less encourage the manufacturers to invest on new emerging technologies. Simple and validated safety tests, as well as clear definitions and standards should be defined in nanotechnology-based food guidelines which may be more updated through scientific experiences in this field. There were also some initial activities regarding regulation of nanotechnology-based products in other American countries, which needs further revisions to be more comprehensive. However, future prospects call for international guidelines and standards to facilitate commercial exchange between different countries.

References

Amenta, V., Aschberger, K., Arena, M., Bouwmeester, H., Botelho Moniz, F., Brandhoff, P., et al. (2015). Regulatory aspects of nanotechnology in the agri/feed/food sector in EU and non-EU countries. *Regulatory Toxicology and Pharmacology*, *73*(1), 463–476.

Bumbudsanpharoke, N., & Ko, S. (2015). Nano-food packaging: an overview of market, migration research, and safety regulations. *Journal of Food Science*, *80*(5), R910–R923.

Canada-United States Regulatory Cooperation Council RCC (2014). Joint Forward Plan. Avaialable from: https://www.whitehouse.gov/sites/default/files/omb/oira/irc/us-canada-rcc-joint-forward-plan.pdf.

Center for Food Safety CFS (2018). Nanotechnology in our food: Interactive database of consumer food products containing nanomaterials. Available from: https://www.center-forfoodsafety.org/nanotechnology-in-food.

Chaturvedi, S., & Dave, P. N. (2020). Application of nanotechnology in foods and beverages. In A. M. Grumezescu, & A. M. Holban (Eds.), *Nanoengineering in the beverage industry* (pp. 137–162). Academic Press.

Dasgupta, N., & Ranjan, S. (2018). Nano-food toxicity and regulations. In N. Dasgupta, & S. Ranjan (Eds.), *An introduction to food grade nanoemulsions* (pp. 151–179). Springer: Singapore.

Delgado-Ramos, G. C. (2014). Nanotechnology in Mexico: Global trends and national implications for policy and regulatory issues. *Technology in Society*, *37*, 4–15.

Duvall, M. N. (2012). *FDA regulation of nanotechnology*. Washington, DC, USA: Beveridge & Diamond, PG.

Environment Canada (2014). Assessment of nanomaterials under the new substances notification regulations (chemicals and polymers). Available from: https://www.ec.gc.ca/subsnouvelles-newsubs/default.asp?lang%C2%BCEn&n%C2%BC53527-F9D-1.

Falkner, R., Breggin, L., Jaspers, N., Pendergrass, J., & Porter, R. (2009). Consumer labelling of nanomaterials in the EU and US: Convergence or divergence? Briefing Paper. London, UK: Chatham House.[online]. Available from: https://www.chathamhouse.org/publications/papers/view/109136.

FDA (2007). Nanotechnology: a report of the U.S. food and drug administration nanotechnology task force.

FDA (2014a). Guidance for industry, assessing the effects of significant manufacturing process changes, including emerging technologies, on the safety and regulatory status of food ingredients and food contact substances, including food ingredients that are color additives. Available from: http://www.fda.gov/Food.

FDA (2014b). Guidance for industry, considering whether an FDA-regulated product involves the application of nanotechnology. Available from: http://www.fda.gov/regulatoryinformation/guidances/ucm257698.htm.

FDA (2017). Regulatory framework for substances intended for use in human food or animal food on the basis of the generally recognized as safe (GRAS) provision of the federal food, drug, and cosmetic act: guidance for industry. Available from: https://www.fda.gov/animal-veterinary/guidance-industry/guidance-number.

Foladori, G. (2013). Nanotechnology policies in Latin America: risks to health and environment. *NanoEthics*, 7(2), 135–147.

Foladori, G., & Lau, E. Z. (2014). The regulation of nanotechnologies in Mexico. *Nanotechnology Law & Business*, *11*, 164.

Hamburg, M. A. (2012). FDA's approach to regulation of products of nanotechnology. *Science*, *336*(6079), 299–300.

He, X., Deng, H., & Hwang, H. M. (2019). The current application of nanotechnology in food and agriculture. *Journal of Food and Drug Analysis*, *27*(1), 1–21.

He, X., & Hwang, H. M. (2016). Nanotechnology in food science: functionality, applicability, and safety assessment. *Journal of Food and Drug Analysis*, *24*(4), 671–681.

Health Canada (2011). Policy statement on Health Canada's working definition for nanomaterial. Available from: http://www.hc-sc.gc.ca/sr-sr/pubs/nano/pol-eng.php.

ISO (2013). Technical Specification 13830, Nanotechnologies—Guidance on voluntary labelling for consumer products containing manufactured nano-objects.

Available from: http://www.iso.org/iso/home/store/catalogue_tc/catalogue_detail.htm?csnumber=54315.

Kay, L., & Shapira, P. (2009). Developing nanotechnology in Latin America. *Journal of Nanoparticle Research*, *11*(2), 259–278.

Park, H. G., & Yeo, M. K. (2016). Nanomaterial regulatory policy for human health and environment. *Molecular and Cellular Toxicology*, *12*(3), 223–236.

Ravichandran, R. (2010). Nanotechnology applications in food and food processing: innovative green approaches opportunities and uncertainties for global market. *International Journal of Green Nanotechnology: Physics and Chemistry*, *1*(2), P72–P96.

Scott, N., Chen, H., & Rutzke, C.J. (2003). Nanoscale Science and Engineering for Agriculture and Food Systems: A Report Submitted to Cooperative State Research, Education and Extension Service, the United States Department of Agriculture : National Planning Workshop, November 18-19, 2002, Washington, DC, USDA.

Wacker, M. G. (2014). Nanotherapeutics—product development along the "nanomaterial" discussion. *Journal of Pharmaceutical Sciences*, *103*(3), 777–784.

CHAPTER 8

Regulatory principles on food nano-particles legislated by international organizations

Özgür Tarhan[a], Atike Ince Yardımcı[b]
[a]Department of Food Engineering, Engineering Faculty, Uşak University, Uşak, Turkey
[b]Technology Transfer Office, Uşak University, Uşak, Turkey

8.1 Introduction

Innovative food products prepared with the incorporation of nanoparticles through nanotechnology methods and tools have garnered great attention worldwide. Organic and inorganic nanoparticles are promising various functionalities for food industry through their unique physicochemical and structural properties. Thus, the number of food products containing nanoscale ingredients, referred to nanofoods, present in market is increasing rapidly and continuously (Chaudhry et al., 2008; Bouwmeester et al., 2014). Engineered nanoparticles blended in foods (e.g., baby and infant foods) directly, food supplements, and packaging materials have been reported in certain databases, for example, European inventories (Sohal et al., 2018). Limited knowledge about the potential risks of nanoparticles ingested, inhaled, or directly exposed can lead to serious consequences. These ultra tiny particles added to food-related matrices may potentially cause circulatory, cardiovascular, and pulmonary diseases due to their ability to pass biological barriers in the body (Bouwmeester et al., 2009).

Besides serious health risks, improperly used nanoparticles can lead to undesirable environmental impacts through high exposure level and widespread implementation. Complete identification of nanoscale materials, assessment of risk factors and toxicity, safe handling principles, exposure rates, informing community, and establishment of regulations are critical subtopics to be taken into consideration when establishing nanofood management strategies.

In the worldwide, many initiatives have been introduced targeting hazard analysis and regulation of nanomaterials used in food and related matrices (Chau et al., 2007; Cushen et al., 2012; Bouwmeester et al., 2014). Both

Safety and Regulatory Issues of Nanoencapsulated Food Ingredients
http://dx.doi.org/10.1016/B978-0-12-815725-1.00008-2

national and international authorities in governmental and non-governmental status are involved in establishing regulatory frameworks (Amenta et al., 2015; Jain et al., 2018). The nation-based or local bodies authorized for that and the related guidance provided by them have been discussed in detail in the previous chapters of this book. International organizations to be discussed here have been mainly supported by governmental sources to conduct nanotechnology researches including safety assessment and regulatory guidance of nanomaterials to be used in the industry.

Codex Alimentarius Commission (CAC), an agency established jointly by Food and Drug Administration (FDA) and World Health Organization (WHO), Organization for Economic Cooperation and Development (OECD), International Standard Organization (ISO), and International Risk Governance Council (IRGC) are the major international bodies directing regulatory issues and guidelines related to nanomaterials in foods. Legislative guidelines and recommendations released by legal international authorities introduce safe and harmonized approaches to assess safety of nanotechnology-based food products. Major principles for regulation of nano products include identification of nanoparticles in terms of size, morphology, surface characteristics, etc., through various analytical methods, assessment of bioabsorption, and cytotoxicity through in vitro and in vivo studies, and degradation in the environment.

Potential risks and hazards associated with the nanomaterials used in food matrices leading to health concerns might be designed out by the regulatory frameworks established with the global consensus. That expectedly provides standardization of the regulatory protocols, thus supports commercial potential of nanofood products and public perception. This chapter will review principles of current regulations launched by international organizations to clarify the mutual approaches in the assessment of nanofood safety.

8.2 Nanoparticle incorporated foods in market and safety issues

Nanotechnology has significant impacts on many industrial areas including agriculture and food at various stages from production to sale. Agricultural raw materials continuously feeding food industry, food formulations and production methodologies, and chemical polymeric materials essentially associated with food manufacturing (e.g., packaging) are naturally involved in whole process and can be designed, improved, and fabricated through innovative technological principles. The benefits of nanotechnology-based food products for consumers and manufacturers are widely reported in terms of bioavailability, food safety, target, and controlled delivery systems (Chen et al., 2006; Bouwmeester et al., 2009).

Many companies are currently manufacturing nano-based food and food-related products for global market (Bumbudsanpharoke & Ko, 2015; Momin & Joshi, 2015). Fundamental targets of using nanotechnology in the food industry are to obtain food formulations fortified with safe macromolecular nanoparticles, innovative, and smart packaging materials supported with functional biopolymeric nanostructures and health promoting dietary supplements enhanced with edible nanoparticles comprising controlled release and target delivery of nutraceuticals (Acosta, 2009; McClements, 2013; Hannon et al., 2015). Depending on those promising opportunities served by nanomaterials for food manufacturers, global market size of nanotechnology yielded food products is continuously increasing as estimated by market research reports (Bumbudsanpharoke & Ko, 2015; Naseer et al., 2018).

Various inventories and databases have declared current food and associated products containing nanoparticles in global market with increasing market size (Bouwmeester et al., 2014, Bazana et al., 2019). Inorganic nanoparticles such as titanium oxide, silicon oxide, and silver, and organic nanoparticles such as carbohydrate, protein, and lipid-based nanostructures incorporated in food materials can be up taken into body continuously through consumption. There might be unforeseen consequences arised from extended and uncontrolled exposure to them.

Nano-sized materials can be integrated in food through various paths such as pesticides and veterinary medicines passing through agricultural livestocks, organic nanoparticles directly added to food for improvement of nutritive, structural, and mechanical properties, and potential migration of nanoparticles from packaging materials (Chaudhry et al., 2008). Once they are ingested via food chain, they can pass through bloodstream and biological barriers and may result in accumulation within tissues and organs (McClements & Xiao, 2012; McCracken et al., 2013; Hannon et al., 2015). Safety issues associated with the use of nanomaterials in food and food contact products have raised public concerns. In fact, in normal daily diet, human can intake various tiny particles ranging from large macromolecules to nanoscale structures with no harm. However, in case of deliberately incorporated organic or inorganic nanoparticles in food and their contact materials can lead to overexposure, allergenity, and toxicity at different levels as they are readily absorbed through digestion system and deposited in tissues (Hillyer & Albrecht, 2001; des Rieux et al., 2006; Hannon et al., 2015; McClements et al., 2016; Sohal et al., 2018).

Those potential risks may have led to public concerns and unwillingness in paying for and consuming nanofood products. A recently published research report presented a nationwide (United States) online survey study targeting to measure consumer acceptance of nanotechnology based food products (Zhou & Hu, 2018). They used canola oil produced from nanoscale modified seeds or packed through a nanotechnaology-derived technique. Their results suggested that consumers were willing to pay less for canola oil obtained using nano-modified seeds or packed with nano-techniques. Moreover, no significant difference was found regarding canola oil enriched with nano-modified drops to improve health. Assessment of the risks properly and setting of the regulatory frameworks have critical importance for safe manufacturing and use of nanofood products and supporting their commercial potential.

8.3 Importance of regulations regarding use of nanomaterials in foods

Nanotechnological applications in food industry can facilitate improvement of functionality, quality, and safety of food products by enhancing nutritional value, developing textural and sensorial properties, extending shelf life, detecting residual substances (e.g., pesticides), and screening pathogens (Dasgupta et al., 2015). Organic and inorganic nanoparticles used for those purposes may possibly lead significant undesirable consequences for health and nature as stated previously. By considering continuous need and interest of market for foods with improved technological features provided by nanotechnology, proper regulations are required to be established for safe handling of nano-based structures in food related processing.

Due to their quite small sizes, nanoparticles may exhibit different physicochemical and structural properties in comparison to their corresponding bulk states, including surface to volume ratio, zeta-potential, conductivity, mechanical strength, and optical properties (Hannon et al., 2015). Identification of the characteristic features, exposure profiles, and potential toxicity of the nanoparticles used in foods are the fundamental requirements to assess the safety of these nanostructures. Currently, potential health risks and hazards of nanomaterials used directly and indirectly in consumable goods have been under investigation. Though there is still uncertainties, consumers, livestocks, and nature should be protected from unforeseen consequences of nanotechnology products by the well-defined regulation schemes.

Potential risks to be arised from uncontrolled diffusion of relatively small-sized nanoparticles through body fluids and biological barriers, and their widespread use, thus over exposure rates and residual effects leading to allergy and toxicity are the critical factors enforcing regulatory legislations (Chau et al., 2007; Cushen et al., 2012; Bajpai et al., 2018). Testing guidelines based on validated analysis methods target the assessment of appropriateness for functionality and application, cytotoxicity to human being and livestocks, and environmental effects due to direct contamination or release through manufacturing wastes. Thus, development and implementation of these procedures crucially helps to achieve safe production and use of the desired nanomaterials in food environments.

Regional regulatory bodies create guidance by considering their own needs, priorities, and preferences based on their cultural and ethical issues. However, international authorities are faced with the challenge of carrying out the balanced regulatory legislations for sustainable agri-food production meeting market needs and protecting consumer and environment from risks and hazards of nanotecnology. Besides, there are some common approaches in the basic principles and applications released by all organizations, commissions, and centers. The most well known international authorities involved in regulation of nanomaterials will be mentioned in the following section.

8.4 International organizations relevant to regulatory frameworks for food nano-particles

Although many countries do not have regulations for production and use of nanoparticles in agro-food applications, some national and international advisory committees are present for providing guidance of safe handling of nanomaterials in foods. Nevertheless, there is still lack of a comprehensive universal guidance for the assessment of safety for nanomaterials in foods. Up to date, list of regulatory bodies and legislations have been introduced by some literature reports (Chau et al., 2007; Amenta et al., 2015; Jain et al., 2018).

The regional level and nation-based regulations are released from governmental or non-governmental organizations including European Union (European Commission (EC), European Food Safety Agency (EFSA)), United States (Food and Drug Administration (FDA), Environmental Protection Agency (EPA)), Australia and New Zealand (Food Standards Australia New Zealand (FASNZ)), Russia (The Federal Service for the

Protection of Consumer Rights and Human Well-Being of the Ministry of Health and Social Development), Japan (Ministry of Health, Labor, and Welfare), and many others, as presented in the previous chapters.

Unfortunately, there are many discrepancies between existing regulations (Mbengue & Charles, 2013). This indicates that global regulatory legislations are needed for international standardization of analysis and treatment of nanotechnology products in the food industry. Though there is not an international institution particularly responsible for the regulation policies regarding nanotechnology in food, there are some efforts by international organizations and initiatives to introduce regulatory guidelines for risk assessment and safe handling of nanoparticles in agro-food/feed industry. CAC jointly founded by Food and Agricultural Organization (FAO) and World Health Organization (WHO), Organization for Economic Development (OECD), International Organization for Standardization (ISO), International Risk Governance Council (IRGC), and some other initiatives including International Life Sciences Institute (ILSI) and Institute of Food Technologists (IFT) are the well-known international organizations exerting for that purpose (Takeuchi et al., 2014). These authorities organize meetings to discuss current developments in nanotechnology, release testing guidance, and legislative documents for identification, safety assessment, and handling of nanoparticles in industrial applications.

Assessment and governance of risks regarding the use of nanomaterials in various industries including agriculture and food via international legislations provide assurance of safety of nanotechnology products, thus expedite their public perception and international trade. Existing international organizations and some of their activities relevant to regulating nanotechnology in foods have been presented in Table 8.1.

8.4.1 Food and Agricultural Organization/World Health Organization/Codex Alimentarius Commission (FAO/WHO CAC)

The Food and Agricultural Organization (FAO) and the World Health Organization (WHO) are two well-known international organizations undertaking risk assessment and regulation of nanotechnology products (van der Meulen et al., 2014). Since it's foundation in 1945, FAO has international efforts in achieving food security and accessibility to high-quality food by the people, especially living in underdeveloped regions. WHO, founded in 1948, coordinate international health within the United Nation (UN) system. The mutual interest regarding food and health resulted in collaborative works in the corresponding area.

Table 8.1 International organizations and some activities relavant to regulating nanotechnology in foods.

Organization	Acronyme	Location	Activities/reports/projects	Considerations/highlights	References
Food and Agricultural Organization/ World Health Organization Codex Alimentarius Comission	FAO/WHO CAC	USA/ America	FAO/WHO Expert Meeting and Report: The application of nanotechnologies in the food and agriculture sectors: potential food safety implications	Needed to develop a common perspective and global guidence needed for governing risk assessment and safety hazards associated with nanotechnology applications in agri-food industry	(Organization, 2010; WHO, 2010)
			FAO/WHO Joint Meeting: Nanotechnologies in Food and Agriculture	Discussing objectives, scope and progress of the initiatives of other international organizations, and future activities	(WHO, 2012)
			Report: State of the art on the initiatives and activities relevant to risk assessment and risk management of nanotechnologies in the food and agriculture sectors	Developments in the application of nanotechnology and the activities of national and international bodies for risk assessment of using nanoparticles in food and agriculture	(Food & Organization, 2013; WHO, 2013)

(*Continued*)

Table 8.1 International organizations and some activities relavant to regulating nanotechnology in foods. (*Cont.*)

Organization	Acronyme	Location	Activities/reports/projects	Considerations/highlights	References
Organization for Economic Co-operation and Development	OECD	France/ Europe	Report: Opportunities and Risks of Nanotechnologies	The Working Party of Nanotechnology and the Working Party of Manufactured Nanomaterials are two relevant working parties for testing and assessment of nanoparticles in industrial applications including food	(OECD, 2005)
			Survey Report: Regulatory Frameworks for Nanotechnology in Foods and Medical Products	Intergovermental discussions for regulatory approaches, legislative regimes, scientific reseaches for establishments of regulations and other activities to be supported by governments	(OECD, 2013)
			Meeting Reports: Series on the Safety of Manufactured Nanomaterials Report No. 59 Report No. 81 Report No. 87 Report No. 89	Round table discussions of safely manufactured nanomaterials including agri/food applications	(OECD, 2015, 2017, 2018, 2019)

Table 8.1 International organizations and some activities relavant to regulating nanotechnology in foods. (*Cont.*)

Organization	Acronyme	Location	Activities/reports/projects	Considerations/highlights	References
			ProSafe project: Science based support for regulation of manufactured nanomaterials	Nanosafety: Nanomaterials identification and characterization, tiered testing, risk assessment, management and governance	(OECD, 2016)
International Organization for Standardization	ISO	Switzerland/ Europe	Nanotechnology standards: ISO Technical Committee 229	Responsibility of developing standards for identification, utilization and safety assessment of nanomaterials	(Benko, 2017)
International Risk Governence Council	IRGC	Switzerland/ Europe	White Paper No 2: Nanotechnology Risk Governance	Inclusion of the societal context in the assessment and management of risks and catagorization of risk-related knowledge as simple, complex, uncertain and ambiguous risks	(Renn & Roco, 2006)
			Report: Risk Governance of Nanotechnology Applications in Food and Cosmetics	The lack of authoritative knowledge gave rise to confusions and speculative discussions in relation to potential risks and safety concerns	(Grobe et al., 2008)

(*Continued*)

Table 8.1 International organizations and some activities relavant to regulating nanotechnology in foods. (*Cont.*)

Organization	Acronyme	Location	Activities/reports/projects	Considerations/highlights	References
International Life Sciences Institute	ILSI	USA/ America	The Task Force on Novel Foods and Nanotechnology	Aimed at developing safety assessment of novel foods and ingredients produced by nanotechnology	(Cockburn et al., 2012)
			NanoRelease Project, NanoRelease Food Additives	A road map for the applications and risk management of nanomaterials in food industry on the basis of identification, detection, quantification and characterization principles	(Szakal, Tsytsikova et al. 2014)
			Workshop: The Outlook and Challenges of Nanotechnology for Food Packaging	Expert reviews on benefits,, consumer acceptance, risk assessment, safety and environmental impact, and regulatory implications	(Wyser et al., 2016)

In 2010, FAO/WHO organized a joint international expert meeting and discussed to develop a common perspective and global guidance needed for governing risk assessment and safety hazards associated with nanotechnology applications in agri-food industry by considering practices of national regulatory programmes (Organization, 2010). Opportunities for the development of innovative food products, smart packaging materials, water treatment, and agricultural improvements promised by nanotechnology were other significant topics discussed. Moreover, the need of international definition and basics of nanotechnology in food was addressed to CAC which is also agreed to be addressed for the use of current risk assessment approaches in engineered nanomaterials in food and agriculture as well. Additionally, the need for further research of risk assessment strategies and test methodologies was emphasized. The second joint international expert meeting of FAO/WHO was organized in 2012 and discussed developments in the application of nanotechnology and the activities of national and international bodies for risk assessment of using nanoparticles in food and agriculture (FAO/WHO, 2012; Food and Agriculture Organization, 2013).

CAC, jointly established by FAO and WHO, stands as an intergovernmental coordinator for formation of food standards undertaken by various governmental or non-governmental institutes (Chau et al., 2007). CAC actively participates in international efforts for the development of safety regulations including the use of emerging technologies. Recently published guideline by CAC for the risk analysis of contaminants used in foods included nanomaterials to be investigated as well (ALIMENTARIUS, 2019).

Up to date, rather than developing strict legislative regulations for risk assessment of nanomaterials in foods, these joint actions have mediated international platforms for discussion of developments in the applications of nanotechnology in agro-food industry presenting best practices for safety assessment of nanoparticles (Mbengue and Charles, 2013). Adaptability of food safety standards defined by CAC for foods and contact materials containing nanoparticles can provide universal measures of standardization for international nanofood market.

8.4.2 Organization for economic co-operation and development (OECD)

Intergovernmental organization, OECD with it's 34-member countries from America, Europe, Asia, and Pacific regions co-ordinate and harmonize economical and industrial policies, discuss mutual concerns, and attempt to solve international problems (OECD, 2015). The organization has specialized committees and working groups composed of member country delegates and dealing with particular topics. The Working Party of Nanotechnology (WPN) and the Working Party of Manufactured Nanomaterials (WPMN) are two relevant working parties for testing and assessment of nanoparticles in industrial applications including food as well. Main activities of OECD working groups are implementing projects regarding risk evaluation approaches, regulatory frameworks and benefits of using nanoparticles in environmentally sustainable manufacturing, producing appropriate test methodologies for safety assessment of nanomaterials, and setting research database contributing safe nanomaterial fabrication (Takeuchi et al., 2014).

OECD working groups published a few reports based on conducted international projects investigating safety assessments and governance of regulatory frameworks regarding production, usage and testing of nanomaterials for food and non-food applications. OECD report published in 2005 was based on opportunities and risks of nanotechnology applications in various industries including food and agriculture and presented risk management in food and related materials (OECD, 2005). Primarily, EU and United States, OECD countries have significant individual efforts to make policy for the safety assessment and regulation of nanotechnology based food and associated products, however discussions have indicated the need for regulatory legislations at international level.

In 2013, OECD presented a survey report on "Regulatory frameworks for nanotechnology in food and medical products" (OECD, 2013). In that survey, many national/regional-based regulatory guidelines belonging OECD countries and relevant scientific research activities were comparatively documented, and responses of delegates indicated intergovernmental discussions for regulatory approaches, legislative regimes, scientific researches for establishments of regulations, and other activities to be supported by governments to encourage investments in nanotechnology. A series of round table discussions were organized and respective reports were published on developments on safely manufactured nanomaterials including agri/food applications (OECD, 2015, 2017, 2019).

The fast rise of nanotechnology applications in consumable products has induced ethical, health, and environmental concerns in OECD countries with regard to potential risks of nano-enabled particles. (Gruère, 2012) has reported a list of OECD countries including Australia, Canada, EU, Japan, etc., with the examples of commercially available nano products and research activities in food and agriculture sectors (Gruère, 2012). In this research, the challenges for companies and investors have been presented based on OECD reports by emphasizing the low scientific knowledge regarding risk assessments including high exposure, bioaccumulation and toxicity issues, and the discouraging factors due to regulatory uncertainties, public perception, and investment fundings.

8.4.3 International organization for standardization (ISO)

The International Organization for Standardization (ISO) is a non-governmental network of national standard institutes from 162 countries with the motivation of internationalization of standards and relevant activities (ISO, 2015). Development of consensus-based and market need-relevant international standards efforts to solve global challenges. A number of ISO technical committees develop standards on varying topics including agriculture and food. ISO Technical Committee 229 established in 2005 have been responsible for standardization in the field of nanotechnologies with the identification and utilization principles of nanomaterials (Takeuchi et al., 2014; Benko, 2017).

Several specific tasks include developing standards for terminology and nomenclature, metrology, and instrumentation with specifications for reference materials, test methodologies, modeling and simulations, and science-based health, safety and environmental practices for the assessment of food and non-food nanomaterials (https://www.iso.org/committee/381983.html). Based on those tasks, working groups (WGs) established under this technical committee with different scopes such as development uniform definition for terminology and nomenclature (WG1), development standards, and test methods for measurement and characterization (WG2), and development of scientific standards for assessment of health, safety, and environmental impacts of nanomaterials (WG3) (Benko, 2017). Besides working groups responsible for developing standards, a number of advisory groups named as task groups and study groups were founded with the charge of exploring novel concepts of nanotechnology and contribute to WGs for rapid progress. Collaborative works established between ISO and other regulatory organizations such as OECD provide effective contribution to develop assured standard documentation with the input of relevant stakeholders.

Scientific principles are the basis for development of ISO standards in nanotechnology concepts. The standards approved after passing all evaluation steps of testing and validity are published and implemented.

8.4.4 International risk governance council (IRGC)

International Risk Governance Council (IRGC) is an independent, non-profit foundation aimed at supporting government, industry, and other organizations by governing global risk issues to protect public health and environment (Renn & Roco, 2006; Grobe et al., 2008). IRGC suggested four progressive stages of nanotechnology shifting from first generation passive nanostructures to forth generation molecular nanosystems used in energy, electronic, chemical, biotechnology and food industries (Renn and Roco, 2006). Two significant reports published by IRGC for governance of risks in food related applications. In the first one, a risk governance framework for nanotechnology applications was introduced on the basis of integrated approach gathering scientific, economic, social, and cultural aspects (Renn and Roco, 2006).

The IRGC framework offered inclusion of the societal context in the assessment and management of risks and categorization of risk-related knowledge as simple, complex, uncertain, and ambiguous risks. This report was also used to shape governance of nanotechnology-based issues in food and cosmetic in general. In 2018, another report of IRGC was published on "Risk governance of nanotechnology in food and cosmetics" (Grobe et al., 2008). Regarding food and relevant applications of nanomaterials, misinformation, and the lack of authoritative knowledge gave rise to confusions and speculative discussions in relation to potential risks and safety concerns.

8.4.5 Other international institutes/organizations

There are some other international institutes associated with various activities such as projects, expertized working groups, meetings and conferences, and publications regarding risk assessment and regulatory principles of nanomaterials in food and related products. International Life Sciences Institute (ILSI), Institute of Food Technologists (IFT) are two well known organizations with recognized activities in regard of safety issues of nanoparticle-incorporated food and contact materials.

ILSI is a non-profit worldwide international organization including many country-based branches with the mission of providing science to improve human health and safeguard environment. It is well recognized with its support for the high quality scientific researches, global conferences and workshops, projects and publications on significant topics including nutrition, food safety, toxicology, risk assessment, and environmental issues. Being one of the primary sponsors, ILSI Research Foundation supported "NanoRelease" project (2010) aiming the promotion of safe development methods and release behavior of nanomaterials incorporated in industrial products on the basis of data, methodology, guidance, standards, and links collected (Takeuchi et al., 2014).

Relevant to this work, NanoRelease Food Additives Project gathered many international experts including scientists from academy, industry, government, and non-government organizations to discuss the applications and risk management of nanomaterials in food industry on the basis of identification, detection, quantification, and characterization principles (Szakal, 2014). An ILSI Europe initiative the "Novel Foods and Nanotechnology Task Force" (2011), with an expert group, focused on developing safety assessment of novel foods and ingredients produced by nanotechnology (Cockburn et al., 2012). Another ILSI activity, the workshop on the "Outlooks and Challenges of Nanotechnology in Food Packaging," facilitated expert reviews on potential opportunities and challenges in terms of benefits, consumer acceptance, risk assessment, safety and environmental impact, and regulatory implications (Wyser et al., 2016).

IFT is a non-profit, worldwide international scientific society of expert members from academy, industry, and government in the field of food science and technology. As a global scientific organization, it supports research activities and provide platforms to discuss the benefits and challenges of innovative technologies and share knowledge and experiences in food area through its worldwide network. IFT journals and conferences facilitated many sessions for sharing advancements of nanotechnology applications and safety assessments in foods (Shahidi, 2006; Bugusu et al., 2007; Clemens and Pressman, 2011). A number of articles presented nanomaterials used in various food matrices, safety issues, and regulatory policies on food nanotechnology (Brody, 2007; Bugusu et al., 2009).

8.5 Regulatory legislations by international organizations

Regulatory policies and guidelines for safe handling and use of nanoparticles in food and agriculture have been implemented at national and international level in worldwide. With the consideration of international efforts, several governmental and non-governmental organizations are involved in regulatory frameworks of emerging technologies including nanotechnology. Though there is not a comprehensive legislation yet, several harmonized approaches based on existing guidelines of food safety standards have been implemented to support commercial potential of nanotechnology derived agro-food products (Amenta et al., 2015). Table 8.2 summarizes regulatory legislations released by international organizations relevant to nanomaterials in foods and contact materials.

Table 8.2 Regulatory legislations released by International Organizations relavant to nanomaterials in foods and contact materials.

Authority	Legislative documents released	Highlighted regulations	References
CAC FAO/WHO	CXG-92-2019	Nanomaterials: Exclusionary contaminant categories in regulatory guideline of contaminants in foods	(Codex, 2019) http://www.fao.org/fao-who-codexalimentarius/codex-texts/all-standards/en/
OECD	ENV/JM/MONO(2006)20	Guidance document for the development of OECD guidelines for the testing of chemicals	(OECD, 2006)
	ENV/JM/MONO(2009)21	Preliminary review of OECD test guidelines for their applicability to manufactured nanomaterials	(OECD, 2009)
	ENV/JM/MONO(2012)40	Guidance on sample preparation and dosimetry for the safety testing on manufactured nanomaterials	(OECD, 2012)
ISO	ISO/TS 27687:2008 *Revised version:* ISO/TS 80004-2:2015	*Basics of Nanotechnology:* Nanotechnologies-Terminology and definitions for nano-objects – Nanoparticle, nanofibre and nanoplate Nanotechnologies-Vocabulary-Part2: Nano-objects	(ISO, 2008)
	ISO/TS 80004-1:2010 *Revised version with the same title:* ISO/TS 80004-1:2015	*Basics of Nanotechnology:* Nanotechnologies-Vocabulary-Part 1: core terms	(ISO, 2010)
	ISO/TR 12888:2008 *Revised version:* ISO/TR 12888:2018	*Health and Safety:* Nanotechnologies-Health and safety practices in occupational settings relevant to nanotechnologies	(ISO, 2008)
	ISO/TS 21975 *(under development)*	*Food Packaging:* Nanotechnologies-Polymeric nanocomposite films for food packaging with barrier properties-Specification of characteristics and measurement methods	(ISO, 2020)

1961 was the establishment year of Codex Alimentarius Commission (CAC) and joint FAO/WHO Food Standards Programme. Afterward, in the joint FAO/WHO conference (1962), CAC was requested to implement Food Standards Programme and create Codex Alimentarius (Depew, 1963), which is a collection of internationally adapted food standards facilitating harmonized international food trade. Since then, CAC has been responsible for the implementation and creation of international food standards and guidelines. Regarding nanotechnology and nanoparticles in foods, there is no particular standard declared for nanoparticle-incorporated foods, however, nanomaterials have been involved in regulatory guideline of contaminants in foods (Codex CXG-92-2019). In this guideline, nanomaterials are considered as exclusionary contaminant categories, thus existing regulatory frameworks, standards, and guidelines are recommended to follow. In fact, in the future Codex Alimentarius is expected to create guidelines of food safety standards adapted for nanotechnology applications, thus results in an internationally recognized legal certainty (Mbengue & Charles, 2013).

OECD publishes a series of guidance documents regarding safety of manufactured materials under its environment, health, and safety publications. These documents introduce recommendations of working groups for existing test guidelines of nanomaterials based on current developments in manufactured nanomaterials independent from the application area. The subdivision of OECD Chemicals Committee, the WPMN mostly concentrated on human health and environmental safety of nanomaterials by involving safety testing and risk assessments. Relevant test guidelines published present scientific and internationally harmonized standard for assessment of risk, exposure, and hazard (OECD, 2009). The WPMN recommended to develop a guidance on sample preparation and dosimetry for tests using manufactured nanomaterials, differently from existing documents (OECD, 2012). Although this guidance is not directly refer the assessment of nanoparticles in foods, some inorganic nanoparticles such as silver to be incorporated in food packaging materials are exposed to testing procedures due to this guideline (Kühnel & Nickel, 2014). Recommendations based on these two guidelines have been taken into consideration to evaluate exposure and hazards, and risk assessment for successfully adaptable regulations for nanoparticles in food related products.

Regarding nanotechnology, the first legislative document of ISO was the technical specification "ISO/TS 27687:2008/ Nanotechnologies-Terminology and definitions for nano-objects-Nanoparticle, nanofiber and nanoplate" published in 2008 (ISO, 2008). In 2015, it was revised and confirmed as "ISO/TS 80004-2:2015/ Nanotechnologies-Vocabulary-Part2: Nano-objects," which lists the terms and definitions related to particles in the field of nanotechnologies on nanotechnology terminology and definitions (ISO, 2015). Within this concept, the other specification published in 2010, "ISO/TS 80004-1:2010/ Nanotechnologies-Vocabulary-Part 1: core terms" (ISO, 2010) was revised in 2015 as "ISO/TS 80004-1:2015" with the same title (Nanotechnologies, 2015) (Ref: ISO, 2015). Both specifications revised in 2018 and 2019, and confirmed to remain current until the next update. Fundamental terms such as "nanoscale" referring size from 1 to 100 nm, and nano-objects, representing nanoparticles including nanofibers, nanorods, and nanowires, and other related basics independent from application area have been described in this specification.

A series of general guidance documents prepared jointly by Technical Committee ISO/TC 229 are currently available by ISO for standardization of manufacturing and characterizing nanoparticles developed for various purposes in different industries. Besides, health and safety practices for occupational settings relevant to manufacturing processes of nanoparticles specified with "ISO/TR 12888:2008" (revised by ISO/TR 12888:2018). This standard does not directly address health and safety issues or practices associated with nanomaterials obtained by standard operations, potential consumer exposures, but helps to prevent safety consequences during production, handling, use, and disposal of manufactured materials in broad applications. Although there is still not a specific standard for nanoparticles used in foods, general specifications are considerable for safety settings of nanoparticles in food and related applications.

Recently introduced another ISO specification, "ISO/TS 21975/Nanotechnologies-Polymeric nanocomposite films for food packaging with barrier properties-Specification of characteristics and measurement methods" is under development (www.iso.org/standard/72330.html). This standard expectedly provides appropriate guidance for handling and safety assessment of nanomaterial incorporated food-packaging materials.

Worldwide several approaches declared by some international authorities, based on current food standards and general regulatory frameworks of nanotechnology provide useful guidance for handling and safety assessment of nanotechnology in food and agriculture applications. Current regulatory schemes, as more general, in nanotechnology applications are adaptable to some extend for food manufacturing as well. However, a number of surveys with the remarks of academia and industrial stakeholders, consumer preferences, and health and environmental concerns clearly indicate the need of a comprehensive international regulatory legislation for safe use of nanomaterials and other nanotechnological approaches in agriculture/feed and food industry. Ongoing discussions at various international platforms shared by all stakeholders and policy makers will further result in development of such specific standards to be applied for food and related products.

8.6 Regulatory principles of food nanoparticles

Regulatory food safety standards have continually been adapted and revised for the presence and safety of nanoparticles in the food industry. Nanoparticle usage in food applications should comply with legislations, recommendations, and guidelines. Otherwise, health risks and diseases caused by nanoparticle insertion could not be prevented. For nanomaterial applications in food-related product, principle physicochemical parameters and techniques to identify and characterize them, their biological and toxicological effects, in vivo and in vitro testing and hazard characterization have been declared in details at FAO and WHO meeting report (FAO/WHO, 2011).

There are two different approaches about proposed definitions of nanomaterials for regulatory purposes. The first one interests in the dimension of materials whether it is at nanometer scale; the second one queries whether the nanomaterial provides a change in the properties of food product (FAO/WHO, 2013). Change in the properties of the food products such as solubility, biodegradability, or digestibility would differentiate nanomaterials and their hazardous effects.

Over the past decade, national and regional food safety organizations have focused on determination of the influences of nanomaterials utilized for foods. Globally accepted test guidelines were reviewed by OECD and found generally suitable for nanomaterial testing.

Besides the direct effects of nanomaterials inserted food, workers in food and feed production and animals eating nanomaterial inserted feed could be exposed to their effects via oral uptake and inhalation of dusts into their lung and skin. Namely, exposure is carried out by the way of inhalation, dermal, and ingestion. Therefore, health risks and hazards of nanomaterials have been under examination. Especially, inorganic nanoparticles should be utilized more carefully because of their potential toxicity. Most commonly used inorganic nanoparticles in foods are silver, gold, carbon, zinc oxide, titanium dioxide, and silica nanoparticles. Exposure risk to these nanomaterials increases as the related applications expand. Toxicity of nanoparticles has currently been under investigation, since there is not enough information yet.

To investigate vital information about health effects of nanoparticles by inhalation, in vivo testing with nanoparticles inserted foods should be performed to assess short or long-term inhalation toxicity (as described in Chapter 4). Concentration, size, and distribution of nanoparticles in the inhalation chamber are significant parameters for these toxicity studies. International Standard suggests inhalation toxicity monitoring with a differential mobility analyzing system (DMAS) and to determine particle number, size, distribution, surface area and estimated mass dose, as well as morphological examination using electron microscopy (EM) equipped with an energy dispersive X-ray analyzer (EDX) for chemical composition (ISO 10810, 2010). Conventional mass dose monitoring and other physicochemical monitoring techniques are also included in this International Standard to conduct for toxicity investigation. Inhalation toxicity results should be supported by nanoparticle particle surface area, mass dose, particle distribution, composition, and dispersion for a comprehensive analysis (Ji et al., 2007; Sung, 2008).

In order to provide safe use of the nanoparticle inserted food products, regulatory guidelines are significant for manufacturers, importers, and consumers as well (Amenta et al., 2015). Although, regulatory guidelines should include fabrication, processing, packaging, and consumption of nanoparticle incorporated food products, there is no specific legislation globally established for the regulation of nanomaterials in food applications. Due to lack of information about the influences of nanoparticle usage in foods on human health and environment, their market potential has also been affected negatively (Bazana et al., 2019).

8.6.1 Aspects of regulation

In 2009, FAO and WHO as the participants of a meeting agreed that nanotechnology methods and nanomaterials provide great benefits and opportunities in many area including food processing and packaging, water treatment, agriculture, and preservation (FAO/WHO, 2010). In 2012, FAO/WHO published a report about activities on risk assessment and management of nanotechnology in food and agriculture at national and international level (FAO/WHO, 2013). After this FAO/WHO meeting, nanostructured ingredients, nanodelivery systems for nutrients, nanosized inorganic and organic food additives, food packaging, nanocoating of food surface and nano filtration, have been built up. Novel food nanomaterials inserted products were developed and penetrated into the market and therefore, national and international regulations required for assessing safe use of nanomaterials.

There is still insufficient knowledge about the toxicity of nanoparticles arisen during production, use, and disposal for human health and nature. Based on EU regulations for the application of nanomaterials in foods and feed, testing rules and required standards have been declared (EFSA, 2011a). Regulations for nanoparticles inserted food products and food packaging, have been carried out by the FDA and EPA in the United States. The permitted maximum nanoparticle levels in food products have been qualified by EU and FDA regulations (Chaudhry et al., 2008).

In general, nanoparticles utilized in foods can be divided into two main types: organic or inorganic nanoparticles. Nanoparticle type has an important influence on gastrointestinal fate and potential toxicity. According to their potential toxicity, inorganic nanoparticles need to have more attention when used in food applications. Many inorganic nanoparticles such as silver, iron oxide, titanium dioxide, silicon dioxide, or zinc oxide are utilized in food products (Pietroiusti et al., 2016). These nanoparticles indicate different surface characteristics and they could be found in either crystalline or amorphous forms (McClements and Xiao, 2017). For example, silver nanoparticle are one of the most common nanoparticle in food, food contact surfaces and packaging materials due to its antimicrobial and antiodorant properties (Pulit-Prociak et al., 2015). It is known that some of nanoparticle used in food storage could migrate into foods and therefore, they could be ingested by humans (Mackevica et al., 2016). Estimated consume amount of silver is 20–80 μg/day for adults. However, there is still insufficient data about its potential toxicity when ingested with foods.

The NanoRelease project of ILSI Research Foundation's Center for Risk Science Innovation and Application aims to promote the safe nanomaterial production and to clarify the release behavior of nanomaterials. Therefore, as a part of the NanoRelease project, related data, proper methods, guidance, standards and links were screened and collected (ILSI, 2011). ILSI Europe started the Novel Foods and Nanotechnology Task Force (ILSI Europe, 2011), which interested in emerging technologies for the safety assessment of foods and food products. OECD has created a database on manufactured nanomaterial safety in terms of data, methods, guidance, standards, and links tracked (ILSI, 2011). Regulatory agencies in the United States and Europe have also introduced policies and guidance for the safe use of nanomaterials in food and related applications by considering various critical aspects (EFSA, 2011a; EFSA, 2011b; EFSA, 2016).

8.6.2 Fundamental principles

As mentioned in the previous section, current and projected nanotechnology applications in the food and agriculture area and risk management approaches were discussed and presented by the FAO and WHO meeting reports (FAO/WHO, 2010; FAO/WHO, 2013). Although the risk assessment has to be carried out in a systematic and consecutive way, indeed, the commercialization of nanomaterial incorporated products continues to pass the development of a precautionary risk management technique widely accepted by the scientific community and enforced by the regulatory bodies.

The risk assessment management is a very complex and dynamic process because the steps included in the process might change with the updating literature. However, it is accepted and published by UN reports that the risk assessment of nanoparticles inserted food and feed products should take into account the following four elements: hazard identification, exposure assessment, hazard characterization, and risk characterization (Codex, 2007; FAO/WHO, 1995a; FAO/WHO, 1997; SSC, 2000). Potential hazard of metal and metal-oxide nanomaterials like TiO_2, Ag, and ZnO nanoparticles could be predicted in hazard identification process. Physicochemical and cytotoxicity properties of nanomaterials are utilized to infer the hazard potential (Sheehan et al., 2018). The dose–response relationship for nanomaterials quantitatively determines the relationship between adverse effects and the material content in a given environment. The relationships between dose and biomarkers are utilized to determine no observed adverse effect levels and human health exposure thresholds. The exposure is commonly assigned through inhalation, ingestion, or dermal routes individually or in combination, or more routes such as environmental compartments such as water, sediment, or soil under different exposure processes. The fourth element, risk characterization covers approximation of the potential for adverse health effects to happen, assessment of uncertainty, and risk information summary (Subramanian, 2018).

The risk assessment procedure for nanomaterials is generally the same with that of engineered materials; however, additional safety concerns should be considered due to nano scale size of these particles (COT, 2005, 2007; SCENIHR, 2007). Toxicological studies indicated that occupational hazards of nanoparticles depend on the production and utilization of nanostructured materials. There has been not enough research regarding the effect of nanoparticles on the gastrointestinal system (FAO/WHO, 2010). However, nowadays some valuable data has been presented based on findings of ongoing researches for in vitro and in vivo studies (McCracken et al., 2013; Sohal et al., 2018).

Food packaging materials and coatings containing nanoparticles are capable of creating antimicrobial, antifungal, antireflective, anticorrosion surfaces to preserve foods. Antibacterial kitchenware, cutting boards and teapots are some of the food preparation surfaces coated with nanoparticles. High barrier nanocoatings are also used to enhance the oxygen barrier properties for food and beverages in order to provide improved shelf life (FAO/WHO, 2010).

The most important advantages using nanomaterials in food packaging materials and containers are the sensing and diagnosing of chemicals and toxicities in food materials. This process called as smart packaging, enhances the quality and freshness of the food (IRGC, 2008). In order to determine and prevent potential hazardous effects of nanoparticles in food and food packaging, physicochemical analysis, and toxicity tests should be performed adequately. Bioabsorption and bioavailability of nanoparticles are significant parameters for safety assessment as well. The risk characterization of nanoparticles combines their hazard effects and exposure risk (Hristozov et al., 2016).

8.6.2.1 Basic information and identification

Functionalities, reactivities, and adverse effects of nanomaterials can change with their size, shape, structure, dose, and environmental conditions. Physical properties of nanoparticles such as morphology, particle size, shape, porosity, agglomeration, and chemical properties such as composition, surface charge, stability are considered as significant parameters should be described for setting regulation procedures for food nanomaterials (Oberdörster et al., 2005). Besides, properties such as solubility and corrodibility are also significant for nanomaterials inserted food application. Therefore, different types of microscopy, spectroscopy, and chromatography techniques are utilized to understand physical and chemical properties of food nanomaterials. These techniques will be emphasized in following sections.

Testing and characterization of nanomaterials provide information about their benefits and potential toxic effects as well. Characterization of nanomaterials in food matrices is more complex than their characterization alone. There are three cases to guide nanomaterial characterization; the case in which a material is being evaluated, the importance of characterization of a specific property in this case, and the feasibility of characterization of the property in a specific case (Oberdörster et al., 2005). While there are many analytical techniques to analyze nanomaterials, there has been not enough study about their in situ characterization. Characterization of chemical and hazardous material in foods is significant for human health.

Nanomaterials in foods and feeds have to be biodegradable in human and animal bodies. If nanomaterials indicate biopersistency, it means these nanomaterials might cause digestibility problem and thus cause adverse effects on health. Therefore, they induce to form foreign regulatory bodies (FAO/WHO, 2010). It is very important to inform policy makers about health risks of nanomaterials. The Flemish service environment and health aimed the assessment of the possible exposure to nanomaterials in products. Some nanomaterial incorporated products found in markets and frequently utilized by consumers, tested for the presence of nanoparticles and evaluation of their exposure. For these products, the exposure during its usage could be investigated by simulating representative exposure conditions and nanoparticles properties such as atomic structure, particle size and surface area, etc. are very important for these toxicological simulations.

Another purpose of the study is to formulate specific policy advises about the potential exposure caused by nanomaterials inserted products and the other results of risks of nanomaterial usage. The NANO Global Risk Assessment (NANOGRA) project purpose is also to state risks of nanomaterials exposure by focusing explosion, ecotoxicological, and toxicological risks. In 2016, a new laboratory was set up with a 20L-sphere to evaluate explosion and ecotoxicological risk determination of nanomaterials. Based on these studies, characterization and testing for the assessment of future exposure of nanoparticle-inserted products will be developed and supported.

8.6.2.2 Testing physical and chemical properties

Different types of microscopy, spectroscopy, and chromatography techniques are utilized to investigate the physical and chemical properties of food materials. Regarding microscopy, electron microscopy utilizing accelerated electron beams for imaging is powerful tool in the analysis of nanomaterials. Organic and inorganic nanoparticles added in foods can be identified through electron microscopy in terms of morphology and microstructure. It gives information about morphological features such as size, shape, appearance, and surface roughness of the nanostructures blended in food matrices.

Scanning electron microscopy (SEM) and transmission electron microscopy (TEM) are the most common electron microscopy methods. SEM uses secondary and backscattered electrons while TEM uses transmitted electrons during imaging process. Therefore, SEM provides information only on the morphology and composition of the sample. However, TEM could provide information about the inner structure and crystallinity. Energy dispersive X-ray spectroscopy (EDX) is an analytical technique to determine the elemental composition of nanoparticles and it is in conjunction with an SEM (Ault et al., 2012). It detects X-rays, which are emitted from the sample during electron bombardment and gives chemical information of the sample surface. Another important microscopy method that gives information about the morphology and topography of nanoparticles is scanning probe microscopy (SPM) techniques.

Atomic force microscopy (AFM) and scanning tunneling microscopy (STM) are main SPM techniques whose working principle is based on the sensing of the interaction between probe and specimen. AFM is generally used in tapping or non-contact mode for the food specimen (Kumar et al., 2017) and the size and shape of nanoparticles are clearly observed in AFM images. STM is a measurement technique which allows to take surface images at atomic level and its working principle is based on quantum tunneling. Its usage for nanoparticles in food applications provides detailed analysis on surface morphology (Pawar et al., 2019). Physical and chemical characterization techniques for nanoparticles incorporated in food matrices are summarized in Table 8.3.

Spectroscopic methods are also useful to detect and analyze organic and inorganic nanoparticles in food matrices. X-ray diffraction (XRD) is one of the most commonly utilized spectroscopic method for nanoparticles. This method uses X-ray to characterize nanomaterials and to investigate crystallographic structure of nanoparticles in food (Chen et al., 2013). Another spectroscopic method to analyze the elemental composition of the material is X-ray photoelectron spectroscopy (XPS), which is a surface-sensitive and quantitative analysis. In this method, X-ray is generated by electron bombardment of a metallic anode. The energy of emitted photons could eject electrons. The elemental composition of the materials could be determined by measuring the kinetic energies of the ejected electrons. Surface chemical composition of nanoparticles and their oxidation states could be characterized by XPS (Longano et al., 2012).

Water contact angle (WCA) test is frequently applied especially for food packaging applications to determine hydrophobicity or hydrophilicity of materials (Al-Naamani et al., 2016). WCA is measured by the angle between the specimen surface and the tangent line at the point of contact of the liquid droplet with the substrate. In WCA, high contact angle means poor wetting and hydrophobic property whereas low contact angle indicated hydrophilic behavior. Zeta potential spectroscopy, which is based on applying an electrical field on colloidal particles in a suspension, gives information about colloidal stability and surface charge of nanoparticles. High

Table 8.3 Physical and chemical characterization techniques for nanoparticles incorporated into food matrices.

Characterization/ test technique	Purpose	Food nanoparticles	References
SEM	To determine morphology, mean particle size, particle size distribution.	TiO_2, ZnO, Ag	(Tiede et al., 2008; Manonmani et al., 2011)
TEM	To determine particle size and crystallinity.	TiO_2, ZnO	(Venkatasubbu et al., 2016)
EDX	To determine chemical composition.	Ag	(Echegoyen & Nerín, 2013)
AFM	To determine surface topography and mechanical strength.	Ag, TiO_2, ZnO	(Tiede et al., 2008; Krishnaraj et al., 2015)
STM	To obtain surface images at atomic level.	CuO, Cu, Ag	(Cárdenas et al., 2009; Jain et al., 2018; Pawar et al., 2019)
XRD	To determine polymorphism and crystallinity.	Carbon, ZnO	(Espitia et al., 2012; Sk et al., 2012)
XPS	To determine surface chemical composition of nanoparticles and their oxidation states.	Chitosan, Au	(Thibault et al., 2008; Aresta et al., 2013)
WCA	To determine hydrophobicity or hydrophylicity.	Chitosan/ZnO	(Al-Naamani et al., 2016)
Zeta Potential Spectroscopy	To determine surface charge.	Chitosan, lechitin/chitosan	(Souza, 2014; Madureira et al., 2015)
TGA	To determine thermal stability.	Zein, chitosan, ZnO	(Espitia et al., 2012; Hosseini et al., 2013; da Rosa et al., 2015)
DSC	To determine melting temperature, melting enthalpy, glass transition temperature.	Chitosan, zein/pectin	(Liu et al., 2011; Hu et al., 2015)
DLS	To determine particle size and distribution.	α-lactalbumin, polystyrene	(Graveland-Bikker et al., 2004; Filipe et al., 2010)
SPR	To confirm the presence of nanoparticles.	Ag, Au	(Velmurugan et al., 2014)

zeta potential value indicates stability between the particles while low zeta potential value indicates the propensity of particles to flocculate (Sharma et al., 2019).

Dynamic light scattering (DLS) method is another important method to determine particle size and distribution of the self-aggregates in foods (Graveland-Bikker et al., 2004; Filipe et al., 2010). In this method, measurements were carried out using a monochromatic light. The scattered light is measured at a constant angle and collected by a photomultiplier. In this method, the hydrodynamic diameter of nanoparticles was calculated by the Stokes–Einstein equation. The presence of nanoparticles could be confirmed by surface plasmon resonance (SPR) method, as well (Tankhiwale and Bajpai, 2012).

Thermal properties of nanoparticles are also very important for foods in processing and sterilization. Thermogravimetric analysis (TGA) and differential scanning calorimetry (DSC) are the most common thermal characterization techniques enabling to measure thermal properties of nanoparticle incorporated food and contact materials. TGA gives weight loss of a sample as a function of temperature and determines thermal stability of the specimen (da Rosa et al., 2015). The DSC thermogram gives information about the melting point and glass transition temperature of nanoparticles (Yi et al., 2015). DSC technique is used to determine thermal behavior of the dried food nanoparticles such as curcumin, zein, and pectin (Hu et al., 2015).

8.6.2.3 Further characterization

While physical and chemical characterization of nanoparticles are crucial in identification of nanoparticle-food complex, additional information such as quantification of radical formation capacity, secondary structure of nanoparticles, impurities, etc. also required to understand their impacts within food matrices. Advanced characterization techniques for better understanding of unique features of nanoparticles and interactions between nanoparticles and food matrix are given in this section. Table 8.4 summarizes advanced characterization techniques for nanoparticles inserted in food complexes.

Mass spectroscopy (MS) is a powerful tool in nanoparticle characterization and quantification at quite low detection limits, even ng/L scale (Bustos et al., 2013). A typical MS system consists of three main particles: ion source, mass analyzer, and detector system. For liquid and solid samples, frequently utilized two ionization techniques are electrospray ionization and matrix-assisted laser ionization. MS could provide both molecular and

Table 8.4 Advanced characterization techniques for nanoparticles inserted in food complexes.

Characterization/ test technique	Purpose	Food nanoparticles	References
MS	To determine chemical composition and core/surface ligand stoichiometries, nanoparticle size, and number concentration.	Iron oxide, silver nitrate, silver	(Lindgren, 2014; Tombuloglu et al., 2020)
ICP-MS	To identify particle size, particle size distributions in the extract, mass and particle concentration.	TiO_2, Ag	(Weir et al., 2012; Loeschner et al., 2013)
HPLC-MS	To separate and identify each component in a mixture.	Indium oxide, silica, Au	(Qi et al., 2014; Yu et al., 2015; Arabi, 2016)
FT-IR	To identify functional groups in nanoparticles.	Gelatin/silver, cerium oxide	(Darroudi et al., 2014; Kanmani & Rhim, 2014)
Raman spectroscopy	To provide information about the phonon modes of the specimen.	Au, Ag	(Giovannozzi et al., 2014; Guo et al., 2015)
UV-visible spectroscopy	To determine size of nanoparticles, colloidal suspension and biomolecules.	Cerium-oxide, Cu	(Longano et al., 2012; Darroudi et al., 2014)
FFF	To determine particle-size distributions.	Silica, starched-based nanoparticles	(Peters et al., 2011; Heroult et al., 2014)
NMR	To define exchange mechanisms of organic molecules at nanoparticle surface and composition of ligand.	CdS, TiO_2	(Ladizhansky et al., 1998; Weir et al., 2012)
CD	Secondary structure analysis	Zein, resrevatrol, coenzyme Q10	(Barrow et al., 1992; Wei et al., 2020)

elemental analysis of nanoparticles. Therefore, besides chemical composition and core/surface ligand stoichiometries, nanoparticle size and number concentration also could be determined by MS technique (Costa-Fernández et al., 2016). It is one of the primary analysis methods for determining the chemical composition of small samples and several of MS techniques have been adapted to analyze single aerosol particles.

Induced coupled plasma (ICP)-MS is a widespread detector for identification of particle size, particle size distributions in the extract, mass and particle concentration (Peters et al., 2014). It provides high sensitivity and robust elemental analysis, especially for metal analysis. Besides, if nanoparticles contain low-level impurities, these could be detected by ICP-MS and detailed composition analysis is provided. MS coupled with liquid (LC-MS) and gas chromatography (GC-MS) are other important MS tools for the detection of nanoparticles in foods. High-performance liquid chromatography (HPLC)/MS is utilized for separation and identification of each component in a mixture. It relies on the separation of the specimen components from each other due to their different interaction degrees with the incorporated nanoparticles (McCullum et al., 2014).

Fourier-Transform Infrared (FT-IR) spectroscopy is based on molecular vibrations. The energy of most molecular vibrations corresponds to that of the infrared region of electromagnetic spectrum and possible functional groups in nanoparticles could be identified by FT-IR (Chitra & Annadurai, 2013). Another important spectroscopy technique is Raman spectroscopy. The working principle of Raman spectroscopy is based on the inelastic scattering of a laser light. The interactions between laser and phonons, molecular vibrations are concluded with a shift in the energy of photons (Colthup, 2012). This shift provides information about the phonon modes of the specimen. Therefore, in food studies, Raman spectroscopy could be utilized to acquire more information about the molecular interactions between the nanoparticle and food material (Hu et al., 2015). UV-visible spectroscopy is based on absorbance or reflectance spectroscopy in UV visible region of electromagnetic spectrum. It is utilized generally for the size determination of nanoparticles, colloidal suspension, and biomolecules (Tiede et al., 2008).

Field-flow fractionation (FFF) (Rambaldi et al., 2011) method, which is based on the different granular or surface properties of nanoparticles, is a hydrodynamic separation technique for nanoparticles. FFF technique also allows to determine particle-size distributions. By combining with specific nanoparticle detectors, coupled separation-detector systems occurs to analyze nanoparticles in environmental or food matrices (Von der Kammer et al., 2011). Solution nuclear magnetic resonance (NMR) spectroscopy defines exchange mechanisms of organic molecules at nanoparticle surface and composition of ligand (Guarino et al., 2012; Hens & Martins, 2013). If detailed information about capping ligand needed, then solid state NMR could provide this information (Abraham et al., 2010).

As an absorption spectroscopy, circular dichroism (CD) technique based on the differential absorption of left-handed and right-handed circularly polarized light. Circularly polarized light could be preferably absorbed in one direction by optically active chiral molecules. The CD spectrum in the far UV region provides a secondary structure analysis (Kelly et al., 2005) and so conformation of the presence of different type of nanoparticles in food matrices and the influence of their secondary structure on food could be analyzed by this method (Wei et al., 2020).

8.6.2.4 Testing for health effects

Nanoparticles used in food applications require to pass some testing procedures to evaluate their safety for health. Toxicological studies show the safety of nanoparticles for successful food applications. Opportunities and potential issues regarding introduction of nanoparticles in food and related products, and safety issues should be examined in detail to comply with the regulation schemes. The cytotoxicity and other undesirable impacts of nanomaterials for human body are still not fully known, however, many research studies have currently been conducted to resolve this issue. Analytical tools and procedures to detect release and toxicity of nanoparticles are discussed in this section. Table 8.5 summarizes safety and biological activity tests for nanoparticles up taken with food materials.

It is very important to determine and understand the release behavior and kinetics of nanoparticles from a food environment for designing nanoparticles safely manufactured and used in food applications. Nano sized particles ingested with food, may affect the interaction with living systems and also carry other foreign substances to blood (Das et al., 2011). In vitro and in vivo release studies provide very important information about the response of body against nanoparticles. Therefore, in vitro and in vivo studies are crucially required as emphasized in guidelines before commercialization of nanoparticle incorporated foods. Main in vitro release methods are sampling and separation, continuous flow, dialysis methods, and combination of methods as well (Souza, 2014).

Table 8.5 Safety and biological activity tests for nanoparticles uptaken with food materials.

Characterization/ test type	Purpose	Food nanoparticles	References
In vitro release test	To understand release properties of nanoparticles	bLf, BC, lutein/ zein, chitosan	(Alishahi et al., 2011; Hu, 2012;Yi et al., 2015; Martins et al., 2016; Zhu et al., 2020
In vivo release test	To understand release properties of nanoparticles	Zein	(Luo et al., 2012)
MTT test	To evaluate cytotoxicity of nanoparticles	Silver, silicon dioxide, ZnO	(Athinarayanan et al., 2014; Choi et al., 2015; Gaillet & Rouanet, 2015)
Antioxidant activity tests	To evaluate antioxidant property of nanoparticules	Sodium caseinate, whey protein isolate and soybean protein isolate	(Yi et al., 2015)
Antibacterial activity tests	To evaluate antibacterial property of nanoparticules	Silver, zinc oxide, chitosan, magnesium oxide	(Cao et al., 2010; Jin & He, 2011; Xie et al., 2011; Sotelo-Boyás, 2017)
MBRT	To evaluate antibacterial property of nanoparticules	Copper oxide	(Pawar et al., 2019)
PCR	To detect foodborn pathogen	Carboxyl modified magnetic nanoparticles	(Yang et al., 2007)

In vitro release studies were carried out to test the safety of many organic and inorganic nanoparticles complexed with food products. Martins et al. studied iron release from bovine lactoferrin (bLf) nanoparticles and in vitro release studies of iron from bLf nanoparticle was achieved in a dialysis membrane (Martins et al., 2016). After dispersion of bLf nanoparticle in the dialysis membrane, they sealed, and the total iron ion concentration after in vitro release studies was determined by inductively coupled plasma optical emission spectrometer (ICP-OES) analysis. Stability, which is a significant parameter for food products, adjusts the range of foods in that nanoparticles could be inserted. bLf nanoparticles were found to be stable during 76 days at 4–60°C and pH 2–11. To immitate gastric stage environment and intestinal stage, proteases could be studied for release in vitro. Beta-carotene (BC) nanoparticles in HCl simulates the gastric stage digestion (Yi et al., 2015) and to determine BC content HPLC could be used an appropriate technique. The release profiles in vitro frequently are plotted according to the cumulative percentage of nanoparticles release. HCl and phosphate buffered saline (PBS) media is also used to simulate stomach and intestine environment in vitro release studies. While the release of vitamin C from chitosan nanoparticles was observed, the release could take place in two different ways, namely diffusion and erosion. To test in vivo release properties, nanoparticles are inserted in diet directly and the amount of nanoparticles in body fluids are evaluated (Alishahi et al., 2011).

Nanoparticles may cause toxicity in cells and so damage cell membranes, organelles, and the nucleus (Wu et al., 2014; Gaillet & Rouanet, 2015). Some reported inorganic nanoparticles producing cytotoxicity are silicon dioxide nanoparticles (Athinarayanan et al., 2014), ZnO nanoparticles (Choi et al., 2015), and silver nanoparticles (Gaillet and Rouanet, 2015). Nanoparticles could be subjected to cell culture tests to investigate whether they have toxic effects on cells. If they do not indicate any toxic effects on cell viability, then these nanoparticles are biocompatible. MTT test could be utilized to evaluate cytotoxicity; cell adhesion test also indicates the cytotoxicity.

The power of bioactive compounds can reduce in aqueous solution depending on their antioxidant activity. Cellular antioxidant activity and chemical antioxidant activity of the nanoparticles should be examined to assess their biological activity and thus functionality in the food complex. Nanoparticles can provide higher the antioxidant activity in comparison to crystalline materials due to their higher surface area, enhanced solubility by protein encapsulation and amorphous structure (Yi et al., 2015). Nanoencapsulation process also enhances cellular antioxidant activity of nanoparticles. Antibacterial activity of nanoparticles could be tested three main methods: disc diffusion method, the colony forming count method, and the minimum inhibitory area method (Cao et al., 2010). Methylene blue reduction test (MBRT) could be also performed to determine antibacterial activity of nanoparticles (Pawar et al., 2019). MBRT is based on the discoloration of methylene blue in milk because of the removal of oxygen and the formation of reducing substances during bacterial activity. This is a primary antimicrobial susceptibility test and its advantage is its limited equipment and materials need, however, it does not provide a quantitative analysis of bacterial amount in a specimen.

8.7 Conclusion

Nanotechnology-derived materials inserted in food products can introduce various opportunities and challenges as well. Potential safety risks associated with these materials mainly due to their tiny size, changeable physicochemical characteristics, intensive exposure, possible toxicity, and environmental deposition, give rise to serious health concerns. In order to assess safety of nanoparticles used in food matrices, many national and international authorities release guidance documents worldwide. Although there is still lack of common international legislations regulating nanoparticles in foods, some significant intergovernmental bodies have been currently involved in following the developments in nanotechnology industry, organizing international meetings, and establishment of recommendations and guidelines. Significant principles as the basics of these international guidelines are identification and characterization of nanomaterials using powerful analytical techniques, describing safe nanoparticle manufacturing methods, evaluation and control of their toxicity, and allergenicity due to high exposure rates through in vitro and in vivo studies. A comprehensive inspection is capable of providing required data for safe manufacturing, use, and consumption of nanomaterial incorporated food products. Global regulatory legislations are expected to be developed soon for the safety assessment of nano-enabled foods and therefore supported international commercialization of these products promising for beneficial opportunities.

References

Abraham, A., et al. (2010). Solid-state NMR study of cysteine on gold nanoparticles. *The Journal of Physical Chemistry C*, *114*(42), 18109–18114.

Acosta, E. (2009). Bioavailability of nanoparticles in nutrient and nutraceutical delivery. *Current Opinion in Colloid and Interface Science*, *14*(1), 3–15.

Al-Naamani, L., et al. (2016). Chitosan-zinc oxide nanoparticle composite coating for active food packaging applications. *Innovative Food Science and Emerging Technologies, 38*, 231–237.

Alimentarius, C. (2019). Guidelines for rapid risk analysis following instances of detection of contaminants in food where there is no regulatory level. Retrieved 15/06/2020.

Alishahi, A., et al. (2011). Shelf life and delivery enhancement of vitamin C using chitosan nanoparticles. *Food Chemistry, 126*(3), 935–940.

Amenta,V., et al. (2015). Regulatory aspects of nanotechnology in the agri/feed/food sector in EU and non-EU countries. *Regulatory Toxicology and Pharmacology, 73*(1), 463–476.

Arabi, M., Ghaedi, M., & Ostovan, A. (2016). Development of dummy molecularly imprinted based on functionalized silica nanoparticles for determination of acrylamide in processed food by matrix solid phase dispersion. *Food chemistry, 210*, 78–84.

Aresta, A., et al. (2013). Development and analytical characterization of vitamin (s)-loaded chitosan nanoparticles for potential food packaging applications. *Journal of Nanoparticle Research, 15*(4), 1592.

Athinarayanan, J., et al. (2014). Presence of nanosilica (E551) in commercial food products: TNF-mediated oxidative stress and altered cell cycle progression in human lung fibroblast cells. *Cell Biology and Toxicology, 30*(2), 89–100.

Ault, A. P., et al. (2012). Single-particle SEM-EDX analysis of iron-containing coarse particulate matter in an urban environment: sources and distribution of iron within Cleveland, Ohio. *Environmental Science and Technology, 46*(8), 4331–4339.

Bajpai,V. K., et al. (2018). Prospects of using nanotechnology for food preservation, safety, and security. *Journal of Food and Drug Analysis, 26*(4), 1201–1214.

Barrow, C. J., et al. (1992). Solution conformations and aggregational properties of synthetic amyloid β-peptides of Alzheimer's disease: analysis of circular dichroism spectra. *Journal of Molecular Biology, 225*(4), 1075–1093.

Bazana, M.T., et al. (2019). Nanoencapsulation of bioactive compounds: challenges and perspectives. *Current Opinion in Food Science, 26*, 47–56.

Benko, H. (2017). ISO Technical Committee 229 Nanotechnologies. *Metrology and Standardization of Nanotechnology: Protocols and Industrial Innovations:*, 259–268.

Bouwmeester, H., et al. (2014). State of the safety assessment and current use of nanomaterials in food and food production. *Trends in Food Science and Technology, 40*(2), 200–210.

Bouwmeester, H., et al. (2009). Review of health safety aspects of nanotechnologies in food production. *Regulatory Toxicology and Pharmacology, 53*(1), 52–62.

Brody, A. L. (2007). Case studies on nanotechnologies for food packaging. *Food Technology, 61*.

Bugusu, B., et al. (2007). International Food Nanoscience Conference, INST Food Technologists 525 West Van Buren, STE 1000, Chicago, IL, United States.

Bugusu, B., et al. (2009). Global regulatory policies on food nanotechnology. *Food Technology (Chicago), 63*(5), 24–28.

Bumbudsanpharoke, N., & Ko, S. (2015). Nano-food packaging: an overview of market, migration research, and safety regulations. *Journal of Food Science, 80*(5), R910–R923.

Bustos, A. R. M., et al. (2013). Mass spectrometry for the characterisation of nanoparticles. *Analytical and Bioanalytical Chemistry, 405*(17), 5637–5643.

Cao, X., et al. (2010). Immobilization of silver nanoparticles onto sulfonated polyethersulfone membranes as antibacterial materials. *Colloids and Surfaces B: Biointerfaces, 81*(2), 555–562.

Cárdenas, G., et al. (2009). Colloidal Cu nanoparticles/chitosan composite film obtained by microwave heating for food package applications. *Polymer Bulletin, 62*(4), 511–524.

Chau, C. -F., et al. (2007). The development of regulations for food nanotechnology. *Trends in Food Science and Technology, 18*(5), 269–280.

Chaudhry, Q., et al. (2008). Applications and implications of nanotechnologies for the food sector. *Food Additives and Contaminants, 25*(3), 241–258.

Chen, H., et al. (2006). Nanotechnology in nutraceuticals and functional foods. *Food Technology (Chicago), 60*(3), 30–36.
Chen, J., et al. (2013). Determination of the domain structure of the 7S and 11S globulins from soy proteins by XRD and FTIR. *Journal of the Science of Food and Agriculture, 93*(7), 1687–1691.
Chitra, K., & Annadurai, G. (2013). Antimicrobial activity of wet chemically engineered spherical shaped ZnO nanoparticles on food borne pathogen. *International Food Research Journal, 20.*(1).
Choi, J., et al. (2015). Toxicity of zinc oxide nanoparticles in rats treated by two different routes: single intravenous injection and single oral administration. *Journal of Toxicology and Environmental Health Part A, 78*(4), 226–243.
Clemens, R., & Pressman, P. (2011). The role of nanotechnology in feeding the World. *Food Technology, 65*(1).
Cockburn, A., et al. (2012). Approaches to the safety assessment of engineered nanomaterials (ENM) in food. *Food and Chemical Toxicology, 50*(6), 2224–2242.
Codex Alimentaarus (2007). Organically Produced Foods. Rome, 2007
Colthup, N. (2012). *Introduction to infrared and Raman spectroscopy*. Elsevier.
Costa-Fernández, J. M., et al. (2016). Mass spectrometry for the characterization and quantification of engineered inorganic nanoparticles. *TrAC Trends in Analytical Chemistry, 84*, 139–148.
COT (2005). COT joint statement on nanomaterial toxicology.
COT (2007). COT report on variability and uncertainty in toxicology.
Cushen, M., et al. (2012). Nanotechnologies in the food industry–Recent developments, risks and regulation. *Trends in Food Science and Technology, 24*(1), 30–46.
D'Souza, S. (2014). A review of in vitro drug release test methods for nano-sized dosage forms. *Advances in Pharmaceutics, 2014*, 12.
da Rosa, C. G., et al. (2015). Characterization and evaluation of physicochemical and antimicrobial properties of zein nanoparticles loaded with phenolics monoterpenes. *Colloids and Surfaces A: Physicochemical and Engineering Aspects, 481*, 337–344.
Darroudi, M., et al. (2014). Food-directed synthesis of cerium oxide nanoparticles and their neurotoxicity effects. *Ceramics International, 40*(5), 7425–7430.
Das, M., et al. (2011). Need for safety of nanoparticles used in food industry. *Journal of Biomedical Nanotechnology*, 7(1), 13–14.
Dasgupta, N., et al. (2015). Nanotechnology in agro-food: from field to plate. *Food Research International, 69*, 381–400.
Depew, F. M. (1963). The Joint FAO-WHO Conference on Food Standards. *Food Drug Cosmetic Law Journal, 18*, 34.
des Rieux, A., et al. (2006). Nanoparticles as potential oral delivery systems of proteins and vaccines: a mechanistic approach. *Journal of Controlled Release, 116*(1), 1–27.
Echegoyen, Y., & Nerín, C. (2013). Nanoparticle release from nano-silver antimicrobial food containers. *Food and Chemical Toxicology, 62*, 16–22.
Espitia, P. J. P., et al. (2012). Zinc oxide nanoparticles: synthesis, antimicrobial activity and food packaging applications. *Food and Bioprocess Technology, 5*(5), 1447–1464.
EFSA (2011a). Comprehensive European Food Consumption Database Comprehensive Database.
EFSA. (2011b). Overview of the procedures currently used at EFSA for the assessment of dietary exposure to different chemical substances. *EFSA Journal, 9*(12), 2490. The European Union summary of soonoses, zoonotic agents and food-borne outbreaks in 2016.
FAO/WHO (1995a). The state of food and agriculture 1995. Rome, 1995.
FAO/WHO (1997). The state of food and agriculture 1997. Rome, 1997.
FAO/WHO (2010). Joint FAO/WHO Expert consultation on the risks and benefits of fish consumption.

FAO/WHO (2011). FAO/WHO guide for application of risk analysis principles and procedures during food safety emergencies. 2011, Rome.

FAO/WHO (2012). Joint FAO/WHO Seminar on Nanotechnologies in food and agriculture. 27 March 2012, Rome.

FAO/WHO (2013). Food Systems For Better Nutrition. 2013, Rome.

Filipe, V., et al. (2010). Critical evaluation of nanoparticle tracking analysis (NTA) by nanosight for the measurement of nanoparticles and protein aggregates. *Pharmaceutical Research*, *27*(5), 796–810.

Food and Agriculture Organization (2013). State of the art on the initiatives and activities relevant to risk assessment and risk management of nanotechnologies in the food and agriculture sectors, FAO.

Gaillet, S., & Rouanet, J. -M. (2015). Silver nanoparticles: their potential toxic effects after oral exposure and underlying mechanisms–a review. *Food and Chemical Toxicology*, *77*, 58–63.

Giovannozzi, A. M., et al. (2014). Rapid and sensitive detection of melamine in milk with gold nanoparticles by Surface Enhanced Raman Scattering. *Food Chemistry*, *159*, 250–256.

Graveland-Bikker, J. F., et al. (2004). Influence of calcium on the self-assembly of partially hydrolyzed α-lactalbumin. *Langmuir*, *20*(16), 6841–6846.

Grobe, A., et al. (2008). Risk governance of nanotechnology applications in food and cosmetics, International Risk Governance Council (IRGC).

Gruère, G. P. (2012). Implications of nanotechnology growth in food and agriculture in OECD countries. *Food Policy*, *37*(2), 191–198.

Guarino, G., et al. (2012). Lanthanide-based NMR:a tool to investigate component distribution in mixed-monolayer-protected nanoparticles. *Journal of the American Chemical Society*, *134*(17), 7200–7203.

Guo, H., et al. (2015). Analysis of silver nanoparticles in antimicrobial products using surface-enhanced Raman spectroscopy (SERS). *Environmental Science & Technology*, *49*(7), 4317–4324.

Hannon, J. C., et al. (2015). Advances and challenges for the use of engineered nanoparticles in food contact materials. *Trends in Food Science & Technology*, *43*(1), 43–62.

Hens, Z., & Martins, J. C. (2013). A solution NMR toolbox for characterizing the surface chemistry of colloidal nanocrystals. *Chemistry of Materials*, *25*(8), 1211–1221.

Heroult, J., et al. (2014). The potential of asymmetric flow field-flow fractionation hyphenated to multiple detectors for the quantification and size estimation of silica nanoparticles in a food matrix. *Analytical and Bioanalytical Chemistry*, *406*(16), 3919–3927.

Hillyer, J. F., & Albrecht, R. M. (2001). Gastrointestinal persorption and tissue distribution of differently sized colloidal gold nanoparticles. *Journal of Pharmaceutical Sciences*, *90*(12), 1927–1936.

Hosseini, S. F., et al. (2013). Two-step method for encapsulation of oregano essential oil in chitosan nanoparticles: preparation, characterization and in vitro release study. *Carbohydrate Polymers*, *95*(1), 50–56.

Hristozov, D., et al. (2016). Frameworks and tools for risk assessment of manufactured nanomaterials. *Environment International*, *95*, 36–53.

Hu, D, Lin, C., Liu, L., Li, S., & Zhao, Y. (2012). Preparation, characterization, and in vitro release investigation of lutein/zein nanoparticles via solution enhanced dispersion by supercritical fluids. *Journal of food engineering*, *109*(3), 545–552.

Hu, K., et al. (2015). Core–shell biopolymer nanoparticle delivery systems: Synthesis and characterization of curcumin fortified zein–pectin nanoparticles. *Food Chemistry*, *182*, 275–281.

ILSI (2011). International Symposium on Health Benefits of Foods -From Emerging Science to Innovative Products. Prague, 5-7 October 2011.

ILSI Europe (2011). ILSI Europe Session at the 11th Europian Nutrition Conference. Madrid, 2011.

IRGC (2008). GLobal risk Governance- Concept and practice using the IRGC framework.

ISO, C. (2008). TS 27687: 2008–Nanotechnologies–Terminology and definitions for nano-objects–Nanoparticle, nanofibre and nanoplate (ISO/TS 27687: 2008). German version: Vornorm DIN CEN ISO/TS 27687.

ISO, I. (2010). TS 80004-1: 2010, Nanotechnologies--Vocabulary--Part 1: Core terms. International Organization for Standardization, Geneva.

ISO (2010). TC 201/SC 7 Electron Spectroscopies ICS: 71.040.40 Chemical Analysis-Surface Chemical Analysis- X-Ray Photoelectron Spectroscopy- Guidelines for Analysis.

ISO (2015). ISO/TS 80004-2: 2015. Nanotechnologies-Vocabulary-Part 2: Nano-objects.

ISO (2020). ISO/TS 21975:2020 Nanotechnologies- Polymeric nanocomposite films for food packaging with barrier properties- Specification of characteristics and measurement methods.

Jain, A., et al. (2018). Nanomaterials in food and agriculture: an overview on their safety concerns and regulatory issues. *Critical Reviews in Food Science and Nutrition*, *58*(2), 297–317.

Ji, J. H., et al. (2007). Twenty-eight-day inhalation toxicity study of silver nanoparticles in Sprague-Dawley rats. *Inhalation Toxicology*, *19*(10), 857–871.

Jin, T., & He, Y. (2011). Antibacterial activities of magnesium oxide (MgO) nanoparticles against foodborne pathogens. *Journal of Nanoparticle Research*, *13*(12), 6877–6885.

Kanmani, P., & Rhim, J. -W. (2014). Physicochemical properties of gelatin/silver nanoparticle antimicrobial composite films. *Food Chemistry*, *148*, 162–169.

Kelly, S. M., et al. (2005). How to study proteins by circular dichroism. *Biochimica et Biophysica Acta (BBA)-Proteins and Proteomics*, *1751*(2), 119–139.

Krishnaraj, C., et al. (2015). Mechanistic aspects of biologically synthesized silver nanoparticles against food-and water-borne microbes. *Bioprocess and Biosystems Engineering*, *38*(10), 1943–1958.

Kumar, M., et al. (2017). Combating food pathogens using sodium benzoate functionalized silver nanoparticles: synthesis, characterization and antimicrobial evaluation. *Journal of Materials Science*, *52*(14), 8568–8575.

Kühnel, D., & Nickel, C (2014). The OECD expert meeting on ecotoxicology and environmental fate—Towards the development of improved OECD guidelines for the testing of nanomaterials. *Science of the Total Environment*, *472*, 347–353.

Ladizhansky, V., et al. (1998). Surface properties of precipitated CdS nanoparticles studied by NMR. *The Journal of Physical Chemistry B*, *102*(43), 8505–8509.

Lindgren, A.L. (2014). The effects of silver nitrate and silver nanoparticles on Chlamydomonas reinhardtii: a proteomic approach. Degree Project, Department of Biology and Environmental Sciences, University of Gothenburg, Germany 9(4): 1–34.

Liu, Y., et al. (2011). Preparation and characterization of α-galactosidase-loaded chitosan nanoparticles for use in foods. *Carbohydrate Polymers*, *83*(3), 1162–1168.

Loeschner, K., et al. (2013). Detection and characterization of silver nanoparticles in chicken meat by asymmetric flow field flow fractionation with detection by conventional or single particle ICP-MS. *Analytical and Bioanalytical Chemistry*, *405*(25), 8185–8195.

Longano, D., et al. (2012). Analytical characterization of laser-generated copper nanoparticles for antibacterial composite food packaging. "*Analytical and Bioanalytical Chemistry*, *403*(4), 1179–1186.

Luo, Y., et al. (2012). Development of zein nanoparticles coated with carboxymethyl chitosan for encapsulation and controlled release of vitamin D3. *Journal of Agricultural and Food Chemistry*, *60*(3), 836–843.

Mackevica, A., et al. (2016). Silver nanoparticle release from commercially available plastic food containers into food simulants. *Journal of Nanoparticle Research*, *18*(1), 5.

Madureira, A. R., et al. (2015). Production of antimicrobial chitosan nanoparticles against food pathogens. *Journal of Food Engineering, 167*, 210–216.

Manonmani, V., et al. (2011). Biosynthesis of Ag nanoparticles for the detection of pathogenic bacteria in food. International Conference on Innovation, Management and Service, IPEDR, IPEDR, vol. 14, IACSIT Press, Singapore.

Martins, J. T., et al. (2016). Lactoferrin-based nanoparticles as a vehicle for iron in food applications–Development and release profile. *Food Research International, 90*, 16–24.

Mbengue, M. M., & Charles, M. (2013). International organizations and nanotechnologies: the challenge of coordination. *Review of European, Comparative & International Environmental Law, 22*(2), 174–185.

McClements, D. J. (2013). Edible lipid nanoparticles: digestion, absorption, and potential toxicity. *Progress in Lipid Research, 52*(4), 409–423.

McClements, D. J., et al. (2016). The role of the food matrix and gastrointestinal tract in the assessment of biological properties of ingested engineered nanomaterials (iENMs): State of the science and knowledge gaps. *NanoImpact, 3*, 47–57.

McClements, D. J., & Xiao, H. (2012). Potential biological fate of ingested nanoemulsions: influence of particle characteristics. *Food & Function, 3*(3), 202–220.

McClements, D. J., & Xiao, H. (2017). Is nano safe in foods? Establishing the factors impacting the gastrointestinal fate and toxicity of organic and inorganic food-grade nanoparticles. *npj Science of Food, 1*(1), 1–13.

McCracken, C., et al. (2013). Minimal intestinal epithelial cell toxicity in response to short- and long-term food-relevant inorganic nanoparticle exposure. *Chemical Research in Toxicology, 26*(10), 1514–1525.

McCullum, C., et al. (2014). Extraction of aflatoxins from liquid foodstuff samples with polydopamine-coated superparamagnetic nanoparticles for HPLC-MS/MS analysis. *Journal of Agricultural and Food Chemistry, 62*(19), 4261–4267.

Momin, J. K., & Joshi, B. (2015). Nanotechnology in foods. In *Nanotechnologies in food and agriculture* (pp. 3–24). Springer.

Nanotechnologies, I. (2015). Vocabulary–Part 1: Core Terms. International Standardization Organisation (ISO), Technical Specification ISO/TS:80004-80001.

Naseer, B., et al. (2018). Importance and health hazards of nanoparticles used in the food industry. *Nanotechnology Reviews*, 7(6), 623–641.

Oberdörster, G., et al. (2005). Principles for characterizing the potential human health effects from exposure to nanomaterials: elements of a screening strategy. *Particle and Fibre Toxicology, 2*(1), 8.

OECD, A. (2005). Opportunities and risks of Nanotechnologies. Report in co-operation with the OECD International Futures Programme.

OECD (2009). Preliminary Review of OECD Test Guidelines for Their Applicability to Manufactured Nanomaterials.

OECD, O. (2012). "Environment, health and safety publications series on the safety of manufactured nanomaterials No. 36. Guidance on sample preparation and dosimetry for the safety testing of manufactured nanomaterials." Organization for Economic Cooperation and Development, Paris, France.

OECD (2013). "Regulatory Frameworks for Nanotechnology in Foods and Medical Products.".

OECD 2017-2019 available at http://www.oecd.org/env/ehs/nanosafety/publications-series-safety-manufactured-nanomaterials.htm.(visited at 14.06.2020)

Organization, W.H. (2010). FAO/WHO Expert Meeting on the Application of Nanotechnologies in the Food and Agriculture Sectors: Potential Food Safety Implications: Meeting Report, World Health Organization.

Pawar, J., et al. (2019). Testing of antibacterial efficacy of CuO nanoparticles by methylene blue reduction test against *Bacillus cereus* responsible for food spoilage and poisoning. *Indian Chemical Engineer, 61*(3), 248–253.

Peters, R., et al. (2011). Identification and characterization of organic nanoparticles in food. *TrAC Trends in Analytical Chemistry, 30*(1), 100–112.

Peters, R. J., et al. (2014). Characterization of titanium dioxide nanoparticles in food products: analytical methods to define nanoparticles. *Journal of Agricultural and Food Chemistry, 62*(27), 6285–6293.

Pietroiusti, A., et al. (2016). New frontiers in nanotoxicology: gut microbiota/microbiome-mediated effects of engineered nanomaterials. *Toxicology and Applied Pharmacology, 299*, 90–95.

Pulit-Prociak, J., et al. (2015). Nanosilver products and toxicity. *Environmental Chemistry Letters, 13*(1), 59–68.

Qi, R., et al. (2014). Rapid identification of synthetic colorants in food samples by using indium oxide nanoparticle-functionalized porous polymer monolith coupled with HPLC-MS/MS. *Analyst, 139*(23), 6168–6177.

Rambaldi, D. C., et al. (2011). Flow field-flow fractionation: recent trends in protein analysis. *Analytical and Bioanalytical Chemistry, 399*(4), 1439–1447.

Renn, O. & M. Roco (2006). "IRGC white paper no. 2: nanotechnology risk governance.

SCENIHR (2007a). The Appropriateness of the Risk Assessment Methodology in Accordance with the Technical Guidance Documents for New and Existing Substances for Assessing the Risks of Nanomaterials. 21-22 June 2007.

Shahidi, F. (2006). Nanotechnology in nutraceuticals and functional foods emerging technology has shown great potential for delivering bioactive compounds in functional foods to improve human health. *Nanotechnology, 60*(3), .

Sharma, S., et al. (2019). Nanostructured materials for food applications: spectroscopy, microscopy and physical properties. *Bioengineering, 6*(1), 26.

Sheehan, B., et al. (2018). Hazard screening methods for nanomaterials: a comparative study. *International Journal of Molecular Sciences, 19*(3), 649.

Sk, M. P., et al. (2012). Presence of amorphous carbon nanoparticles in food caramels. *Scientific Reports, 2*(1), 1–5.

Sohal, I. S., et al. (2018). Ingested engineered nanomaterials: state of science in nanotoxicity testing and future research needs. *Particle and Fibre Toxicology, 15*(1), 29.

Sotelo-Boyás, M.E., Correa-Pacheco, Z.N., Bautista-Baños, S., & Corona-Rangel, M.L. (2017). Physicochemical characterization of chitosan nanoparticles and nanocapsules incorporated with lime essential oil and their antibacterial activity against food-borne pathogens. *LWT*, 77, 15–20.

Subramanian, K.S., & Rajkishore, S.K. (2018). Regulatory framework for nanomaterials in agri-food systems. *Nanomaterials: Ecotoxicity, Safety, and Public Perception*. Cham: Springer, pp. 319-342.

Sung, J.H., Ji, J.H., Yoon, J.U., Kim, D.S., Song, M.Y., Jeong, J., & Yu, I.J. (2008). Lung function changes in Sprague-Dawley rats after prolonged inhalation exposure to silver nanoparticles. *Inhalation toxicology, 20*(6), 567–574.

Szakal, C., Tsytsikova, L., Carlander, D., & Duncan, T.V. (2014). Measurement methods for the oral uptake of engineered nanomaterials from human dietary sources: summary and outlook. *Comprehensive Reviews in Food Science and Food Safety, 13*(4), 669–678.

Takeuchi, M.T., et al. (2014). State of the art on the initiatives and activities relevant to risk assessment and risk management of nanotechnologies in the food and agriculture sectors. *Food Research International, 64*, 976–981.

Tankhiwale, R., & Bajpai, S. (2012). Preparation, characterization and antibacterial applications of ZnO-nanoparticles coated polyethylene films for food packaging. *Colloids and Surfaces B: Biointerfaces, 90*, 16–20.

Tiede, K., et al. (2008). Detection and characterization of engineered nanoparticles in food and the environment. *Food Additives and Contaminants, 25*(7), 795–821.

Thibault, S., et al. (2008). Gold nanoparticles and a glucose oxidase based biosensor: an attempt to follow-up aging by XPS. *Microchimica Acta, 163*(3–4), 211–217.

Tombuloglu, H., et al. (2020). Iron oxide nanoparticles translocate in pumpkin and alter the phloem sap metabolites related to oil metabolism. *Scientia Horticulturae, 265*, 109223.
van der Meulen, B., et al. (2014). *Governing nano foods: principles-based responsive regulation: EF-FoST critical reviews# 3.* Academic Press.
Velmurugan, P., et al. (2014). Green synthesis of silver and gold nanoparticles using *Zingiber officinale* root extract and antibacterial activity of silver nanoparticles against food pathogens. *Bioprocess and Biosystems Engineering, 37*(10), 1935–1943.
Venkatasubbu, G. D., et al. (2016). Toxicity mechanism of titanium dioxide and zinc oxide nanoparticles against food pathogens. *Colloids and Surfaces B: Biointerfaces, 148*, 600–606.
Von der Kammer, F., et al. (2011). Separation and characterization of nanoparticles in complex food and environmental samples by field-flow fractionation. *TrAC Trends in Analytical Chemistry, 30*(3), 425–436.
Wei, Y., et al. (2020). Fabrication, characterization and in vitro digestion of food grade complex nanoparticles for co-delivery of resveratrol and coenzyme Q10. *Food Hydrocolloids, 105*, 105791.
Weir, A., et al. (2012). Titanium dioxide nanoparticles in food and personal care products. *Environmental Science & Technology, 46*(4), 2242–2250.
WHO (2010). World health statistics 2010.
WHO (2012). World health statistics 2012.
WHO (2013). Wrold Health report 2013: Research for universal health coverage.
Wu, H., et al. (2014). Reactive oxygen species-related activities of nano-iron metal and nano-iron oxides. *Journal of Food and Drug Analysis, 22*(1), 86–94.
Wyser, Y., et al. (2016). Outlook and challenges of nanotechnologies for food packaging. *Packaging Technology and Science, 29*(12), 615–648.
Xie, Y., et al. (2011). Antibacterial activity and mechanism of action of zinc oxide nanoparticles against *Campylobacter jejuni*. *Applied and Environmental Microbiology*, 77(7), 2325–2331.
Yang, H., et al. (2007). Rapid detection of Listeria monocytogenes by nanoparticle-based immunomagnetic separation and real-time PCR. *International Journal of Food Microbiology, 118*(2), 132–138.
Yi, J., et al. (2015). Beta-carotene encapsulated in food protein nanoparticles reduces peroxyl radical oxidation in Caco-2 cells. *Food Hydrocolloids, 43*, 31–40.
Yu, Q., et al. (2015). Gold nanoparticles-based lateral flow immunoassay with silver staining for simultaneous detection of fumonisin B1 and deoxynivalenol. *Food Control, 54*, 347–352.
Zhou, G., & Hu, W. (2018). Public acceptance of and willingness-to-pay for nanofoods in the US. *Food Control, 89*, 219–226.
Zhu, N., et al. (2020). A novel coronavirus from patients with pneumonia in China, 2019. *New England Journal of Medicine* 382(8): 727–733.

Index

Note: Page numbers followed by "f" indicate figures, "t" indicate tables.

M

N

S

T

U

V

W

X

Z

FSC
www.fsc.org
MIX
Papier | Fördert
gute Waldnutzung
FSC® C083411

Kolibri 360 GmbH
Ferdinand-Jühlke-Straße 7
99095 Erfurt, Deutschland
produktsicherheit@kolibri360.de